GOVERNMENT SERIES

Energy: Ethanol

The Production and Use of Biofuels, Biodiesel, and Ethanol; Agriculture-Based Renewable Energy Production Including Corn and Sugar; the Ethanol "Blend Wall"; Renewable Fuel Standard (RFS and RFS2); Cellulosic Biofuels; 2007 Energy Bill; 2008 Farm Bill; Food and Livestock Feed Price Inflation; Caribbean Basin Initiative; and U.S.-Brazil Energy Cooperation

Compiled by TheCapitol.Net

Authors: By Brent D. Yacobucci, Randy Schnepf, Salvatore Lazzari, Megan Stubbs, Fred Sissine, Remy Jurenas, Scott A. Malcolm, Marcel Aillery, Marca Weinberg, Kelsi Bracmort, Tom Capehart, Joe Richardson, Geoffrey S. Becker, and Clare Ribando Seelk

For over 30 years, TheCapitol.Net and its predecessor, Congressional Quarterly Executive Conferences, have been training professionals from government, military, business, and NGOs on the dynamics and operations of the legislative and executive branches and how to work with them.

Our training and publications include congressional operations, legislative and budget process, communication and advocacy, media and public relations, research, business etiquette, and more.

TheCapitol.Net is a non-partisan firm.

Our publications and courses, written and taught by *current* Washington insiders who are all independent subject matter experts, show how Washington works.™ Our products and services can be found on our web site at *<www.TheCapitol.Net>*.

Additional copies of *Energy: Ethanol* can be ordered online: *<www.GovernmentSeries.com>*.

Design and production by Zaccarine Design, Inc., Evanston, IL; 847-864-3994.

∞ The paper used in this publication exceeds the requirements of the American National Standard for Information Sciences—Permanence of Paper for Printed Library Materials, ANSI Z39.48-1992.

PO Box 25706
Alexandria, VA 22313-5706
703-739-3790 Toll free: 1-877-228-5086
<www.TheCapitol.Net>

v 1

Energy: Ethanol, softbound:
ISBN: 158733-191-8
ISBN 13: 978-1-58733-191-6

Summary Table of Contents

Table of Contents

Introduction

Energy: Ethanol

The Production and Use of Biofuels, Biodiesel, and Ethanol; Agriculture-Based Renewable Energy Production Including Corn and Sugar; the Ethanol "Blend Wall"; Renewable Fuel Standard (RFS and RFS2); Cellulosic Biofuels; 2007 Energy Bill; 2008 Farm Bill; Food and Livestock Feed Price Inflation; Caribbean Basin Initiative; and U.S.-Brazil Energy Cooperation

Biofuels have grown significantly in the past few years as a component of U.S. motor fuel supply. Current U.S. biofuels supply relies primarily on ethanol produced from Midwest corn. Today, ethanol is blended in more than half of all U.S. gasoline (at the 10% level or lower in most cases). Federal policy has played a key role in the emergence of the U.S. biofuels industry in general, and the corn ethanol industry in particular. U.S. biofuels production is supported by federal and state policies that include minimum usage requirements, blending and production tax credits, an import tariff to limit importation of foreign-produced ethanol, loans and loan guarantees to facilitate the development of biofuels production and distribution infrastructure, and research grants.

Since the late 1970s, U.S. policy makers at both the federal and state levels have enacted a variety of incentives, regulations, and programs to encourage the production and use of agriculture-based renewable energy. Motivations cited for these legislative initiatives include energy security concerns, reduction in greenhouse gas emissions, and raising domestic demand for U.S.-produced farm products.

Agricultural households and rural communities have responded to these government incentives and have expanded their production of renewable energy, primarily in the form of biofuels and wind power, every year since 1996.

Ethanol and biodiesel, the two most widely used biofuels, receive significant government support under federal law in the form of mandated fuel use, tax incentives, loan and grant programs, and certain regulatory requirements.

Ethanol plays a key role in policy discussions about energy, agriculture, taxes, and the environment. In the United States it is mostly made from corn; in other countries it is often made from cane sugar. Fuel ethanol is generally blended in gasoline to reduce emissions, increase octane, and extend gasoline stock.

U.S. policy to expand the production of biofuel for domestic energy use has significant implications for agriculture and resource use. While ongoing research and development investment may radically alter the way biofuel is produced in the future, for now, corn-based ethanol continues to account for most biofuel production. As corn ethanol production increases, so does the production of corn. The effect on agricultural commodity markets has been national, but commodity production adjustments, and resulting environmental consequences, vary across regions. Changes in the crop sector have also affected the cost of feed for livestock producers.

Links to Internet resources are available on the book's web site at <*www.TCNEthanol.com*>.

Biofuels: Ethanol and Biodiesel *Explained*

Basics

"Biofuels" are transportation fuels like ethanol and biodiesel that are made from biomass materials. These fuels are usually blended with the petroleum fuels — gasoline and diesel fuel, but they can also be used on their own. Using ethanol or biodiesel means we don't burn quite as much fossil fuel. Ethanol and biodiesel are usually more expensive than the fossil fuels that they replace, but they are also cleaner-burning fuels, producing fewer air pollutants.

What Is Ethanol?

Ethanol is an alcohol fuel made from the sugars found in grains, such as:

- Corn
- Sorghum
- Barley

USDA research geneticists study switchgrass as a source of ethanol.

Photo Credit: Brett Hampton, USDA Agricultural Research Sevice (Public Domain)

Other sources of sugars to produce ethanol include:

- Potato skins
- Rice
- Sugar cane
- Sugar beets
- Yard clippings
- Bark
- Switchgrass

Most of the ethanol used in the United States today is distilled from corn. Scientists are working on cheaper ways to make ethanol by using all parts of plants and trees rather than just the grain. Farmers are experimenting with "woody crops," mostly small poplar trees and switchgrass, to see if they can be grown cheaply and abundantly.

Ethanol Is Blended With Gasoline

About 99% of the ethanol produced in the United States is used to make "E10" or "gasohol," a mixture of 10% ethanol and 90% gasoline. Any gasoline powered engine can use E10, but only specially made vehicles can run on E85, a fuel that is 85% ethanol and 15% gasoline.

What Is Biodiesel?

Biodiesel is a fuel made from vegetable oils, fats, or greases — such as recycled restaurant grease. Biodiesel fuel can be used in diesel engines without changing them. It is the fastest growing alternative fuel in the United States. Biodiesel, a renewable fuel, is safe, biodegradable, and produces lower levels of most air pollutants than petroleum-based products.

Fuel Ethanol Statistics

Data for 2008 except where noted

U.S. Production	9.2 billion gallons 220 million barrels
U.S. Net Imports	12.3 million barrels
U.S. Consumption	9.6 billion gallons 229 million barrels

Biodiesel Basics

Data for 2008 except where noted

U.S. Production	0.7 billion gallons 16 million barrels
U.S. Net Exports	8.6 million barrels
U.S. Consumption	0.3 billion gallons 7.6 million barrels

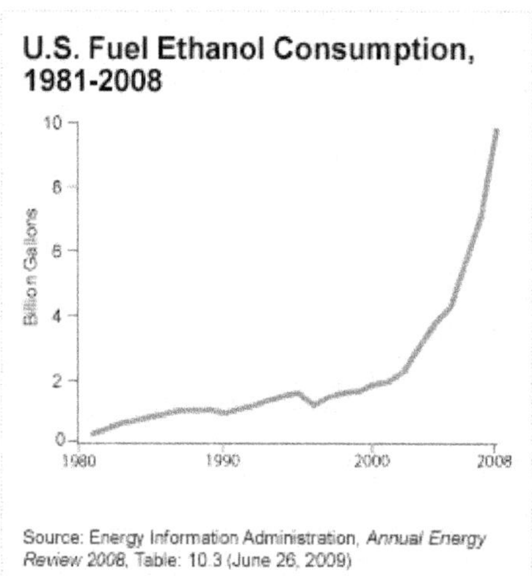

Source: U.S. Energy Information Administration , June, 2009

U.S. Department of Energy - Energy Efficiency and Renewable Energy

Alternative Fuels and Advanced Vehicles Data Center - Ethanol

Ethanol Basics

Ethanol is a renewable fuel made from various plant materials, which collectively are called "biomass." Ethanol contains the same chemical compound (C_2H_5OH) found in alcoholic beverages. Nearly half of U.S. gasoline contains ethanol in a low-level blend to oxygenate the fuel and reduce air pollution. Ethanol is also increasingly available in E85, an alternative fuel that can be used in flexible fuel vehicles. Studies have estimated that ethanol and other biofuels could replace 30% or more of U.S. gasoline demand by 2030.

Several steps are required to make ethanol available as a vehicle fuel—see the supply chain diagram below. Biomass feedstocks are grown, then various logistical systems are used to collect and transport them to ethanol production facilities. After ethanol is produced at the facilities, a distribution network supplies ethanol-gasoline blends to fueling stations for use by drivers.

This page lists basic ethanol topics. To learn more, select one of the following links or select a step in the ethanol supply chain diagram.

What is Ethanol?

Ethanol (CH_3CH_2OH; also known as ethyl alcohol, grain alcohol, and EtOH) is a clear, colorless liquid. Its molecules contain a hydroxyl group (-OH) bonded to a carbon atom. To learn more, see the fuel properties table on the E85 Specifications page. Ethanol is made of the same chemical compound—and it is the same renewable biofuel—whether it is produced from starch- and sugar-based feedstocks such as corn grain (as it primarily is in the United States) and sugar cane (as it primarily is in Brazil) or from cellulosic feedstocks.

Making ethanol from cellulosic feedstocks—such as grass, wood, crop residues, or old newspapers—is more challenging than using starch or sugars. These materials must first be broken down into their component sugars for subsequent fermentation to ethanol in a process called biochemical conversion. Cellulosic feedstocks also can be converted into ethanol using heat and chemicals in a process called thermochemical conversion. Cellulosic ethanol conversion processes are a major focus of U.S. Department of Energy research.

Ethanol works well in internal combustion engines. In fact, Henry Ford and other early automakers thought ethanol would be the world's primary fuel before gasoline became so readily available. A gallon of pure ethanol (E100) contains 34% less energy than a gallon of gasoline.

Ethanol is a high-octane fuel. Octane helps prevent engine knocking and is extremely important in engines designed to operate at a higher compression ratio, so they generate more power. These engines tend to be found in high-performance vehicles. Low-level blends of ethanol, such as E10 (10% ethanol, 90% gasoline), generally have a higher octane rating than unleaded gasoline. Low-octane gasoline can be blended with 10% ethanol to attain the standard 87 octane requirement. Ethanol is the main component in E85, a high-level blend of 85% ethanol and 15% gasoline.

Ethanol Benefits

Ethanol is a renewable, largely domestic transportation fuel. Whether used in low-level blends, such as E10 (10% ethanol, 90% gasoline), or in E85 (85% ethanol, 15% gasoline), ethanol helps reduce imported oil and greenhouse gas emissions. Its use also supports the U.S. agricultural sector. This page describes the benefits of ethanol. For additional details, visit the U.S. Department of Energy Biomass Program's Benefits page.

Increasing Energy Security

About two-thirds of U.S. petroleum demand is in the transportation sector. Sixty percent of U.S. petroleum is currently imported. Depending heavily on foreign petroleum supplies puts the United States at risk for trade deficits, supply disruption, and price changes. Ethanol, on the other hand, is almost entirely produced from domestic crops today. Its use, and that of other alternative fuels, can displace a significant amount of imported petroleum.

Fueling the Economy

Ethanol production is a new industry that is creating jobs in rural areas where employment opportunities are needed. The Renewable Fuels Association's Ethanol Industry Outlook 2009 Report (PDF 4 MB) calculated that in 2008 the ethanol industry added more than $65 billion to gross domestic product and supported the creation of more than 494,000 jobs. A recent report claims there is an economic return on investment of nearly five to one for each dollar spent in the form of the federal tax incentive for ethanol use (PDF 83 KB). Download Adobe Reader. For additional information on the economic benefits of ethanol, see Local Economic Impact of Ethanol Plants (Excel 21 KB) and U.S. Ethanol Plant Ownership and Capacity (Excel 22 KB).

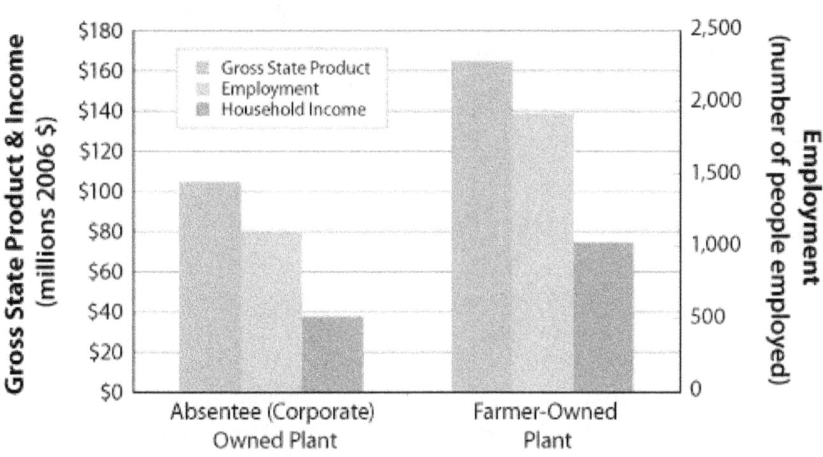

Source: National Corn Growers Association, 2006 (Excel 21 KB)

Reducing Greenhouse Gas Emissions

The carbon dioxide released when ethanol is burned is balanced by the carbon dioxide captured when the crops are grown to make ethanol. This differs from petroleum, which is made from plants that grew millions of years ago. According to Argonne National Laboratory, on a life-cycle analysis basis, corn-based ethanol production and use reduces greenhouse gas emissions (GHGs) by up to 52% compared to gasoline production and use. Cellulosic ethanol use could reduce GHGs by as much as 86%. For more information, see the Greenhouse Gas Emissions and Energy Balance sections.

To learn more about fuel economy, GHG scores, and air pollution scores for E85-capable flexible fuel vehicles (FFVs), visit the U.S. Department of Energy/U.S. Environmental Protection Agency's Fuel Economy Guide.

Providing Convenience, Availability, and Flexibility

Low-level ethanol blends such as E10 already constitute much of the gasoline sold in the United States. Low-level blends require no special fueling equipment and can be used in any gasoline-powered vehicle. E85 fueling equipment is slightly different than petroleum fueling equipment, but the costs are similar. In some cases, it is possible to convert existing petroleum equipment to handle E85. FFVs designed to run on E85 are becoming more common each model year, and FFVs are typically available as standard equipment with little or no incremental cost. Also, because FFVs can be fueled with gasoline as well as E85, drivers have the flexibility to travel outside of areas served by E85 fueling stations.

Protecting the Environment

Ethanol is biodegradable and, if spilled, poses much less of a threat than petroleum to surface and ground water. After the sinking of the Bow Mariner off the Virginia coast in February 2004, U.S. Coast Guard officials noted the cargo of 3.2 million gallons of industrial ethanol had dissipated quickly and did not pose an environmental threat to humans or marine life.

Ethanol Market Penetration

Domestically produced ethanol has the potential to displace a significant amount of U.S. petroleum consumption. A joint U.S. Department of Energy (DOE)/U.S. Department of Agriculture (USDA) study found that 1.3 billion tons of biomass—predominantly cellulosic feedstocks—could be produced for biofuel production in the United States annually with only modest changes in farming practices (see the report PDF 2.8 MB). Download Adobe Reader. This quantity of feedstocks could be used to produce enough biofuels—mostly ethanol—to satisfy about one third of current U.S. petroleum demand. The potential could be even larger if technology is developed to take advantage of additional forms of biomass such as algae.

The potential of biomass-derived fuels is key to U.S. petroleum displacement goals. The Energy Policy Act of 2005 established a nationwide renewable fuels standard requiring use of 7.5 billion gallons of renewable fuel by 2012. The Energy Independence and Security Act of 2007 boosted this renewable fuels standard substantially, requiring 36 billion gallons of annual renewable fuel use by 2022. Of this requirement, 21 billion gallons must be "advanced biofuels"—fuels that cut greenhouse gas emissions by at least 50%—including 16 billion gallons of cellulosic biofuels. Visit the Renewable Fuels Association Web site for more information about the Renewable Fuels Standard.

DOE and USDA are leading a multi-agency effort to achieve these ambitious goals. The National Biofuels Action Plan (PDF 4.9 MB) outlines the collaborative strategy.

Download Adobe Reader. For additional details, visit the DOE Biomass Program, USDA Energy, and Biomass Research & Development Initiative Web sites.

Ethanol primarily replaces gasoline as a fuel for passenger vehicles. The table below shows the penetration of ethanol, flexible-fuel vehicles (FFVs), and E85 stations into the gasoline market. As of 2007, about 3% of gasoline-vehicle fuel consumption—4.6 billion gasoline gallon equivalents (GGEs)—was attributed to ethanol, which was produced almost entirely from corn and consumed predominantly in the form of low-level blends.

Penetration of Ethanol, Flexible Fuel Vehicles, and E85 Fueling Stations into the U.S. Market

	Year	Total	Ethanol, FFV, E85	Market penetration
Fuel consumption (million gasoline gallon equivalents)	2007	136,932[1]	4,643[2]	3.4%
Passenger vehicles	2006	234,524,720[3]	5,494,258[4]	2.3%
Stations	2007	164,000[5]	1,208[6]	0.7%

[1] Total motor gasoline consumption (including ethanol blended into motor gasoline) from EIA Annual Energy Review 2007, Table 5.13c (preliminary), converted to gasoline gallon equivalents
[2] Ethanol consumption from EIA Annual Energy Review 2007, Table 10.3 (preliminary), converted to gasoline gallon equivalents; also see AFDC U.S. Total Production and Consumption of Ethanol (Excel 25 KB)
[3] Registered passenger cars and other 2-axle 4-tire vehicles from Bureau of Transportation Statistics National Transportation Statistics 2008, Table 1-11
[4] Flexible fuel vehicles from AFDC Light-Duty E85 FFVs in Use (Excel 22 KB)
[5] Approximate public retail gasoline station count from National Petroleum News; also see AFDC Public Retail Gasoline Stations by State and Year (Excel 25 KB)
[6] E85 stations from AFDC U.S. Alternative Fueling Stations by Fuel Type (Excel 28 KB)

The table above gives fuel, vehicle, and station characteristics for 2006 and 2007, the most recent years for which all the data are available. Ethanol, FFVs, and E85 stations have a higher market penetration today as their growth rates outpace the overall rates of growth in U.S. fuel consumption, passenger vehicles, and retail fueling stations (see graph below). Visit the DOE Biomass Program's Biofuels Data page for the latest values of several market penetration indicators.

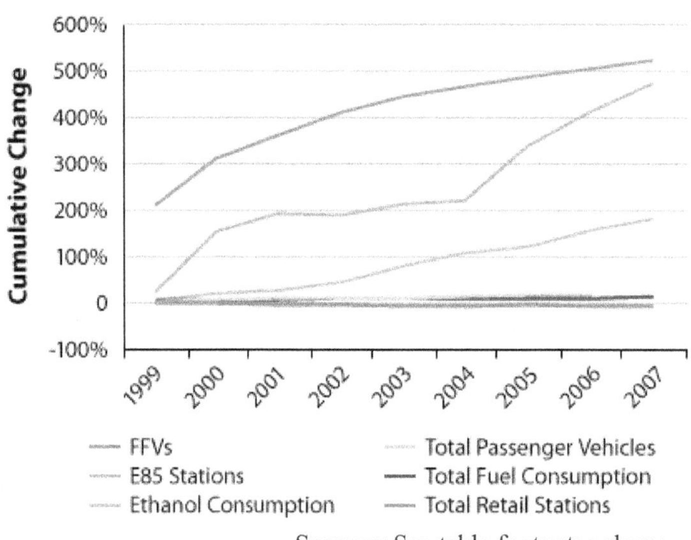

Sources: See table footnotes above

The AFDC's Data, Analysis & Trends section contains a wealth of information on the market penetration of ethanol and other alternative fuels, vehicles, and infrastructure. The following pages contain particularly useful ethanol-related information:

- Vehicles
- Fuels
- Infrastructure
- Biomass Resources
- Geographic

Also see the Energy Information Administration's Renewable & Alternative Fuels page for data related to ethanol and other alternative fuels. For more information about the potential of ethanol, see the AFDC's Benefits, Energy Balance, and Greenhouse Gas Emissions pages.

Ethanol Energy Balance

Ethanol - The complete Energy Lifecycle Picture Brochure (PDF 4 MB) Download Adobe Reader.

Ethanol in the United States is mainly produced from the starch in corn grain. Some studies have suggested that corn-based ethanol has a negative energy balance. However, a preponderance of recent studies using updated data about corn production methods demonstrates a positive energy balance for corn ethanol.

In addition, once the technology to produce cellulosic ethanol becomes widely available, the energy lifecycle balance of ethanol will improve. That's because it will be produced using less fossil fuel and more energy-efficient feedstocks, such as fast-growing trees, corn stover, grain straw, switchgrass, forest product residues, and municipal waste. Cellulosic ethanol also produces lower levels of greenhouse gas emissions. Learn how Ethanol's Lifecycle Energy Balance relates to emissions.

For more information on the energy balance of ethanol, see the U.S. Department of Energy Biomass Program's Ethanol Myths and Facts and download the following documents. Download Adobe Reader.

- Argonne National Laboratory's GREET Model
- DOE response to article, "Use of U.S. Croplands for Biofuels Increases Greenhouse Gases through Emissions from Land Use Change" (PDF 35 KB)
- Life-Cycle Energy Use and Greenhouse Gas Emission Implications of Brazilian Sugarcane Ethanol Simulated with the GREET Model, abstract (PDF 90 KB)
- Energy and Environmental Aspects of Using Corn Stover for Fuel Ethanol (PDF 1.7 MB)
- The 2001 Net Energy Balance of Corn Ethanol (PDF 30 KB)
- The Energy Balance of Corn Ethanol: An Update (PDF 169 KB)

Ethanol Greenhouse Gas Emissions

Using ethanol as a vehicle fuel provides local and global benefits—reducing emissions of harmful pollutants and greenhouse gases.

Highlights of GAO-09-446, a report to congressional requesters

August 2009

BIOFUELS

Potential Effects and Challenges of Required Increases in Production and Use

Why GAO Did This Study

In December 2007, the Congress expanded the renewable fuel standard (RFS), which requires rising use of ethanol and other biofuels, from 9 billion gallons in 2008 to 36 billion gallons in 2022. To meet the RFS, the Departments of Agriculture (USDA) and Energy (DOE) are developing advanced biofuels that use cellulosic feedstocks, such as corn stover and switchgrass. The Environmental Protection Agency (EPA) administers the RFS.

This report examines, among other things, (1) the effects of increased biofuels production on U.S. agriculture, environment, and greenhouse gas emissions; (2) federal support for domestic biofuels production; and (3) key challenges in meeting the RFS. GAO extensively reviewed scientific studies, interviewed experts and agency officials, and visited five DOE and USDA laboratories.

What GAO Recommends

GAO suggests that the Congress consider requiring EPA to develop a strategy to assess lifecycle environmental effects of increased biofuels production and whether revisions are needed to the VEETC. GAO also recommends that EPA, DOE, and USDA develop a coordinated approach for addressing uncertainties in lifecycle greenhouse gas analysis and give priority to R&D that addresses future blend wall issues. DOE, USDA, and EPA generally agreed with the recommendations.

View GAO-09-446 or key components. For more information, contact Patricia Dalton at (202) 512-3841 or daltonp@gao.gov.

What GAO Found

To meet the RFS, domestic biofuels production must increase significantly, with uncertain effects for agriculture and the environment. For agriculture, many experts said that biofuels production has contributed to crop price increases as well as increases in prices of livestock and poultry feed and, to a lesser extent, food. They believe that this trend may continue as the RFS expands. For the environment, many experts believe that increased biofuels production could impair water quality—by increasing fertilizer runoff and soil erosion—and also reduce water availability, degrade air and soil quality, and adversely affect wildlife habitat; however, the extent of these effects is uncertain and could be mitigated by such factors as improved crop yields, feedstock selection, use of conservation techniques, and improvements in biorefinery processing. Except for lifecycle greenhouse gas emissions, EPA is currently not required by statute to assess environmental effects to determine what biofuels are eligible for inclusion in the RFS. Many researchers told GAO there is general agreement on the approach for measuring the direct effects of biofuels production on lifecycle greenhouse gas emissions but disagreement about how to estimate the indirect effects on global land use change, which EPA is required to assess in determining RFS compliance. In particular, researchers disagree about what nonagricultural lands will be converted to sustain world food production to replace land used to grow biofuels crops.

The Volumetric Ethanol Excise Tax Credit (VEETC), a 45-cent per gallon federal tax credit, was established to support the domestic ethanol industry. Unless crude oil prices rise significantly, the VEETC is not expected to stimulate ethanol consumption beyond the level the RFS specifies this year. The VEETC also may no longer be needed to stimulate conventional corn ethanol production because the domestic industry has matured, its processing is well understood, and its capacity is already near the effective RFS limit of 15 billion gallons per year for conventional ethanol. A separate $1.01 tax credit is available for producing advanced cellulosic biofuels.

The nation faces several key challenges in expanding biofuels production to achieve the RFS's 36-billion-gallon requirement in 2022. For example, farmers face risks in transitioning to cellulosic biofuels production and are uncertain whether growing switchgrass will eventually be profitable. USDA's new Biomass Crop Assistance Program may help mitigate these risks by providing payments to farmers through multi-year contracts. In addition, U.S. ethanol use is approaching the so-called blend wall—the amount of ethanol that most U.S. vehicles can use, given EPA's 10 percent limit on the ethanol content in gasoline. Research has been initiated on the long-term effects of using 15 percent or 20 percent ethanol blends, but expanding the use of 85 percent ethanol blends will require substantial new investment because ethanol is too corrosive for the petroleum distribution infrastructure and most vehicles. Alternatively, further R&D on biorefinery processing technologies might lead to price-competitive biofuels that are compatible with the existing petroleum distribution and storage infrastructure and the current fleet of U.S. vehicles.

United States Government Accountability Office

Order Code RL33928

CRS Report for Congress

Ethanol and Biofuels: Agriculture, Infrastructure, and Market Constraints Related to Expanded Production

March 16, 2007

Brent D. Yacobucci
Specialist in Energy Policy
Resources, Science, and Industry Division

Randy Schnepf
Specialist in Agricultural Policy
Resources, Science, and Industry Division

Prepared for Members and Committees of Congress

Ethanol and Biofuels: Agriculture, Infrastructure, and Market Constraints Related to Expanded Production

Summary

High petroleum and gasoline prices, concerns over global climate change, and the desire to promote domestic rural economies have greatly increased interest in biofuels as an alternative to petroleum in the U.S. transportation sector. Biofuels, most notably corn-based ethanol, have grown significantly in the past few years as a component of U.S. motor fuel supply. Ethanol, the most commonly used biofuel, is blended in nearly half of all U.S. gasoline (at the 10% level or lower in most cases). However, current biofuel supply represents less than 4% of total gasoline demand.

While recent proposals have set the goal of significantly expanding biofuel supply in the coming decades, questions remain about the ability of the U.S. biofuel industry to meet rapidly increasing demand. Current U.S. biofuel supply relies almost exclusively on ethanol produced from Midwest corn. In 2006, 17% of the U.S. corn crop was used for ethanol production. To meet some of the higher ethanol production goals would require more corn than the United States currently produces, if all of the envisioned ethanol was made from corn.

Due to the concerns with significant expansion in corn-based ethanol supply, interest has grown in expanding the market for biodiesel produced from soybeans and other oil crops. However, a significant increase in U.S. biofuels would likely require a movement away from food and grain crops. Other biofuel feedstock sources, including cellulosic biomass, are promising, but technological barriers make their future uncertain.

Issues facing the U.S. biofuels industry include potential agricultural "feedstock" supplies, and the associated market and environmental effects of a major shift in U.S. agricultural production; the energy supply needed to grow feedstocks and process them into fuel; and barriers to expanded infrastructure needed to deliver more and more biofuels to the market. This report outlines some of the current supply issues facing biofuels industries, including the limitations on agricultural feedstocks, infrastructure constraints, energy supply for biofuel production, and fuel price uncertainties.

Contents

List of Tables

Ethanol and Biofuels: Agriculture, Infrastructure, and Market Constraints Related to Expanded Production

Introduction

High petroleum and gasoline prices, concerns over global climate change, and the desire to promote domestic rural economies have raised interest in biofuels as an alternative to petroleum in the U.S. transportation sector. Biofuels, most notably corn-based ethanol, have grown significantly in the past few years as a component of U.S. motor fuels. For example, nearly half of all U.S. gasoline contains some ethanol (mostly blended at the 10% level or lower). However, current supply represents only about 3.6% of annual gasoline demand on a volume basis, and only about 2.4% on an energy basis. In 2006, the United States consumed roughly 5 billion gallons of biofuels (mostly ethanol); this 5 billion gallons was blended into roughly 65 billion gallons of gasoline. Total annual gasoline consumption is roughly 140 billion gallons.

Recent proposals, including President Bush's goal in his 2007 State of the Union Address, aim to significantly expand biofuel supply in the coming decades. The President's goal would be to expand consumption from 5 billion gallons in 2007 to 35 billion gallons in 2017. While this proposal included not just biofuels but alternative fuels (including fuels from coal or natural gas), it would likely mean a significant growth in biofuels production over the next 10 years. Other legislative proposals would require significant expansion of biofuels production in the coming decades; some proposals would require 30 billion gallons of biofuels alone by 2030 or 60 billion gallons by 2050.

Current U.S. biofuel supply relies almost exclusively on ethanol produced from Midwest corn. Other fuels that play a smaller role include ethanol from Brazilian sugar, biodiesel from U.S. soybeans, and ethanol from U.S. sorghum. A significant increase in U.S. biofuels would likely require a movement away from food and grain crops. For example, U.S. ethanol production in 2006 consumed roughly 17% of the U.S. corn crop. If only corn is used, expanding ethanol production to 35 billion gallons would require more corn than the United States currently produces, which would be infeasible. Corn (and other grains) have myriad other uses, and such a shift would have drastic consequences for most agricultural markets, including grains (since corn would compete with other grains for land), livestock (since the cost of animal feed would likely increase), and land (since total harvested acreage would likely increase). In addition to agricultural effects, such an increase in corn-based ethanol would likely affect fuel costs (since biofuels tend to be more expensive than petroleum fuels), energy supply (natural gas is a key input into corn production), and the environment (since the expansion of corn-based ethanol production raises many environmental questions). These constraints are discussed below

Due to the concerns with significant expansion in corn-based ethanol supply, interest has grown in expanding the market for biodiesel (a diesel substitute produced from vegetable and animal oils) and spurring the development of motor fuels produced from cellulosic materials (including grasses, trees, and agricultural and municipal wastes). However, all of these technologies are currently even more expensive than corn-based ethanol.

In addition to expanding domestic production of biofuels, there is some interest in expanding imports of sugar-based ethanol from Brazil and other countries. However, ethanol from Brazil is currently subject to a 54 cent-per-gallon tariff that in most years is a significant barrier to direct Brazilian imports. Some Brazilian ethanol can be brought into the United States duty free if it is dehydrated (reprocessed) in Caribbean Basin Initiative (CBI) countries. Up to 7% of the U.S. ethanol market could be supplied duty-free in this fashion, although historically ethanol dehydrated in CBI countries has only represented about 2% of the total U.S. market.

Any fuel produced from biological materials (e.g., food crops, agricultural residues, municipal waste) is generally referred to as a "biofuel." More specifically, the term generally refers to liquid transportation fuels. The most significant biofuel in the United States is ethanol produced from U.S. corn.[1] Approximately 4.9 billion gallons of ethanol were produced in the United States in 2006, mostly from corn. Other domestic feedstocks for ethanol include grain sorghum and sweet sorghum; imported ethanol (600 million gallons in 2006) is usually produced from sugar cane in Brazil. Ethanol is generally blended into gasoline at the 10% level or lower (E10). Ethanol can be used in purer forms such as E85 (85% ethanol and 15% gasoline) in vehicles specially designed for its use, although E85 represents less than 1% of U.S. ethanol consumption.

After ethanol, biodiesel is the next most significant biofuel in the United States, although 2006 U.S. production is estimated at only about 100 million gallons. Biodiesel is a diesel fuel substitute produced from vegetable and animal oils (mainly soybean oil in the United States), as well as recycled cooking grease. Other biofuels with the potential to play a role in the U.S. market include ethanol and diesel fuel substitutes produced from various biomass feedstocks containing cellulose. However, these "cellulosic" biofuels are currently prohibitively expensive relative to conventional ethanol and biodiesel. Other potential biofuels include other alcohols (e.g., methanol and butanol produced from biomass).

This report outlines some of the current supply issues facing biofuels industries, including supply limitations of agricultural feedstocks, infrastructure limitations, energy supply for fuel conversion, and fuel prices.

---

[1] For more information on ethanol, see CRS Report RL33290, *Fuel Ethanol: Background and Public Policy Issues*, by Brent D. Yacobucci.

Table 1. U.S. Production of Biofuels from Various Feedstocks

Fuel	Feedstock	U.S. Production in 2006
Ethanol	Corn	4.9 billion gallons
	Sorghum	less than 100 million gallons
	Cane sugar	No production (600 million gallons imported from Brazil and Caribbean countries)
	Cellulose	No production (one demonstration plant in Canada)
Biodiesel	Soybean oil	approximately 90 million gallons
	Other vegetable oils	less than 10 million gallons
	Recycled grease	less than 10 million gallons
	Cellulose	No production
Methanol	Cellulose	No production
Butanol	Cellulose, other biomass	No production

Sources: Renewable Fuels Association; National Biodiesel Board; CRS analysis.

Issues with Corn-Based Ethanol Supply

Overview of Long-Run Corn Ethanol Supply Issues

The U.S. ethanol industry has shown rapid growth in recent years with national production increasing from 1.8 billion gallons in 2001 to 4.9 billion in 2006. This rapid growth — which is projected to continue for the foreseeable future — has important consequences for U.S. and international fuel, feed, and food markets.

Corn accounts for about 98% of the feedstocks used in ethanol production in the United States. USDA estimates that 2.15 billion bushels of corn (or 20% of the 2006 corn crop) will be used to produce ethanol during the September 2006 to August 2007 corn marketing year.[2] As of February 2, 2007, existing U.S. ethanol plant capacity was a reported 5.6 billion gallons per year (BGPY), with an additional capacity of 6.1 BGPY under construction.[3] Thus, total annual U.S. ethanol production capacity in existence or under construction as of February 2, 2007, was 11.7 billion gallons. This production capacity is well in excess of the 7.5 billion gallon supply required in 2012 by the Renewable Fuel Standard (Energy Policy Act

[2] USDA, WAOB, *World Agricultural Supply and Demand Estimates (WASDE) Report*, Jan. 12, 2007, Washington; available at [http://www.usda.gov/oce/].

[3] See Renewable Fuels Association, *Industry Statistics*, at [http://www.ethanolrfa.org/industry/statistics/].

of 2005 [P.L. 109-58]). The current pace of plant construction suggests that annual corn-for-ethanol use will likely require more than 3 billion bushels in 2007 and approach, or possibly exceed, 4 billion bushels in 2008.

The ethanol-driven surge in corn demand has fueled a sharp rise in corn prices. For example, the futures contract for March 2007 corn on the Chicago Board of Trade, rose from $2.50 per bushel in September 2006 to a contract high of over $4.16 per bushel in January 2007 (a rise of 66%). This sharp rise in corn prices owes its origins largely to increasing corn demand spurred by the rapid expansion of corn-based ethanol production capacity in the United States since mid-2006. The rapid growth in ethanol capacity has been fueled by both strong energy prices and a variety of government incentives, regulations, and programs. Major federal incentives include a tax credit of $0.51 to fuel blenders for every gallon of ethanol blended with gasoline; a Renewable Fuel Standard (RFS) that mandates a renewable fuels blending requirement for gasoline suppliers that grows annually from 4 billion gallons in 2006 to 7.5 billion gallons in 2012; and a 54¢ per gallon most-favored-nation duty on most imported ethanol.[4] A recent survey of federal and state government subsidies in support of ethanol production reported that the total annual federal support fell somewhere in the range of $5.1 to $6.8 billion per year.[5]

Market participants, economists, and biofuels skeptics have begun to question the need for continued large federal incentives in support of ethanol production, particularly when the sector would have been profitable during much of 2006 without such subsidies.[6] Their concerns focus on the potential for widespread unintended consequences that might result from excessive federal incentives adding to the rapid expansion of ethanol production capacity and the demand for corn to feed future ethanol production. These questions extend to issues concerning the ability of the gasoline-marketing infrastructure to accommodate more ethanol in fuel, the likelihood of modifications in engine design, the environmental impacts, and other considerations.

Agricultural Issues

Rapidly expanding corn-based ethanol production could have significant consequences for traditional U.S. agricultural crop production. As corn prices rise, so too does the incentive to expand corn production either by expanding onto more marginal soil environments or by altering the traditional corn-soybean rotation that dominates Corn Belt agriculture. This would crowd out other field crops, primarily soybeans, and other agricultural activities. Large-scale shifts in agricultural production activities will likely have important regional economic consequences that

---

[4] For more information on incentives (both tax and non-tax) for ethanol, see CRS Report RL33572, *Biofuels Incentives: A Summary of Federal Programs*, by Brent D. Yacobucci.

[5] Koplow, Doug. *Biofuels — At What Cost? Government Support for Ethanol and Biodiesel in the United States*, Global Subsidies Initiative of the International Institute for Sustainable Development, Geneva, Switzerland, October 2006; available at [http://www.globalsubsidies.org].

[6] Chris Hurt, Wally Tyner, and Otto Doering, Department of Agricultural Economics, Purdue University, *Economics of Ethanol*, December 2006, West Lafayette, IN.

have yet to be fully explored or understood. Further, corn production is among the most energy-intensive of the major field crops. An expansion of corn area would likely have important and unwanted environmental consequences due to the resulting increase in fertilizer and chemical use and soil erosion. The National Corn Growers Association estimates that U.S. corn-based ethanol production could expand to between 12.8 and 17.8 billion gallons by 2015 without significantly affecting agricultural markets.[7] However, as noted above, other evidence suggests effects are already being felt in the current run-up in production.

Feed Markets. Prolonged higher corn prices could have significant consequences for traditional feed markets and the livestock industries that depend on those feed markets. Corn has traditionally represented about 57% of feed concentrates and processed feedstuffs fed to animals in the United States.[8] As corn-based ethanol production increases, so do total corn demand and corn prices. Dedicating an increasing share of the U.S. corn harvest to ethanol production will likely lead to higher prices for all grains and oilseeds that compete for the same land, resulting in higher feed costs for cattle, hog, and poultry producers. In addition, supply distortions are likely to develop in protein-meal markets related to expanding production of the ethanol processing by-product Distiller's Dried Grains (DDG), which averages about 30% protein content and can substitute in certain feed and meal markets.[9] While DDG use would substitute for some of the lost feed value of corn used in ethanol processing, about 66% of the original weight of corn is consumed in producing ethanol and is no longer available for feed. Furthermore, not all livestock species are well adapted to dramatically increased consumption of DDG in their rations — dairy cattle appear to be best suited to expanding DDG's share in feed rations; poultry and pork are much less able to adapt. Also, DDG must be dried before it can be transported long distances, adding to feed costs. There may be some potential for large-scale livestock producers to relocate near new feed sources, but such relocation would likely have important regional economic effects.

Exports. The United States is the world's leading producer and exporter of corn. Increased use of corn for ethanol production could diminish U.S. capacity for exports. Since 1980 U.S. corn production has accounted for over 40% of world production, while U.S. corn exports have represented nearly a 66% share of world corn trade during the past decade. In the 2006/2007 marketing year, the United States is expected to export about 21% of its corn production.[10] Higher corn prices would likely result in lost export sales. It is unclear what type of market adjustments would occur in global feed markets, since several different grains and feedstuffs are

---

[7] National Corn Growers Association, *How Much Ethanol Can Come From Corn?*, November 9, 2006, Washington.

[8] USDA, ERS, *Feed Situation and Outlook Yearbook*, FDS-2003, Apr. 2003, Washington.

[9] For a discussion of potential feed market effects due to growing ethanol production, see Bob Kohlmeyer, "The Other Side of Ethanol's Bonanza," *Ag Perspectives* (World Perspectives, Inc.), Dec. 14, 2004; and R. Wisner and P. Baumel, "Ethanol, Exports, and Livestock: Will There be Enough Corn to Supply Future Needs?," *Feedstuffs*, no. 30, vol. 76, July 26, 2004.

[10] USDA, WAOB, *WASDE Report*, Jan. 12, 2007; available at [http://www.usda.gov/oce/].

relatively close substitutes. Price-sensitive corn importers may quickly switch to alternative, cheaper sources of feed, depending on the availability of supplies and the adaptability of animal rations. In contrast, less price-sensitive corn importers, such as Japan and Taiwan, may choose to pay a higher price in an attempt to bid the corn away from ethanol plants. There could be significant economic effects to U.S. grain companies and to the U.S. agricultural sector if ethanol-induced higher corn prices caused a sustained reshaping of international grain trade.

Food vs. Fuel. A sustained rise in grain prices driven by ethanol feedstock demand could lead to higher U.S. and world food prices. Most corn grown in the United States is used for animal feed. Higher feed costs ultimately lead to higher meat prices. The feed-price effect will first translate into higher prices for poultry and hogs which are less able to use alternate feedstuffs. Dairy and beef cattle are more versatile in their ability to shift to alternate feed sources, but eventually a sustained rise in corn prices will push their feed costs upward as well. The price of corn is also linked to the price of other grains, including those destined for food markets, through competition in the feed marketplace and in the producer's planting choices for limited acreage. The price run-up in the U.S. corn market has already spilled over into price increases in the markets for soybeans and soybean oil.

Since food costs represent a relatively small share of consumer spending for most U.S. households, the price run-up is relatively easily absorbed in the short run. However, the situation is very different for lower-income households, as well as in many foreign markets, where food expenses can represent a larger portion of the household budget. Due to trade linkages, the increase in U.S. corn prices has become a concern for international markets, as well. In January, Mexico experienced riots following a nearly 30% price increase for tortillas, the country's dietary staple. In China, where corn is also an important food source, the government has recently put a halt to its planned ethanol plant expansion due to the threat it poses to the country's food security. Similarly, humanitarian groups have expressed concern for the potential difficulties that higher grain prices imply for developing countries that are net food importers.

Energy Supply Issues

Ethanol is not a primary energy source. Energy stored in biological material (through photosynthesis) must be converted into a more useful, portable fuel. This conversion requires energy. The amount and types of energy used to produce ethanol, and the feedstocks for ethanol production, are of key concern. Because of the input energy requirements, the energy and environmental benefits of corn ethanol may be limited.

Energy Balance. A frequent argument for the use of ethanol as a motor fuel is that it reduces U.S. reliance on oil imports, making the U.S. less vulnerable to a fuel embargo of the sort that occurred in the 1970s. However, while corn ethanol use displaces petroleum, its overall effect on total energy consumption is less clear. To analyze the net energy consumption of ethanol, the entire fuel cycle must be considered. The fuel cycle consists of all inputs and processes involved in the development, delivery and final use of the fuel. For corn-based ethanol, these inputs include the energy needed to produce fertilizers, operate farm equipment, transport corn, convert corn to ethanol, and distribute the final product. Some studies find a significant positive energy balance of 1.5 or greater — in other words, the energy contained in a gallon of corn ethanol is 50% higher than the amount of energy needed to produce and distribute it. However, other studies suggest that the amount of energy needed to produce ethanol is roughly equal to the amount of energy obtained from its combustion. A review of research studies on ethanol's energy balance and greenhouse gas emissions found that most studies give corn-based ethanol a slight positive energy balance of about 1.2.[11]

Natural Gas Demand. As ethanol production increases, the energy needed to process the corn into ethanol, which is derived primarily from natural gas in the United States, can be expected to increase. For example, if the entire 4.9 billion gallons of ethanol produced in 2006 used natural gas as a processing fuel, it would have required an estimated 240 to 290 billion cubic feet (cu. ft.) of natural gas.[12] If the entire 2006 corn crop of 10.5 billion bushels were converted into ethanol, the energy requirements would be equivalent to approximately 1.4 to 1.7 trillion cu. ft. of natural gas. This would have represented about 6% to 8% of total U.S. natural gas consumption, which was an estimated 22.2 trillion cu. ft. in 2005.[13] The United States has been a net importer of natural gas since the early 1980s. Because natural gas is used extensively in electricity production in the United States, a significant increase in its use as a processing fuel in the production of ethanol would likely increase prices and imports of natural gas.

Energy Security. Despite the fact that ethanol displaces gasoline, the benefits to energy security from corn-based ethanol are not certain. As was stated above, while roughly 20% of the U.S. corn crop is used for ethanol, ethanol only accounts for approximately 2.4% of gasoline consumption on an energy equivalent basis.[14] The import share of U.S. petroleum consumption was estimated at 54% in 2004, and is expected to grow to 70% by 2025.[15] Further, as long as ethanol remains dependent on the U.S. corn supply, any threats to this supply (such as drought), or increases in corn prices, would negatively affect the supply and/or cost of ethanol. In fact, that happened when high corn prices caused by strong export demand in 1995 contributed to an 18% decline in ethanol production between 1995 and 1996.

---

[11] Alexander E. Farrell, Richard J. Plevin, Brian T. Turner, Andrew D. Jones, Michael O'Hare, and Daniel M. Kammen, "Ethanol Can Contribute to Energy and Environmental Goals," *Science*, Jan. 27, 2006, pp. 506-508.

[12] CRS calculations based on energy usage rates of 49,733 Btu/gal of ethanol from Shapouri (2004), roughly 60,000 Btu/gal from Farrell (2006). Hosein Shapouri and Andrew McAloon, USDA, Office of the Chief Economist, *The 2001 Net Energy Balance of Corn-Ethanol*, 2004, Washington; Farrell, op. cit.

[13] U.S. Department of Energy (DOE), Energy Information Administration (EIA), *Annual Energy Outlook 2006 with Projections to 2030*, Table 1-Total Energy Supply and Disposition Summary, Washington; at [http://www.eia.doe.gov/oiaf/aeo/index.html].

[14] By volume, ethanol accounted for approximately 3.6% of gasoline consumption in the United States in 2006.

[15] DOE, EIA, *Annual Energy Outlook 2004 with Projections to 2025*, Washington.

Further, expanding corn-based ethanol production to levels needed to significantly promote U.S. energy security is likely to be infeasible. If the entire 2006 U.S. corn crop of 10.5 billion bushels were used as ethanol feedstock, the resultant 28 billion gallons of ethanol (18.9 billion gasoline-equivalent gallons (GEG)) would represent about 13.4% of estimated national gasoline use of approximately 141 billion gallons.[16] In 2006, an estimated 71 million acres of corn were harvested. Nearly 137 million acres would be needed to produce enough corn (20.5 billion bushels) and resulting ethanol (56.4 billion gallons or 37.8 billion GEG) to substitute for roughly 20% of petroleum imports.[17] Since 1950, the U.S. corn-harvested acreage has never reached 76 million acres. Thus, barring a drastic realignment of U.S. field crop production patterns, corn-based ethanol's potential as a petroleum import substitute appears to be limited by crop area constraints, among other factors.[18]

Infrastructure and Distribution Issues

In addition to the above concerns about raw material supply for ethanol production (both corn and energy), there are additional issues involving ethanol distribution and infrastructure. Expanding ethanol production will likely strain existing supply infrastructure. Further, expansion of ethanol use beyond certain levels will require investment in entirely new infrastructure that would be necessary to handle a higher and higher percentage of ethanol in gasoline.

Distribution Issues. Ethanol-blended gasoline tends to separate in pipelines. Further, ethanol is corrosive and may damage existing pipelines. Therefore, unlike petroleum products, ethanol and ethanol blended gasoline cannot be shipped by pipeline in the United States. Another issue with pipeline transportation is that corn ethanol must be moved from rural areas in the Midwest to more populated areas, which are often located along the coasts. This shipment is in the opposite direction of existing pipeline transportation, which moves gasoline from refiners along the coast to other coastal cities and into the interior of the country. While some studies have concluded that shipping ethanol or ethanol-blended gasoline via pipeline could be feasible, no major U.S. pipeline has made the investments to allow such shipments.[19]

---

[16] Based on USDA's Jan. 12, 2007, *World Agricultural Supply and Demand Estimates (WASDE) Report*, and using comparable conversion rates.

[17] This represents roughly half of gasoline's share of imported petroleum. However, petroleum imports are primarily unrefined crude oil, which is then refined into a variety of products. CRS calculations assume corn yields of 150 bushel per acre and an ethanol yield of 2.75 gal/bu.

[18] Two recent articles by economists at Iowa State University examine the potential for obtaining a 10 million acre expansion in corn planting: Bruce Babcock and D. A. Hennessy, "Getting More Corn Acres From the Corn Belt"; and Chad E. Hart, "Feeding the Ethanol Boom: Where Will the Corn Come From?" *Iowa Ag Review*, Vol. 12, No. 4, Fall 2006.

[19] Some small, proprietary ethanol pipelines do exist. American Petroleum Institute, *Shipping Ethanol Through Pipelines*. Available at [http://www.api.org/aboutoilgas/sectors /pipeline/upload/pipelineethanolshipment-2.doc]

CRS-9

Thus, the current distribution system for ethanol is dependent on rail cars, tanker trucks, and barges. These deliver ethanol to fuel terminals where it is blended with gasoline before shipment via tanker truck to gasoline retailers. However, these transport modes lead to higher prices than pipeline transport, and the supply of current shipping options (especially rail cars) is limited. For example, according to industry estimates, the number of ethanol carloads has tripled between 2001 and 2006, and the number is expected to increase by another 30% in 2007.[20] A significant increase in corn-based ethanol production would further strain this tight transport situation.

Because of these distribution issues, some pipeline operators are seeking ways to make their systems compatible with ethanol or ethanol-blended gasoline. These modifications could include coating the interior of pipelines with epoxy or some other, corrosion-resistant material. Another potential strategy could be to replace all susceptible pipeline components with newer, hardier components. However, even if such modifications are technically possible, they will likely be expensive, and could further increase ethanol transportation costs.

Higher-Level Ethanol Blends. One key benefit of gasoline-ethanol blends up to 10% ethanol is that they are compatible with existing vehicles and infrastructure (e.g., fuel tanks, retail pumps, etc.). All automakers that produce cars and light trucks for the U.S. market warranty their vehicles to run on gasoline with up to 10% ethanol (E10). As a major producer of ethanol for its domestic market, Brazil has a mandate that all of its gasoline contain 20-25% ethanol. For the United States to move to E20 (20% ethanol, 80% gasoline), it may be that few (if any) modifications would need to be made to existing vehicles and infrastructure. Vehicle testing, however, would be necessary to determine whether new vehicle parts would be required, or if existing vehicles are compatible with E20. Similar testing would be necessary for terminal tanks, tanker trucks, retail tanks, pumps, etc.

There is also interest in expanding the use of E85 (85% ethanol, 15% gasoline). Current E85 consumption represents only approximately 1% of ethanol consumption in the United States. A key reason for the relatively low consumption of E85 is that relatively few vehicles operate on E85. The National Ethanol Vehicle Coalition estimates that there are approximately six million E85-capable vehicles on U.S. roads,[21] as compared to approximately 230 million gasoline- and diesel-fueled vehicles.[22] Most E85-capable vehicles are "flexible fuel vehicles" or FFVs. An FFV can operate on any mixture of gasoline and between 0% and 85% ethanol. However, owners of a large majority of the FFVs on U.S. roads choose to fuel them exclusively

---

[20] Ilan Brat and Daniel Machalaba, "Can Ethanol Get a Ticket to Ride?," *The Wall Street Journal*, Feb. 1, 2007, p. B1.

[21] National Ethanol Vehicle Coalition, *Frequently Asked Questions*, accessed February 3, 2006 [http://www.e85fuel.com/e85101/faq.php].

[22] Federal Highway Administration, *Highway Statistics 2003*, November 2004, Washington.

CRS-10

with gasoline, largely due to higher per-mile fuel cost[23] and lower availability of E85.

However, E85 capacity is expanding rapidly, with the number of E85 stations roughly doubling between February 2006 and February 2007. But those stations still represent less than 1% of U.S. gasoline retailers. Further expansion will require significant investments, especially at the retail level. If a new E85 pump and underground tank are necessary, they can cost as much as $100,000 to $200,000 to install.[24] However, if existing equipment can be used with little modification, the cost could be less than $10,000.

Sugar Ethanol

Excluding feedstock costs, producing ethanol from sugar cane can be less costly than producing it from corn. This is because the starch in corn must first be broken down into sugar, before it can be fermented, adding costly processing. Further, sugar cane waste (bagasse) can be burned to provide process energy for the ethanol plant, reducing associated energy costs, and improving sugar ethanol's energy balance relative to corn ethanol. Using sugar cane, Brazil produces nearly as much ethanol as the United States, but its passenger vehicle fuel demand is roughly 90% lower.

Brazil's success at integrating sugar ethanol into the fuel supply has stimulated interest in adopting Brazilian practices in the United States. However, differences in market prices for sugar in the two countries cause the economics of sugar-based ethanol to differ significantly. USDA concluded that producing sugar cane ethanol in the United States would be more than twice as costly as U.S. corn ethanol and nearly three times as costly as Brazilian sugar ethanol.[25] Feedstock costs accounted for most of this price differential. Therefore, the USDA study shows that while sugar ethanol may be a positive energy strategy in other countries, it may not be economical in the United States.

Because of the above cost advantage, it is expected that any sugar ethanol used in the United States will continue to be Brazilian ethanol reprocessed in Caribbean Basin Initiative (CBI) countries or imported directly from Brazil. Up to 7% of the U.S. ethanol market can be met duty-free with Brazilian ethanol dehydrated in CBI countries.[26] While only about 2% of the U.S. market historically has been supplied in this manner, there is growing interest among investors in expanding CBI dehydration capacity to approach the 7% quota. Further, high ethanol prices in recent

---

[23] Ethanol has a lower energy content than gasoline per gallon. Therefore, FFVs tend to have lower fuel economy when operating on E85. For the use of E85 to be economical, the pump price for E85 must be low enough to make up for the decreased fuel economy relative to gasoline. Generally, to have equivalent per-mile costs, E85 must cost 20% to 30% less per gallon at the pump than gasoline.

[24] David Sedgwick, *Automotive News*, January 29, 2007. p. 112.

[25] USDA, *The Economic Feasibility of Ethanol Production from Sugar in the United States*, July 2006, Washington.

[26] For more information, see CRS Report RS21930, *Ethanol Imports and the Caribbean Basin Initiative*, by Brent D. Yacobucci.

years (mostly brought on by the elimination of MTBE, another gasoline blending component) have led to a significant increase in direct imports from Brazil despite the tariff.

Biodiesel

Biodiesel is a diesel fuel substitute produced from agricultural products. While the term is generally used to refer to fuel produced from vegetable oils (such as soybean oil or palm oil), it can also be produced from animal fats and recycled cooking grease.[27] Further, research is underway on gasification and other processes to convert cellulose and biomass waste products into synthetic diesel fuel. Biodiesel has the key advantage of being largely compatible with existing infrastructure and vehicles. Most diesel engines can run on low-percentage blends of biodiesel and conventional diesel, and most new diesel engines can likely tolerate significantly higher percentage blends.[28] Further, although there is little experience with transporting biodiesel in pipelines, the fuel may be more compatible with existing infrastructure than ethanol.

Currently, U.S. biodiesel production and consumption are significantly lower than ethanol: approximately 100 million gallons of biodiesel were produced in 2006, compared to roughly 5 billion gallons of ethanol. However, in relative terms, biodiesel production has been expanding rapidly, with U.S. production roughly quadrupling between 2004 and 2006. Like ethanol, there are limits on the amount of biodiesel that could be produced from existing agricultural products. If all U.S. oilseed production, available animal fats, and recycled grease were used for biodiesel production, only about 4 billion gallons of biodiesel could be produced annually — less than current U.S. production of corn-based ethanol.[29]

Despite the limitations on current feedstock supply for biodiesel, like ethanol there is the potential in the future to produce a significantly greater amount of biodiesel from cellulosic materials.

Cellulosic Biofuels

Ethanol and biodiesel produced from cellulosic feedstocks, such as prairie grasses and fast-growing trees, have the potential to improve the energy and environmental effects of U.S. biofuels. Further, moving away from feed and food crops to dedicated energy crops could avoid some of the agricultural supply and price concerns discussed above. Although research is ongoing, there are no demonstration- or commercial-scale cellulosic biofuel plants in the United States at this time, and

---

[27] For more information on biodiesel, see CRS Report RL32712, *Agriculture-Based Renewable Energy Production*, by Randy Schnepf.

[28] Currently, few engine and vehicle manufacturers warranty their engines at levels higher than B5 (5% biodiesel, 95% conventional diesel), but research and testing are ongoing.

[29] Randy Schnepf, *Agriculture-Based Renewable Energy Production*, CRS Report RL32712, January 8, 2007.

there is only one demonstration-scale plant in Canada.[30] A major barrier to cellulosic fuel production is that production costs still remain significantly higher than for corn ethanol or other alternative fuels. Currently, various production processes are prohibitively expensive, including physical, chemical, enzymatic, and microbial treatment and conversion of these feedstocks into motor fuel.

A key potential benefit of energy crops (e.g., prairie grasses, fast-growing trees) is that they can be grown without the need for chemicals. Reducing or eliminating the need for chemical fertilizers would address one of the largest energy inputs for corn-based ethanol production. Using biomass to power a biofuel production plant could further reduce fossil fuel inputs. Improving the net energy balance of ethanol would also reduce net fuel cycle greenhouse gas emissions.

However, increases in per-acre yields would be required to make energy crops for fuel production economically competitive. Questions remain whether high yields can be achieved without the use of fertilizers and pesticides. Another question is whether there is sufficient feedstock supply available. USDA estimates that by 2030 1.3 billion tons of biomass could be available for bioenergy production (including electricity from biomass, and fuels from corn and cellulose).[31] From that, enough biofuels could be produced to replace roughly 70 billion gallons of gasoline per year. However, this projection assumes significant increases in per-acre yields and, according to USDA, should be seen as an upper bound on what is possible. Further, new harvesting machinery would need to be developed to guarantee an economic supply of cellulosic feedstocks.[32]

In addition to the above concerns, other potential environmental drawbacks associated with cellulosic fuels must be addressed, such as the potential for soil erosion, runoff, and the spread of invasive species.

Conclusion

There is continuing interest in expanding the U.S. biofuel industry as a strategy for promoting energy security and environmental goals. However, there are limits to the amount of biofuels that can be produced and questions about the net energy and environmental benefits they would provide. Further, rapid expansion of biofuel production may have many unintended and undesirable consequences for agricultural

---

[30] However, on February 28, 2007, DOE announced $385 million in grant funding for six cellulosic ethanol plants in six states. If operational, combined capacity of these six plants would be 130 million gallons per year. DOE, *DOE Selects Six Cellulosic Ethanol Plants for Up to $385 Million in Federal Funding*, February 28, 2007, Washington.

[31] Oak Ridge National Laboratory for DOE and USDA, *Biomass as a Feedstock for a Bioenergy and Bioproducts Industry: The Technical Feasibility of a Billion-Ton Annual Supply*, April 2005, Oak Ridge, TN.

[32] For example, the study assumes roughly 400 million tons of biomass from agricultural residues. To economically supply those residues to biofuel producers, farm equipment manufacturers would likely need to develop one-pass harvesters that could collect and separate crops and crop residues at the same time.

CRS-13

commodity costs, fossil energy use, and environmental degradation. As policies are implemented to promote ever-increasing use of biofuels, the goal of replacing petroleum use with agricultural products must be weighed against these other potential consequences.

Order Code RL32712

CRS Report for Congress

Agriculture-Based Renewable Energy Production

Updated October 16, 2007

Randy Schnepf
Specialist in Agricultural Policy
Resources, Science, and Industry Division

Prepared for Members and Committees of Congress

Agriculture-Based Renewable Energy Production

Summary

Since the late 1970s, U.S. policy makers at both the federal and state levels have enacted a variety of incentives, regulations, and programs to encourage the production and use of agriculture-based renewable energy. Motivations cited for these legislative initiatives include energy security concerns, reduction in greenhouse gas emissions, and raising domestic demand for U.S.-produced farm products.

Agricultural households and rural communities have responded to these government incentives and have expanded their production of renewable energy, primarily in the form of biofuels and wind power, every year since 1996. The production of ethanol (the primary biofuel produced by the agricultural sector) has risen from about 175 million gallons in 1980 to nearly 4.9 billion gallons per year in 2006. The U.S. ethanol production capacity has also been expanding rapidly, particularly since mid-2006, with important implications for the food and fuel sectors. Current ethanol production capacity is 5.6 billion gallons per year (February 28, 2007), with another 6.2 billion gallons of capacity under construction and potentially online by late 2008. Biodiesel production is at a much smaller level, but has also shown growth rising from 0.5 million gallons in 1999 to an estimated 386 million gallons in 2006. Wind energy systems production capacity has also grown rapidly, rising from 1,706 megawatts in 1997 to an estimated 12,633 megawatts by July 1, 2007. Despite this rapid growth, agriculture- and rural-based energy production accounted for only about 0.7% of total U.S. energy consumption in 2006.

Key points that emerge from this report are:

- substantial federal and state programs and incentives have facilitated development of agriculture's renewable energy production capacity;
- rising fossil fuel prices improve renewable energy's market competitiveness, whereas higher costs for feedstock and plant operating fuel (e.g., natural gas) dampen profitability;
- technological improvements for biofuel production (e.g., cellulosic conversion) enhance its economic competitiveness with fossil fuels;
- farm-based energy production is unlikely to substantially reduce the nation's dependence on petroleum imports unless there is a significant decline in energy consumption; and
- ethanol-driven higher corn prices have raised concerns from corn users over rising food and feed costs, as well as the potential for increased soil erosion and chemical usage from substantially expanded corn production.

This report provides background information on farm-based energy production and how this fits into the national energy-use picture. It briefly reviews the primary agriculture-based renewable energy types and issues of concern associated with their production, particularly their economic and energy efficiencies and long-run supply. Finally, this report examines the major legislation related to farm-based energy production and use. This report will be updated as events warrant.

Contents

List of Figures

List of Tables

Agriculture-Based Renewable Energy Production

Introduction

Agriculture's role as a consumer of energy is well known.[1] However, under the encouragement of expanding government support the U.S. agricultural sector also is developing a capacity to produce energy, primarily as renewable biofuels and wind power. Farm-based energy production — biofuels and wind-generated electricity — has grown rapidly in recent years, but still remains small relative to total national energy needs. In 2006, ethanol, biodiesel, and wind provided 0.7% of U.S. energy consumption (**Table 1**). Ethanol accounted for about 78% of agriculture-based energy production in 2006; wind energy systems for 19%; and biodiesel for 3%.

Historically, fossil-fuel-based energy has been less expensive to produce and use than energy from renewable sources.[2] However, since the late 1970s, U.S. policy makers at both the federal and state levels have enacted a variety of incentives, regulations, and programs to encourage the production and use of cleaner, renewable agriculture-based energy.[3] These programs have proven critical to the economic success of rural renewable energy production. The benefits to rural economies and to the environment contrast with the generally higher costs, and have led to numerous proponents as well as critics of the government subsidies that underwrite agriculture-based renewable energy production.

Proponents of government support for agriculture-based renewable energy have cited national energy security, reduction in greenhouse gas emissions, and raising domestic demand for U.S.-produced farm products as viable justification.[4] In addition, proponents argue that rural, agriculture-based energy production can enhance rural incomes and employment opportunities, while encouraging greater value-added for U.S. agricultural commodities.[5]

---

[1] For more information on energy use by the agricultural sector, see CRS Report RL32677, *Energy Use in Agriculture: Background and Issues*, by Randy Schnepf.

[2] Excluding the costs of externalities associated with burning fossil fuels such as air pollution, environmental degradation, and illness and disease linked to emissions.

[3] See section on "Public Laws That Support Agriculture-Based Energy Production and Use," below, for a listing of major laws supporting farm-based renewable energy production.

[4] For examples of proponent policy positions, see the Renewable Fuels Association (RFA) at [http://www.ethanolrfa.org], the National Corn Growers Association (NCGA) at [http://www.ncga.com/ethanol/main/index.htm], and the American Soybean Association (ASA) at [http://www.soygrowers.com/policy/].

[5] Several studies have analyzed the positive gains to commodity prices, farm incomes, and (continued...)

CRS-2

In contrast, petroleum industry critics of biofuel subsidies argue that technological advances such as seismography, drilling, and extraction continue to expand the fossil-fuel resource base, which has traditionally been cheaper and more accessible than biofuel supplies.[6] Other critics argue that current biofuel production strategies can only be economically competitive with existing fossil fuels in the absence of subsidies if significant improvements in existing technologies are made or new technologies are developed.[7] Until such technological breakthroughs are achieved, critics contend that the subsidies distort energy market incentives and divert research funds from the development of other potential renewable energy sources, such as solar or geothermal, that offer potentially cleaner, more bountiful alternatives. Still others question the rationale behind policies that promote biofuels for energy security. These critics question whether the United States could ever produce sufficient feedstock of either starches, sugars, or vegetable oils to permit biofuel production to meaningfully offset petroleum imports.[8] Finally, there are those who argue that the focus on development of alternative energy sources undermines efforts to conserve and reduce the nation's energy dependence.

The economics underlying agriculture-based renewable energy production include decisions concerning capital investment, plant or turbine location (relative to feedstock supplies and by-product markets or power grids), production technology, and product marketing and distribution, as well as federal and state production incentives and usage mandates.[9] Several additional criteria may be used for comparing different fuels, including performance, emissions, safety, and infrastructure needs.[10] This report will discuss and compare agriculture-based energy production of ethanol, biodiesel, and wind energy based on three criteria:

---

[5] (...continued)
rural employment attributable to increased government support for biofuel production. For examples, see the "For More Information" section at the end of this report.

[6] For example, see Elizabeth Ames Jones, "Energy Security 101," editorial, *The Washington Post*, October 9, 2007.

[7] Advocates of this position include free-market proponents such as the Cato Institute, and federal budget watchdog groups such as Citizens Against Government Waste and Taxpayers for Common Sense.

[8] For example, see James and Stephen Eaves, "Is Ethanol the 'Energy Security' Solution?" editorial, *Washingtonpost.com*, October 3, 2007; or R. Wisner and P. Baumel, "Ethanol, Exports, and Livestock: Will There be Enough Corn to Supply Future Needs?," *Feedstuffs*, no. 30, vol. 76, July 26, 2004.

[9] For more information on the economics underlying the capital investment decision see D. Tiffany and V. Eidman. *Factors Associated with Success of Fuel Ethanol Producers*, Dept of Appl. Econ., Univ. of Minnesota, Staff Paper P03-7, August 2003; hereafter referred to as Tiffany and Eidman (2003). For a discussion of ethanol plant location economics see B. Babcock and C. Hart, "Do Ethanol/Livestock Synergies Presage Increased Iowa Cattle Numbers?" *Iowa Ag Review*, Vol. 12 No. 2, Spring 2006.

[10] For more information on these additional criteria and others, see CRS Report RL30758, *Alternative Transportation Fuels and Vehicles: Energy, Environment, and Development Issues*, by Brent Yacobucci. For information concerning greenhouse gas emissions associated with ethanol use, see CRS Report RL33290, *Fuel Ethanol: Background and Public Policy Issues* by Brent Yacobucci.

CRS-3

- **Economic Efficiency** compares the price of agriculture-based renewable energy with the price of competing energy sources, primarily fossil fuels.
- **Energy Efficiency** compares energy output from agriculture-based renewable energy relative to the fossil energy used to produce it.
- **Long-Run Supply Issues** consider supply and demand factors that are likely to influence the growth of agriculture-based energy production.

Table 1. U.S. Energy Production and Consumption, 2006

Energy source	Production & Imports		Consumption	
	Quadrillion Btu	% of total	Quadrillion Btu	% of total
Total	**100.7**	**100.0%**	**100.7**	**100.0%**
Fossil Fuels	**56.1**	**79.5%**	**86.1**	**85.4%**
Petroleum and products	13.2	13.1%	40.6	40.3%
Coal	23.8	23.6%	23.1	23.0%
Natural Gas	19.1	19.0%	22.4	22.2%
Nuclear	**8.2**	**8.1%**	**8.2**	**8.1%**
Renewables	**6.8**	**6.7%**	**6.8**	**6.7%**
Hydroelectric power	2.8	2.8%	3.0	3.0%
Biomass	3.0	3.0%	2.4	2.4%
Wood, waste, other	2.4	2.4%	2.4	2.4%
Ethanol	0.6	0.9%	0.6	0.6%
Biodiesel	0.0	0.0%	0.0	0.0%
Geothermal	0.4	0.3%	0.4	0.3%
Solar	0.1	0.1%	0.2	0.2%
Wind	0.2	0.2%	0.1	0.1%
Total Domestic Production	**71.1**	**70.6%**		
Net Imports	**29.8**	**29.6%**		

Source: For ethanol data: Renewable Fuels Association, [http://www.ethanolrfa.org]; for biodiesel data: National Biodiesel Board, [http://www.biodiesel.org]; for all other data: DOE, Energy Information Agency (EIA), Annual Energy Outlook 2007 (early release), [http://www.eia.doe.gov/oiaf/aeo/index.html].

Agriculture's Share of Energy Production

In 2006, the major agriculture-produced energy source — ethanol — accounted for about 0.6% of total U.S. energy consumption (see **Table 1**) In addition, the agricultural sector produced other types of renewable energy — biodiesel, wind, and methane from anaerobic digesters and non-traditional biomass — although their production volume remains small relative to ethanol's.

Table 2. Energy and Price Comparisons for Alternate Fuels, July 2007

Fuel type	Unit	Btu's per unit[a]	National Avg. Price: $ per unit	GEG[b]	National Avg. Price: $ per GEG
Gasoline: regular	gallon	115,400	$3.03	1.00	$3.03
Ethanol (E85)[c]	gallon	81,630	$2.63	0.71	$3.70
Diesel fuel	gallon	128,700	$2.96	1.11	$2.67
Biodiesel (B20)	gallon	126,940	$2.96	1.10	$2.69
Biodiesel (B100)	gallon	117,093	$3.27	1.01	$3.24
Propane	gallon	83,500	$2.58	0.72	$3.58
Compressed Natural Gas[d]	1,000 ft.^{3}	960,000	$2.09	1.00	$2.09
Natural Gas[e]	1,000 ft.^{3}	1,030,000	$7.58	8.24	$0.92
Biogas	1,000 ft.^{3}	10 x (% of methane)[f]	na	na	na
Electricity[g]	kilowatt-hour	3,413	5.73¢	na	na

Source: Prices and conversion rates (unless otherwise cited in a footnote below) are for July 2007, from DOE, EIA, *Clean Cities Alternative Fuel Price Report*, July 2007; available at [http://www.eere.energy.gov/afdc/resources/pricereport/price_report.html].

na = not applicable.

a. Conversion rates for petroleum-based fuels and electricity are from DOE, *Alternative Fuel Price Report*, July 2007, p. 14. A Btu (British thermal unit) is a measure of the heat content of a fuel and indicates the amount of energy contained in the fuel. Because energy sources vary by form (gas, liquid, or solid) and energy content, the use of Btu's provides a common benchmark for various types of energy.

b. GEG = gasoline equivalent gallon. The GEG allows for comparison across different forms — gas, liquid, kilowatt, etc. It is derived from the Btu content by first converting each fuel's units to Btu's, then dividing each fuel's Btu unit rate by gasoline's Btu unit rate of 115,400, and finally multiplying each fuel's price by the resulting ratio.

c. 100% ethanol has an energy content of 75,670 Btu per gallon (see table source, p. 14).

d. Compressed natural gas (CNG) is generally stored under pressure at between 2,000 to 3,500 pounds per square inch (psi). The energy content varies with the pressure. Conversion data is from DOE, *Alternative Fuel Price Report*, July 2007, p. 14.

e. Natural Gas prices, $ per 1,000 cu. ft., are industrial prices for the month of July 2007, from DOE, EIA, available at [http://tonto.eia.doe.gov/dnav/ng/ng_pri_sum_dcu_nus_m.htm].

f. When burned, biogas yields about 10 Btu per percentage of methane composition. For example, 65% methane yields 650 Btu per cubic foot or 650,000 per 1,000 cu. ft.

g. Prices are for total industry electricity rates per kilowatt-hour for 2005; from DOE, EIA, available at [http://www.eia.doe.gov/cneaf/electricity/epa/epat7p4.html].

Renewable energy sources must compete with a large number of conventional petroleum-based fuels in the marketplace (see **Table 2**). However, an expanding list of federal and state incentives, regulations, and programs that were enacted over the past decade have helped to encourage more diversity in renewable energy production and use. In late September 2006, the House Agriculture Committee expressed its support for the continued expansion of energy production from renewable sources when it reported favorably a resolution (H.Con.Res. 424) that expressed the sense of Congress that, "not later than January 1, 2025, the agricultural, forestry, and working land of the United States should provide from renewable resources not less than 25% of the total energy consumed in the United States."[11]

Agriculture-Based Biofuels

Biofuels are liquid fuels produced from biomass. Types of biofuels include ethanol, biodiesel, methanol, and reformulated gasoline components; however, the two principal biofuels are ethanol and biodiesel.[12] The Biomass Research and Development Act of 2000 (P.L. 106-224; Title III) defines biomass as "any organic matter that is available on a renewable or recurring basis, including agricultural crops and trees, wood and wood wastes and residues, plants (including aquatic plants), grasses, residues, fibers, and animal wastes, municipal wastes, and other waste materials."

Biofuels are primarily used as transportation fuels for cars, trucks, buses, airplanes, and trains. As a result, their principal competitors are gasoline and diesel fuel. Unlike fossil fuels, which have a fixed resource base that declines with use, biofuels are produced from renewable feedstock. Despite rapid growth in recent years (as discussed below), the two major biofuels — ethanol and biodiesel — still account for very small shares of U.S. motor-vehicle fuel consumption (**Figure 1**).

Under most circumstances biofuels are more environmentally friendly (in terms of emissions of toxins, volatile organic compounds, and greenhouse gases) than petroleum products. Supporters of biofuels emphasize that biofuel plants generate value-added economic activity that increases demand for local feedstock, which raises commodity prices, farm incomes, and rural employment.

Ethanol

Ethanol, or ethyl alcohol, is an alcohol made by fermenting and distilling simple sugars.[13] As a result, ethanol can be produced from any biological feedstock that contains appreciable amounts of sugar or materials that can be converted into sugar such as starch or cellulose. Sugar beets and sugar cane are examples of feedstock that contain sugar. Corn contains starch that can relatively easily be converted into sugar.

---

[11] The resolution was also referred to the House Energy and Commerce Committee and the House Resources Committee. Only the Agriculture Committee acted upon it. No further action was taken on H.Con.Res. 424 by the 109th Congress. The same bill has been reintroduced in the 110th Congress in the House (Rep. Peterson) as H.Con.Res. 25 and in the Senate (Sen. Salazar) as S.Con.Res. 3.

[12] For more information on alternative fuels, see CRS Report RL30758, *Alternative Transportation Fuels and Vehicles: Energy, Environment, and Development Issues*, by Brent D. Yacobucci. See also DOE, National Renewable Energy Laboratory (NREL), *Biomass Energy Basics*, available at [http://www.nrel.gov/learning/re_biomass.html].

[13] For more information, see CRS Report RL33290, *Fuel Ethanol: Background and Public Policy Issues*, by Brent D. Yacobucci.

In the United States corn is the principal ingredient used in the production of ethanol; in Brazil, sugar cane is the primary feedstock. Trees and grasses are made up of a significant percentage of cellulose which can also be converted to sugar, although with more difficulty than required to convert starch. In recent years, researchers have begun experimenting with the possibility of growing hybrid grass and tree crops explicitly for ethanol production. In addition, sorghum and potatoes, as well as crop residue and animal waste, are potential feedstocks.

Figure 1. U.S. Motor Vehicle Fuel Use, 2006

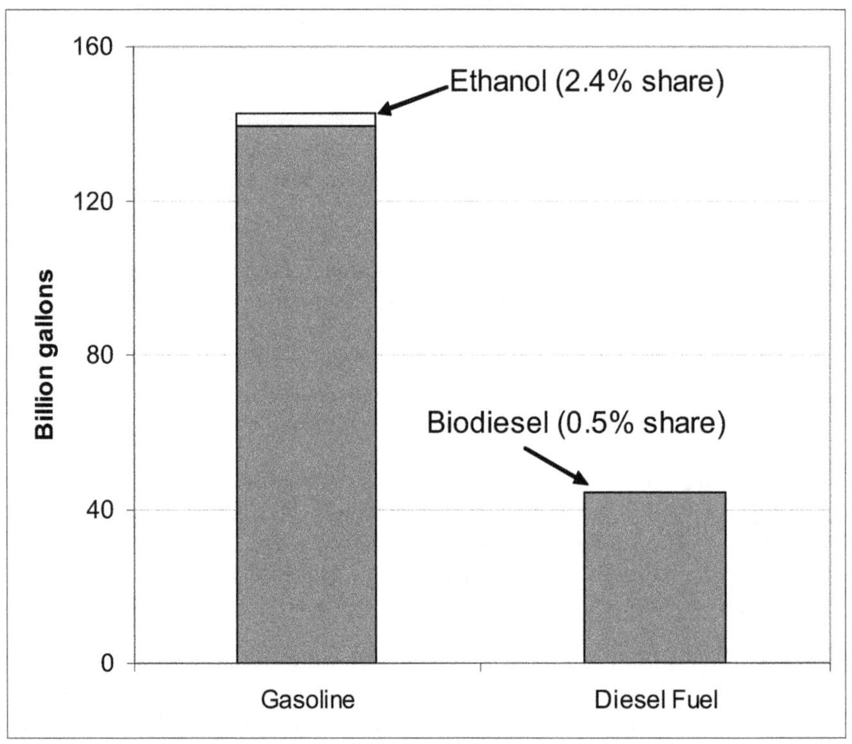

Source: Motor vehicle fuel use: U.S. DOE, EIA, *2007 Annual Outlook*: Ethanol use: Renewable Fuels Association; Biodiesel use: National Biodiesel Board.
Note: Ethanol is in gasoline-equivalent gallons (GEG).

Ethanol production has shown rapid growth in the United States since 2001 and expectations are for this trend to continue to at least 2010 (**Figure 3**). In 2005, the United States surpassed Brazil as the world's leading producer of ethanol. Several events contributed to the historical growth of U.S. ethanol production: the energy crises of the early and late 1970s; a partial exemption from the motor fuels excise tax (legislated as part of the Energy Tax Act of 1978); ethanol's emergence as a gasoline oxygenate; and provisions of the Clean Air Act Amendments of 1990 that favored ethanol blending with gasoline.[14] Ethanol production is projected to continue growing

[14] USDA, Office of Energy Policy and New Uses, *The Energy Balance of Corn Ethanol: An* (continued...)

rapidly through at least 2010 on the strength of both the extension of existing and the addition of new government incentives including a per gallon tax credit of $0.51, a Renewable Fuels Standard (RFS) of 7.5 billion gallons by 2012, and $0.51 per gallon tariff on most imported ethanol.[15]

U.S. ethanol production presently is underway or planned in 28 states based primarily around the central and western Corn Belt, where corn supplies are most plentiful (see **Table 3**).[16] Corn accounts for about 99% of the feedstock used in ethanol production in the United States. As of October 1, 2007, existing U.S. ethanol plant capacity was a reported 6.9 billion gallons per year (BGPY), with an additional planned capacity of 7.6 BGPY under construction (as either new plants or expansion of existing plants). Thus, total annual U.S. ethanol production capacity in existence or under construction is nearly 14.5 billion gallons, well in excess of the 7.5 billion gallon RFS mandated for 2012 (**Figure 3**).

---

[14] (...continued)
Update, AER-813, by Hosein Shapouri, James A. Duffield, and Michael Wang, July 2002.

[15] For more information, see CRS Report RL33572, *Biofuels Incentives: A Summary of Federal Programs* by Brent D. Yacobucci.

[16] The source for this data is "Ethanol Plant List," *The Ethanol Monitor*; published by Oil Intelligence Link, Inc., Editor & Publisher: Tom Waterman; The Ethanol Monitor©2007 October 1, 2007. An additional reference with slightly different reported data is Renewable Fuels Association, *Industry Statistics*, at [http://www.ethanolrfa.org/industry/statistics/].

Table 3. Ethanol Production Capacity by State, October 1, 2007

Rank	State	Total planned capacity Million gal/yr	%	Currently operating Million gal/yr	%	Under construction Million gal./yr.
1	Iowa	3,399	23%	1,979	29%	1,420
2	Nebraska	2,153	15%	955	14%	1,198
3	South Dakota	1188	8%	750	11%	438
4	Illinois	1,180	8%	834	12%	346
5	Minnesota	1,142	8%	499	7%	643
6	Indiana	977	7%	382	6%	595
7	Kansas	643	4%	253	4%	390
8	Ohio	551	4%	3	0%	548
9	Wisconsin	534	4%	284	4%	250
10	Texas	470	3%	0	0%	470
11	North Dakota	354	2%	134	2%	220
12	Michigan	257	2%	157	2%	100
13	California	253	2%	79	1%	174
14	Tennessee	205	1%	67	1%	138
15	New York	164	1%	0	0%	164
	Others	1,008	7%	488	4%	520
	U.S. Total	14,476	100%	6,863	100%	7,614

Source: "Ethanol Plant List," *The Ethanol Monitor*; published by Oil Intelligence Link, Inc., Editor & Publisher: Tom Waterman; The Ethanol Monitor©2007 October 1, 2007.

Ethanol Pricing Issues. From a national perspective, marketing channels, pricing arrangements, and distribution networks are still evolving in an attempt to keep up with the ethanol industry's rapid growth in production and the federally mandated use requirements. These circumstances can contribute to substantial price volatility. For example, in early 2006, several market circumstances combined to push ethanol prices to levels substantially above gasoline prices (see **Figure 2**). In May 2006, the spot market price per gallon for ethanol reached $3.75 in Chicago and $4.50 in New York, while the monthly average ethanol rack price, f.o.b. Omaha, reached $3.58 in June 2006.

These price surges generated considerable concern among consumers regarding possible price manipulation in the marketplace and the reliability of ethanol as a fuel source. However, a review of the circumstances suggests that two market phenomena appear to be the behind the rise in ethanol prices and the ethanol-to-gasoline price disparity — the general price rise in petroleum and natural gas markets,[17] and the elimination of the oxygen requirement for reformulated gasoline (legislated by the Energy Policy Act of 2005, P.L. 109-58), which resulted in a rapid shift from MTBE

[17] For more information see CRS Report RL32530, *World Oil Demand and its Effect on Oil Prices*, by Robert Pirog.

CRS-9

to ethanol by the automotive fuel industry and pushed near-term demand substantially above available ethanol supplies.[18]

Figure 2. Ethanol Versus Gasoline Prices, 2000-2007

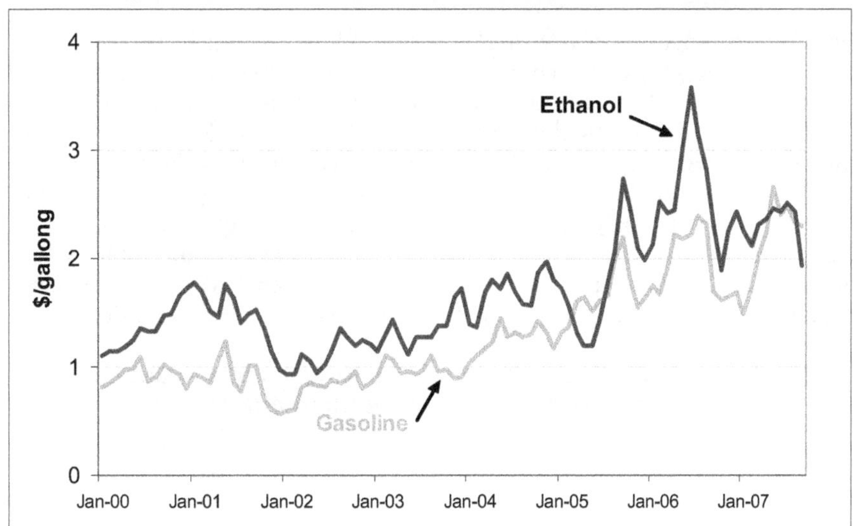

Source: Ethanol and unleaded gasoline rack prices per gallon. F.O.B. Omaha Ethanol Board, Lincoln, NE. Nebraska Energy Office, Lincoln, NE.

Most dry-mill ethanol plants typically employ one or more of three pricing strategies for marketing their ethanol production: sell at the rack price to nearby refinery and fuel blending sites; forward contract at a fixed price for future delivery; and forward contract where the ethanol price is based on a monthly futures contract price (e.g., the wholesale ethanol contract at either the Chicago or New York Boards of Trade or the wholesale gasoline contract at the New York Mercantile Exchange) plus a per-gallon premium.[19] Because a large portion of ethanol is sold under forward contract, the market is vulnerable to near-term, temporary price rises when demand exceeds available non-contracted supplies — as was the case in late 2005 and 2006 when the MTBE-phase-out-induced demand surged above existing supplies while the ethanol industry was already operating near full capacity.

By May 2007 ethanol prices had fallen below the Omaha gasoline rack price; by September 2007 they were below $2 per gallon and any concerns of price manipulation had long since dissipated. The ethanol-to-gasoline price disparity is expected to diminish gradually as more ethanol production capacity comes online.

[18] For more information on the MTBE phase-out, see CRS Report RL33564, *Alternative Fuels and Advanced Technology Vehicles: Issues in Congress*, by Brent Yacobucci.

[19] Tiffany and Eidman (2003), p. 20.

Corn-Based Ethanol. USDA estimated that 1.6 billion bushels of corn (or 14.4% of total U.S. corn production) from the 2005 corn crop were used to produce ethanol during the 2005/06 (September-August) corn marketing year.[20] Ethanol's

share of corn production expanded to 20.2% (or 2.125 billion bushels) in 2006/07, and is projected to reach 24.8% and 3.3 billion bushels in 2007/08.[21] In its annual baseline projections (February 2007), USDA projected that U.S. ethanol production would reach 11 billion gallons and use 30% (3.9 billion bushels) of the corn crop. However, the rapid expansion of ethanol production capacity in the later half of 2006 outpaced USDA projections. In August 2007, the Food and Agricultural Policy Research Institute (FAPRI) — using more recent data — projected that by 2010 U.S. ethanol production would reach 13.7 billion gallons and use 36.2% (4.9 billion bushels) of the U.S. corn crop (see **Figure 3**).[22] Even FAPRI's more recent projections appear to be behind the current pace of plant capacity expansion.

Figure 3. U.S. Ethanol Production: Actual and Projected, Versus the Renewable Fuels Standard (RFS)

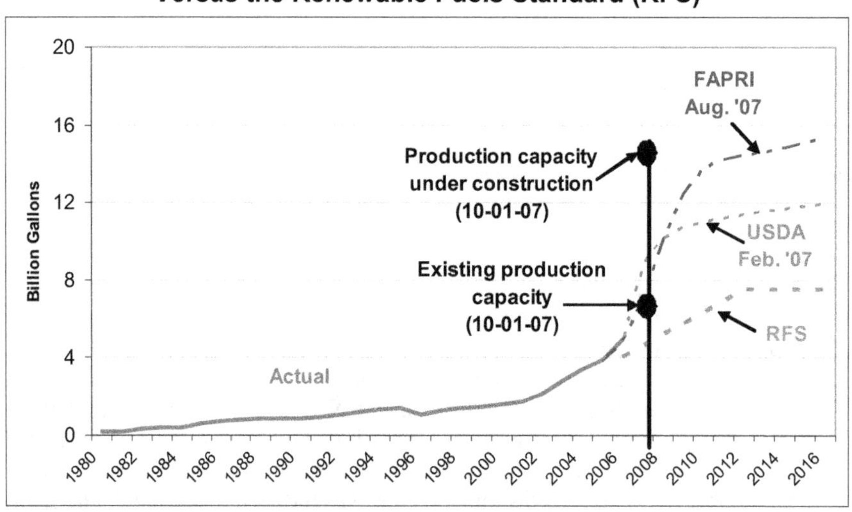

Source: 1980-2006, Renewable Fuels Association (RFA); current and planned capacity are from *The Ethanol Monitor;* projections are from FAPRI's August 2007 Baseline Update, and USDA's *USDA Agr. Projections to 2016*, Feb. 2007.

Despite its rapid growth, ethanol production represents a minor part of U.S. gasoline consumption (**Figure 1**). In calendar 2006, U.S. ethanol production of 4.9 billion gallons accounted for about a 2.4% projected share of national gasoline use

[20] Corn use for ethanol: USDA, World Agricultural Outlook Board, *World Agricultural Supply and Demand Estimates*, September 12, 2007.

[21] Ibid.

[22] FAPRI, August Baseline Update for U.S. Agricultural Markets, FAPRI-UMC Report #28-076, University of Missouri.

(3.3 billion gasoline-equivalent gallons (GEG) out of an estimated 140.3 billion gallons).[23]

Economic Efficiency. Apart from government incentives, the economics underlying corn-based ethanol's market competitiveness hinge primarily on the following factors:

- the price of feedstock, primarily corn;[24]
- the price of the processing fuel, primarily natural gas or electricity, used at the ethanol plant;
- the cost of transporting feedstock to the ethanol plant and transporting the finished ethanol to the user;
- the price of feedstock co-products (for dry-milled corn: distillers dried grains (DDGs); for wet-milled corn: corn gluten feed, corn gluten meal, and corn oil); and
- the price of gasoline, ethanol's main competitor in the marketplace.

Higher prices for corn, processing fuel, and transportation hurt ethanol's market competitiveness, while higher prices for corn by-products and gasoline improve ethanol's competitiveness in the marketplace. Using 2002 data (see **Table 4**), USDA estimated that the average production cost for a gallon of ethanol was $0.958 when corn prices averaged about $2.32 per bushel and natural gas cost about $4.10 per 1,000 cubic feet (mcf). Feedstock costs are the largest expense item in the production of ethanol, representing about 57% of total ethanol production costs (net of by-product credits obtained by selling the DDGs and carbon dioxide) or about $0.55 per gallon. Each $1.00 increase in the price of corn raises the per gallon production cost of ethanol by about $0.36 per gallon ($0.54 per GEG).[25]

Processing fuel (usually natural gas) is the second largest expense representing about 14% of total costs or about $0.14 per gallon. Natural gas prices have risen substantially since 2002 (see **Figure 8**). However, because of its smaller cost share, each $1.00 increase in the price of natural gas only raises the per gallon production cost of ethanol by about $0.034 per gallon ($0.051 per GEG).

---

[23] Based on a conversion rate of 0.67 GEG per gallon of ethanol.

[24] Note that the price for corn and other major program crops grown in the United States are directly influenced by the various federal farm programs which have been shown to encourage over-production during periods of low prices by muting market signals.

[25] Based on CRS simulations of an ethanol dry mill spreadsheet model developed by Tiffany and Eidman (2003).

Table 4. Ethanol Dry Mill Cost of Production Estimates, 2002

Item	Unit	Value	Share
PRICES[a]			
Ethanol (rack, f.o.b. Omaha)	$/gal.	$1.12	
Corn (average farm price received)	$/bushel	$2.32	
Distiller's Dried Grain (Lawrenceburg, IN)	$/short ton	$82.44	
COSTS[b]			
Feedstock (Corn, Sorghum, or Other)	$/gal.	$0.803	
By-Product Credit	$/gal.	$0.258	
Distiller's Dried Grain	$/gal.	$0.252	
Carbon Dioxide	$/gal.	$0.006	
Net Feedstock Costs	$/gal.	$0.545	56.9%
Total Processing Costs	$/gal.	$0.413	43.1%
Processing Fuel Costs	$/gal.	$0.136	14.1%
Chemical Costs	$/gal.	$0.102	10.6%
Labor, Maintenance, & Repair Costs	$/gal.	$0.091	9.5%
Administrative & Miscellaneous Costs	$/gal.	$0.048	5.0%
Electricity Costs	$/gal.	$0.037	3.9%
Total Processing Costs & Net Feedstock Costs	$/gal.	$0.958	100.0%

[a]Prices are for 2002.
[b]All costs are in 2002 dollars.

Source: Ethanol prices from Nebraska Ethanol Board, Lincoln, NE. Nebraska Energy Office, Lincoln, NE; Corn and DDGS prices from ERS, USDA; Natural Gas prices from DOE/EIA; Ethanol cost of production data from Hosein Shapouri and Paul Gallagher, *USDA's 2002 Ethanol Cost-of-Production Survey*, AER 841, USDA, Office of Energy Policy and New Uses, July 2005.

These ethanol production costs ignore capital costs (e.g., depreciation, interest charges, return on equity, etc.), which may play a significant role depending on market conditions. Capital costs for a 40 million gallon per year ethanol plant with an initial capital investment of $60 million (of which 60% is debt financed) have been estimated at roughly $0.14 per gallon assuming a 12% rate of return on equity.[26]

Because ethanol delivers only about 67% of the energy of a gallon of gasoline, the 2002 cost of production in gasoline equivalent gallons is $1.43. However, the federal tax credit (see below) of $0.51 per gallon of pure (100%) ethanol is a direct offset to the production costs. To date, ethanol has been used at low blend ratios (5% or 10%) with gasoline, functioning primarily either as an oxygenate or as a fuel extender. At higher blend ratios (e.g., 85% ethanol), ethanol competes directly with gasoline as a motor fuel.

[26] Ibid.

Since corn is the largest expense in the production of ethanol, the relative relationship of corn to ethanol prices provides a strong indicator of the ethanol industry's well-being (see **Figure 4**). From mid-2005 through mid-2006, the general trend was clearly in ethanol's favor, as average monthly ethanol rack prices (f.o.b. Omaha) surged above the $2.00 per gallon level while corn prices fluctuated around the $2.00 per bushel level. Since each bushel of corn yields approximately 2.75 gallons of ethanol, the profitability of ethanol production escalated rapidly with the increase in ethanol prices. Since mid-2006, ethanol prices have fallen back near the $2.00 per gallon level while corn prices have risen sharply. By November 2006, corn prices had surged to $3.50 per bushel or higher in most cash markets, while nearby futures contracts were trading near $4.00 per bushel.

Figure 4. Corn Versus Ethanol Prices, 2000-2007

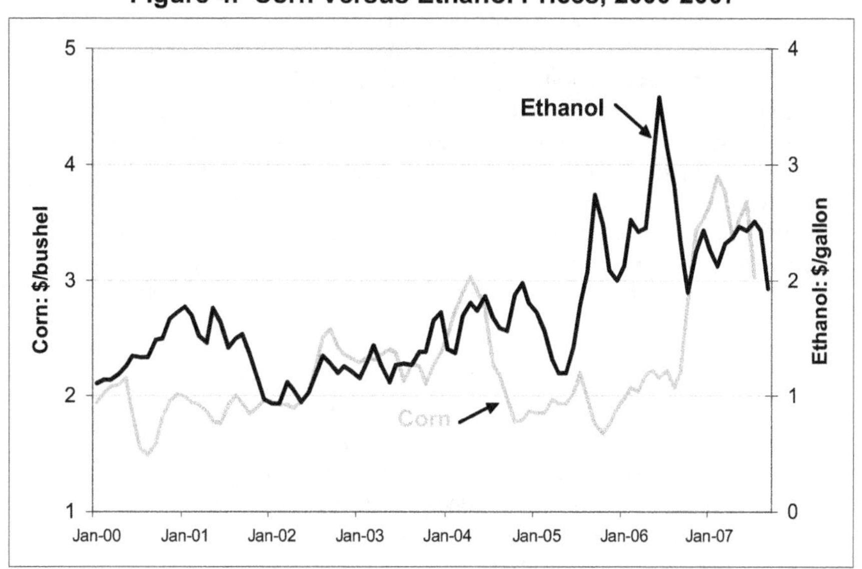

Source: Prices are monthly averages: Corn, No.2, yellow, Central Illinois; USDA, AMS; Ethanol are rack, f.o.b. Omaha, Nebraska Ethanol Board, Lincoln, NE. Nebraska Energy Office, Lincoln, NE.

The price relationship that persisted between ethanol, corn, and natural gas during late 2005 and through much of 2006, coupled with the federal production tax credit (PTC) of 51¢ per gallon of pure ethanol, represented a period of enormous profitability for ethanol producers and helps to explain the surge in ethanol production capacity since late 2005. For example, a model simulation based on prices of $2.50 per gallon for ethanol, $2.20/bushel for corn, and $6.00/mcf for natural gas (as existed during the summer of 2006) suggests that a 40 million gallon-per-year ethanol plant with initial capital of $60 million (of which 60% is debt financed) would be able to recover its entire capital investment in substantially less than a year.[27] When ethanol prices are lowered to $1.80, while corn prices are raised to $4.50, the simulation

[27] Ibid. Note, the results of these scenarios are merely suggestive of an average plant's situation and are not intended to imply uniformity in profitability across all ethanol plants.

model suggests that the ethanol plant still remains profitable. After removing the ethanol PTC of 51¢ per gallon from the simulation, the ethanol plant's per unit profitability falls to zero with corn prices of about $3.80 per bushel.

Government Support. Federal subsidies have played an important role in encouraging investment in the U.S. ethanol industry. The Energy Tax Act of 1978 first established a partial exemption for ethanol fuel from federal fuel excise taxes.[28] In addition to the partial excise tax exemption, certain income tax credits are available for motor fuels containing biomass alcohol. However, the different tax credits are coordinated such that the same biofuel cannot be claimed for both income and excise tax purposes. The primary federal incentives include:[29]

- a production tax credit of 51¢ per gallon of pure (100%) ethanol — the tax incentive was extended through 2010 and converted to a tax credit from a partial tax exemption of the federal excise tax under the American Jobs Creation Act of 2004 (P.L. 108-357);
- a small producer income tax credit (26 U.S.C. 40) of 10¢ per gal. for the first 15 million gal. of production for ethanol producers whose total output does not exceed 60 million gal. of ethanol per year;
- a Renewable Fuels Standard (RFS) (Energy Policy Act of 2005, P.L. 109-58) that mandates renewable fuels blending requirements for fuel suppliers — 4 billion gallons of renewable fuels must be blended into gasoline in 2006; the blending requirement grows annually until 7.5 billion gallons in 2012; and
- a 54¢ per gallon most-favored-nation tariff on most imported ethanol (extended through December 2008 by a provision in P.L. 109-432).

Also important was USDA's now-expired Bioenergy Program (7 U.S.C. 8108), which provided incentive payments (contingent on annual appropriations) on year-to-year production increases of renewable energy during the FY2001 to FY2006 period. Indirectly, other federal programs support ethanol production by requiring federal agencies to give preference to biobased products in purchasing fuels and other supplies and by providing incentives for research on renewable fuels. Also, several states have their own incentives, regulations, and programs in support of renewable fuel research, production, and consumption that supplement or exceed federal incentives.

Energy Efficiency. The net energy balance (NEB) of a fuel can be expressed as a ratio of the energy produced from a production process relative to the energy used in the production process. An output/input ratio of 1.0 implies that energy output equals energy input. The critical factors underlying ethanol's energy efficiency or NEB include:

---

[28] For a legislative history of federal ethanol incentives, see GAO, *Tax Incentives for Petroleum and Ethanol Fuels*, RCED-00-301R, September 25, 2000.

[29] For more information on federal incentives for biofuel production, see CRS Report RL33572, *Biofuels Incentives: A Summary of Federal Programs*, by Brent D. Yacobucci, or see section on "Public Laws That Support Agriculture-Based Energy Production and Use," later in this report.

CRS-15

- corn yields per acre (higher yields for a given level of inputs improves ethanol's energy efficiency);
- the energy efficiency of corn production, including the energy embodied in inputs such as fuels, fertilizers, pesticides, seed corn, and cultivation practices;
- the energy efficiency of the corn-to-ethanol production process — clean burning natural gas is the primary processing fuel for most ethanol plants, but several plants (including an increasing number of new plants) are designed to use coal; and
- the energy value of corn by-products, which act as an offset by substituting for the energy needed to produce market counterparts.

Over the past decade, technical improvements in the production of agricultural inputs (particularly nitrogen fertilizer) and ethanol, coupled with higher corn yields per acre and stable or lower input needs, appear to have raised ethanol's NEB. About 82% of the corn used for ethanol is processed by "dry" milling (a grinding process) and about 18% is processed by "wet" milling plants (a chemical extraction process).[30] All new plants under construction or coming online are expected to dry mill corn into ethanol, thus the dry milling share will continue to rise for the foreseeable future.

In 2004, USDA reported that, assuming "best production practices and state of the art processing technology," the NEB of corn-ethanol (based on 2001 data) was a positive 1.67 — that is, 67% more energy was returned from a gallon of ethanol than was used in its production.[31] Other researchers have found much lower NEB values under less optimistic assumptions, leading to some dispute over corn-to-ethanol's representative NEB.[32] A review (Farrel *et al.*, 2006) of several major corn-to-ethanol NEB analyses found that, when by-products are properly accounted for, the corn-to-ethanol process has a positive NEB (i.e., greater than 1.0) and that the NEB is improving with technology.[33] This result was confirmed by another comprehensive study (Hill *et al.*, 2006) that found a NEB of 1.25 for corn ethanol.[34] However, these studies clearly imply that inefficient processes for producing corn (e.g., excessive

---

[30] Dry milling and wet milling production shares are from the Renewable Fuels Association, *Ethanol Industry Outlook 2007.* According to USDA, dry milling is more energy efficient than wet milling, particularly when corn co-products are considered. These ethanol yield rates have been improving gradually overtime with technological improvements in the efficiency of ethanol processing from corn.

[31] H. Shapouri, J. Duffield, and M. Wang, *New Estimates of the Energy Balance of Corn Ethanol*, presented at 2004 Corn Utilization & Technology Conference of the Corn Refiners Association, June 7-9, 2004, Indianapolis, IN (hereafter cited as Shapouri (2004)).

[32] For example, Prof. David Pimentel, Cornell Univ., College of Agr. and Life Sciences, has researched and published extensive criticisms of corn-based ethanol production.

[33] Alexander E. Farrel, Richard J. Pleven, Brian T. Turner, Andrew D. Jones, Michael O'Hare, and Daniel M. Kammon, "Ethanol Can Contribute to Energy and Environmental Goals," *Science*, vol. 311 (January 27, 2006), pp. 506-508.

[34] Hill, J., E. Nelson, D. Tilman, S. Polasky, and D. Tiffany. "Environmental, economic, and energetic costs and benefits of biodiesel and ethanol biofuels," *Proceedings of the National Academy of Sciences*, Vol. 103, No. 30, July 25, 2006, 11206-11210.

reliance on chemicals and fertilizer or bad tillage practices) or for processing ethanol (e.g., coal-based processing), or extensive trucking of either the feedstock or the finished ethanol long distances to plant or consumer, can result in a NEB significantly less than 1.0. In other words, not all ethanol production processes have a positive energy balance.

Long-Run Supply Issues. The sharp rise in corn prices that has occurred since July 2006 owes its origins largely to the rapid expansion of corn-based ethanol production capacity that has occurred in the United States since 2004.[35] With 6.9 billion gallons of annual ethanol production capacity currently online (October 1, 2007) and another 7.6 billion gallons of capacity under construction and potentially online by early 2009, the U.S. ethanol sector is projected to need over 4.2 billion bushels of corn as feedstock in 2008/09 to service this capacity. This would be an 95% increase from the 2.15 billion bushels of corn projected as ethanol feedstock in 2006/07. Such a strong jump in corn demand is highly unusual and has already fueled substantially higher prices — for both current market prices (Figure 4) as well as for long-run projected prices. In its February 2007 baseline report, USDA projects U.S. farm-gate prices to remain in the $3.30 to $3.50 per bushel range through 2016.[36]

Questions Emerge Surrounding Further Subsidy-Fueled Corn Ethanol Expansion. Market participants, economists, and biofuels skeptics have begun to question the need for continued large federal incentives in support of ethanol production, particularly when the sector would have been profitable during much of 2006 without such subsidies. Their concerns focus on the potential for widespread unintended consequences that might result from excessive federal incentives adding to the rapid expansion of ethanol production capacity and the demand for corn to feed future ethanol production.[37] Such consequences include a rapid expansion of corn area (crowding out other field crops and agricultural activities) and the likelihood of both expanded fertilizer and chemical use and increased soil erosion. Growth in corn-for-ethanol use would reduce both exports and domestic feed use unless accompanied by offsetting growth in domestic production.

Rapidly Expanding Corn Planting. As corn prices rise, so too does the incentive to expand corn production (whether by expanding onto more marginal soil environments or by altering the traditional corn-soybean rotation that dominates Corn Belt agriculture), crowding out other field crops, primarily soybeans, and other agricultural activities. Large-scale shifts in agricultural production activities will likely have important regional economic consequences that have yet to be fully explored or understood. Further, corn production is among the most energy-intensive of the major field crops. An expansion of corn area would likely have important and unwanted

---

[35] International market factors such as the failure of the 2006 Australian wheat and barley crops added psychological momentum to the corn price runup; however, ample U.S. feed grain supplies at the time of the rising corn price (fall 2006) strongly imply that future corn demand attributable to the rapid surge in investment in U.S. ethanol production capacity is the principal factor behind higher corn prices.

[36] *USDA Agricultural Projections to 2016*, OCE-2007-1, USDA, February 2007; available at [http://www.ers.usda.gov/Briefing/Baseline/].

[37] For a list or related articles, see the Reference Section entitled, "Consequences of Expanded Agriculture-Based Biofuel Production" at the end of this report.

environmental consequences due to the resulting increase in fertilizer and chemical use and soil erosion.

Domestic Feed Market Distortions. Corn traditionally represents about 57% of feed concentrates and processed feedstuffs fed to animals in the United States.[38] As corn-based ethanol production increases, so does total corn demand and corn prices. Dedicating an increasing share of the U.S. corn harvest to ethanol production will likely lead to higher prices for all grains and oilseeds that compete for the same land, resulting in higher feed costs for cattle, hog, and poultry producers. In addition, distortions are likely to develop in protein-meal markets related to expanding production of the ethanol processing by-product Distiller's Dried Grains (DDG), which averages about 30% protein content and can substitute in certain feed and meal markets.[39] While DDG use would substitute for some of the lost feed value of corn used in ethanol processing, about 66% of the original weight of corn is consumed in producing ethanol and is no longer available for feed. Furthermore, not all livestock species are well adapted to dramatically increased consumption of DDG in their rations — dairy cattle appear to be best suited to expanding DDG's share in feed rations; poultry and pork are much less able to adapt. Also, DDG must be dried before it can be transported long distances. Will large-scale movements of livestock production occur to relocate near new feed sources? Such a relocation would likely have important regional economic effects.

Domestic and International Food Markets. Most corn grown in the United States is used for animal feed. Higher feed costs ultimately lead to higher meat prices. The feed-price effect will first translate into higher prices for poultry and hogs, which are less able to use alternate feedstuffs. Dairy and beef cattle are more versatile in their ability to shift to alternate feed sources, but eventually a sustained rise in corn prices will push their feed costs upward as well. The price of corn is also linked to the price of other grains, including those destined for food markets, through their competition in both the feed marketplace and the producer's planting choices for his (or her) limited acreage. The price runup in the U.S. corn market has already clearly spilled over into the market for soybeans (and soybean oil).

Since food costs represent a relatively small share of consumer spending in the United States, the price runup is more easily absorbed in the short run. However, the situation is very different for lower-income households as well as in many foreign markets, where food expenses can represent a substantial portion of the household budget. This is becoming a concern since, because of trade linkages, the increase in U.S. corn prices has carried into international markets as well. In January 2007, Mexico experienced riots following a nearly 30% price increase for tortillas, the country's dietary staple. In China, where corn is also an important food source, the government has recently put a halt to its planned ethanol plant expansion due to the

---

[38] USDA, ERS, *Feed Situation and Outlook Yearbook*, FDS-2003, April 2003.

[39] For a discussion of potential feed market effects due to growing ethanol production, see Bob Kohlmeyer, "The Other Side of Ethanol's Bonanza," *Ag Perspectives* (World Perspectives, Inc.), December 14, 2004; and R. Wisner and P. Baumel, "Ethanol, Exports, and Livestock: Will There be Enough Corn to Supply Future Needs?," *Feedstuffs*, no. 30, vol. 76, July 26, 2004.

threat it poses to the country's food security. Similarly, humanitarian groups have expressed concern for the potential difficulties that higher grain prices imply for net-food-importing developing countries.

U.S. Corn Exports. The United States is the world's leading producer and exporter of corn. Since 1980 U.S. corn production has accounted for over 40% of world production, while U.S. corn exports have represented nearly a 66% share of world trade during the past decade. In 2006/2007, the United States exported about 20% of its corn production.[40] Higher corn prices would likely result in lost export sales. It is unclear what type of market adjustments would occur in global feed markets, since several different grains and feedstuffs are relatively close substitutes. Price-sensitive corn importers may quickly switch to alternate, cheaper sources of energy depending on the availability of supplies and the adaptability of animal rations. In contrast, less price-sensitive corn importers, such as Japan and Taiwan, may choose to pay a higher price in an attempt to bid the corn away from ethanol plants. There could be significant economic effects to U.S. grain companies and to the U.S. agricultural sector if ethanol-induced higher corn prices cause a sustained reshaping of international grain trade.

Ethanol Processing Energy Needs. As ethanol production increases, the energy needed to process the corn into ethanol (derived primarily from natural gas in the United States) can be expected to increase. For example, if the entire 4.9 billion gallons of ethanol produced in 2006 used natural gas as a processing fuel, it would have required an estimated 243 billion cu. ft. of natural gas.[41] The energy needed to process the entire 2006 corn crop of 10.5 billion bushels into ethanol would be approximately 1.4 trillion cubic feet of natural gas. Total U.S. natural gas consumption was an estimated 22.2 trillion cu. ft. in 2005.[42] The United States has been a net importer of natural gas since the early 1980s. Because natural gas is used extensively in electricity production in the United States, a significant increase in its use as a processing fuel in the production of ethanol would likely result in increases of both prices and imports of natural gas.

Ethanol as a Substitute for Imported Fuel. Despite improving energy efficiency, the ability for domestic ethanol production to measurably substitute for petroleum imports is questionable, particularly when U.S. ethanol production depends almost entirely on corn as the primary feedstock. The import share of U.S. petroleum consumption was estimated at 65% in 2004, and is expected to grow to 71% by 2030.[43] Presently, ethanol production accounts for less than 3% of U.S. gasoline consumption while using about 20% of the U.S. corn production. If the entire 2006 U.S. corn crop of 10.5 billion bushels were used as ethanol feedstock, the resultant 28 billion gallons of ethanol (18.9 billion gasoline-equivalent gallons (GEG)) would

---

[40] USDA, WAOB, *WASDE Report*, October 12, 2007; available at [http://www.usda.gov/oce/].

[41] CRS calculations based on Shapouri (2004) energy usage rates of 49,733 Btu/gal of ethanol.

[42] DOE, EIA, *Annual Energy Outlook 2007.*

[43] Ibid.

represent about 13.4% of estimated national gasoline use of approximately 140 billion gallons.[44] In 2006, an estimated 71 million acres of corn were harvested. Nearly 137 million acres would be needed to produce enough corn (20.5 billion bushels) and subsequent ethanol (56.4 billion gallons or 37.8 billion GEG) to substitute for 50% of petroleum imports.[45] Since 1950, U.S. corn harvested acres have never reached 76 million acres. Thus, barring a drastic realignment of U.S. field crop production patterns, corn-based ethanol's potential as a petroleum import substitute appears to be limited by a crop area constraint.[46]

These supply issues suggest that corn's long-run potential as an ethanol feedstock is somewhat limited. The Department of Energy (DOE) suggests that the ability to produce ethanol from low-cost biomass will ultimately be the key to making it competitive as a gasoline additive.[47]

In light of these growing concerns, particularly as relates to livestock feed markets, the Nebraska Cattlemen (NC), at their annual convention on November 30, 2006, adopted two resolutions relating to federal policy intervention in the U.S. ethanol sector that are perhaps indicative of the looming tradeoff between feed and fuel and the types of policy options that will likely be debated in the coming months:[48]

> First Resolution: NC support a transition to a market-based approach for the usage and production of ethanol and are opposed to any additional federal or state mandates for ethanol usage and/or production.
>
> Second Resolution: NC favor the implementation of a variable import levy to prevent the price of oil, and its derivatives from dropping below long-term equilibrium prices. This should be the sole incentive for the development of alternative energy facilities in the United States.

Similarly, the National Cattlemen's Beef Association (NCBA), at their industry convention on February 3, 2007, approved an interim policy calling for: a phase-out of government incentives for ethanol production and an end to the 54 cent per gallon

---

[44] Based on USDA's February 9, 2007, *World Agricultural Supply and Demand Estimates (WASDE) Report*, and using comparable conversion rates.

[45] CRS calculations — which assume corn yields of 150 bushel per acre and an ethanol yield of 2.75 gal/bu. — are for gasoline only. Petroleum imports are primarily unrefined crude oil, which is then refined into a variety of products.

[46] Two recent articles by economists at Iowa State University examine the potential for obtaining a 10 million acre expansion in corn planting: Bruce Babcock and D. A. Hennessy, "Getting More Corn Acres From the Corn Belt"; and Chad E. Hart, "Feeding the Ethanol Boom: Where Will the Corn Come From?" *Iowa Ag Review*, Vol. 12, No. 4, Fall 2006.

[47] DOE, EIA, "Outlook for Biomass Ethanol Production and Demand," by Joseph DiPardo, July 30, 2002, available at [http://www.eia.doe.gov/oiaf/analysispaper/biomass.html]; hereafter referred to as DiPardo (2002).

[48] "Nebraska Cattlemen Adopts Ethanol Policy," News Release, December 1, 2006; available at [http://www.nebraskacattlemen.org/home/News/NewsReleases/tabid/116/articleType/ArticleView/articleId/142/Nebraska-Cattlemen-Adopts-Ethanol-Policy.aspx].

tariff on imported ethanol; a transition to a market-based approach to renewable energy production; and greater policy emphasis on transitioning from corn-based to cellulosic ethanol. The NCBA also announced support for a segmentation of the RFS whereby different biofuels would be given a specific portion of the RFS rather than letting it be filled on a first-come, first-serve basis. This would involve carving out a specific portion for cellulosic ethanol and biodiesel.

Ethanol from Cellulosic Biomass Crops.[49] Besides corn, several other agricultural products are viable feedstock and appear to offer attractive long-term supply potential — particularly cellulose-based feedstock such as prairie grasses and fast-growing woody crops such as hybrid poplar and willow trees, as well as waste biomass materials — logging residues, wood processing mill residues, urban wood wastes, and selected agricultural residues such as sugar cane bagasse and rice straw.[50] In particular, native prairie grasses such as switchgrass appear to offer large potential as a cellulosic feedstock because they thrive on marginal lands (as well as on prime cropland) and need little water and no fertilizer.[51]

The main impediment to the development of a cellulose-based ethanol industry is the state of cellulosic conversion technology (i.e., the process of gasifying cellulose-based feedstock or converting them into fermentable sugars). Currently, cellulosic conversion technology is rudimentary and expensive. In 2002, the DOE estimated that the cost of producing ethanol from cellulose was between $1.15 and $1.43 per gallon in 1998 dollars ($1.43 and $1.78 per gallon in current January 2007 dollars).[52] This compares with USDA's estimated cost of producing corn-based ethanol in 2002 of $0.958 per gallon ($1.08 per gal. in current January 2007 dollars).[53]

The projected high cost of production coupled with the uncertainty surrounding the commercial application of a new technology has inhibited commercial investments into cellulosic ethanol production. In addition, the logistics of harvesting, transporting, and storing large volumes of bulky cellulosic material for daily processing at a central plant remain daunting. As of October 2007, no commercial cellulose-to-ethanol facilities are in operation in the United States, although plans to

---

[49] For more information on biomass from non-traditional crops as a renewable energy, see the DOE, EERE, Biomass Program, "Biomass Feedstock," at [http://www1.eere.energy.gov/biomass/biomass_feedstocks.html]. See also, *Ethanol From Cellulose: A General Review*, P.C.Badger, Purdue University, Center for New Crops and Plant Products at [http://www.hort.purdue.edu/newcrop/ncnu02/v5-017.html].

[50] USDA and DOE. *Biomass as Feedstock for a Bioenergy and Bioproducts Industry: The Technical Feasibility of a Billion-Ton Annual Supply*, April 2005; available at [http://feedstockreview.ornl.gov/pdf/billion_ton_vision.pdf]; referred to hereafter as the Billion Ton Study (2005).

[51] Hill, Jason. *Overcoming Barriers to Biofuels: Energy from Diverse Prairie Biomass*, presentation to staff of House Committee on Agriculture, Dept of Applied Economics and Dept. of Ecology, Evolution, and Behavior, University of Minnesota, February 26, 2007.

[52] DiPardo, Joseph. *Outlook for Biomass Ethanol Production and Demand*, DOE, undated: available at [http://www.eia.doe.gov/oiaf/analysispaper/biomass.html].

[53] Shapouri (2004).

build several facilities are underway.[54] Private sector investment received a substantial federal policy boost on February 28, 2007, when the DOE announced the awarding of up to $385 million in cost-share funding for the construction of six cellulosic ethanol plant projects.[55] When fully operational, the six plants are expected to produce up to 130 million gallons per year of cellulosic ethanol. The combined cost-share plus federal funding for the six projects represents total planned investment of more than $1.2 billion. The six companies (and their proposed funding levels) are:

- Abengoa Bioenergy Biomass of Kansas, LLC of Chesterfield, Missouri (up to $76 million). The proposed plant will be located in Kansas. The plant will produce 11.4 million gallons of ethanol annually and enough energy to power the facility, with any excess energy being used to power the adjacent corn dry grind mill. The plant will use 700 tons per day of corn stover, wheat straw, milo stubble, switchgrass, and other feedstock.

- ALICO, Inc. of LaBelle, Florida (up to $33 million). The proposed plant will be in LaBelle, Florida. The plant will produce 13.9 million gallons of ethanol a year and 6,255 kilowatts of electric power, as well as 8.8 tons of hydrogen and 50 tons of ammonia per day. For feedstock, the plant will use 770 tons per day of yard, wood, and vegetative wastes and eventually energy cane.

- BlueFire Ethanol, Inc. of Irvine, California (up to $40 million). The proposed plant will be in Southern California. The plant will be sited on an existing landfill and produce about 19 million gallons of ethanol a year. As feedstock, the plant would use 700 tons per day of sorted green waste and wood waste from landfills.

- POET (originally Broin Companies) of Sioux Falls, South Dakota (up to $80 million). The plant is in Emmetsburg, Iowa, and after expansion, it will produce 125 million gallons of ethanol per year, of which roughly 25% will be cellulosic ethanol. For feedstock in the production of cellulosic ethanol, the plant expects to use 842 tons per day of corn fiber, cobs, and stalks.

- Iogen Biorefinery Partners, LLC, of Arlington, Virginia (up to $80 million). The proposed plant will be built in Shelley, Idaho, and will produce 18 million gallons of ethanol annually. The plant will use 700 tons per day of agricultural residues including wheat straw, barley straw, corn stover, switchgrass, and rice straw as feedstocks.

- Range Fuels (formerly Kergy Inc.) of Broomfield, Colorado (up to $76 million). The proposed plant will be constructed in Soperton, Georgia. The plant will produce about 40 million gallons of ethanol

[54] Perkins, Jerry. "New Crops Could Fuel New Wave of Ethanol," *Des Moines Register*, February 25, 2007.

[55] For more information see the DOE news release at [http://www.doe.gov/news/4827.htm].

> per year and 9 million gallons per year of methanol. As feedstock, the plant will use 1,200 tons per day of wood residues and wood based energy crops.

Celunol Corporation has a pilot cellulosic plant in Jennings, Louisiana, but is currently building a demonstration plant at the same location that will use sugar cane residues (called bagasse), as is done in Brazil, to fuel its ethanol production.[56] Celunol says that it will use the demonstration plant to train its plant operators in anticipation of a commercial-scale plant scheduled for construction starting in late 2008.

Economic Efficiency. The conversion of cellulosic feedstock to ethanol parallels the corn conversion process, except that the cellulose must first be converted to fermentable sugars. As a result, the key factors underlying cellulosic-based ethanol's price competitiveness are similar to those of corn-based ethanol, with the addition of the cost of cellulosic conversion. Cellulosic feedstock are significantly less expensive than corn; however, at present they are more costly to convert to ethanol because of the extensive processing required. Currently, cellulosic conversion is done using either dilute or concentrated acid hydrolysis — both processes are prohibitively expensive. However, the DOE suggests that enzymatic hydrolysis, which processes cellulose into sugar using cellulase enzymes, offers both processing advantages as well as the greatest potential for cost reductions. Current cost estimates of cellulase enzymes range from 30¢ to 50¢ per gallon of ethanol.[57] The DOE is also studying thermal hydrolysis as a potentially more cost-effective method for processing cellulose into sugar. Iogen — a Canadian firm with a pilot-scale cellulosic ethanol plant in Ottawa, Canada, and one of the six companies receiving a DOE award for construction of a commercial-scale ethanol plant (see above) — uses recombinant DNA-produced enzymes to break apart cellulose to produce sugar for fermentation into ethanol.

Based on the state of existing technologies and their potential for improvement, the DOE estimates that improvements to enzymatic hydrolysis could eventually bring the cost to less than 5¢ per gallon, but this may still be a decade or more away. Were this to happen, then the significantly lower cost of cellulosic feedstock would make cellulosic-based ethanol dramatically less expensive than corn-based ethanol and gasoline at current prices. Both the DOE and USDA are funding research to improve cellulosic conversion as well as to breed higher yielding cellulosic crops. In 1978, the DOE established the Bioenergy Feedstock Development Program (BFDP) at the Oak Ridge National Laboratory. The BFDP is engaged in the development of new crops and cropping systems that can be used as dedicated bioenergy feedstock. Some of the crops showing good cellulosic production per acre with strong potential for further gains include fast-growing trees (e.g., hybrid poplars and willows), shrubs, and grasses (e.g., switchgrass).

Government Support. Although no commercial cellulosic ethanol production has occurred yet in the United States, several federal laws support the development

---

[56] For more information on Celunol, visit: [http://www.celunol.com/].

[57] DOE, EERE, *Biomass Program*, "Cellulase Enzyme Research," available at [http://www1.eere.energy.gov/biomass/cellulase_enzyme.html].

of cellulose-based ethanol in the United States. These include various provisions under the Biomass Research and Development Act of 2000, two provisions (Section 2101 and Section 9008) of the 2002 farm bill (P.L. 107-171), and several provisions of the Energy Policy Act of 2005 (EPACT; P.L. 109-58).[58] The specifics of these provisions are discussed later in the report (see "Public Laws That Support Agriculture-Based Energy Production and Use," below). In addition to coordinating the activities of USDA and DOE, these provisions provide competitive grants, loans, and loan guarantees in support of research, education, extension, production, and market development of cellulosic biomass-based ethanol. In addition, both the Senate and House of passed new energy bills this year that include several provisions supportive of cellulosic ethanol production.[59]

In addition to existing legislation, the 2007 farm bill is likely to retain an energy title with updated and/or expanded provisions concerning agriculture-based renewable energy.[60] On July 27, 2007, the House approved a new farm bill — the Farm, Nutrition, and Bioenergy Act of 2007 (H.R. 2419) — which includes an energy title (Title IX). H.R. 2419, as amended and passed by the House, expands and extends several provisions from the energy title of the enacted 2002 farm bill with substantial increases in funding and a heightened focus on developing cellulosic ethanol production. A key departure from current farm-bill related energy provisions is that most new funding would be directed away from corn-starch-based ethanol production and towards either cellulosic-based biofuels production or to new as-yet-undeveloped technologies with some type of agricultural linkage. The Senate Agriculture Committee (SAC) is expected to mark up its version of a 2007 farm bill in late-October.

President Bush has mentioned renewable energy in his past two State of the Union (SOU) speeches. In his 2006 SOU, President Bush introduced the notion of "switchgrass" as a potential energy source and announced the "Advanced Energy Initiative," which included a goal of making cellulosic ethanol cost competitive with corn-based ethanol by 2012. In his 2007 SOU, President Bush announced his "20 in 10" plan, which calls for reducing U.S. gasoline consumption by 20% in 10 years (i.e., by 2017).

Energy Efficiency. The use of cellulosic biomass in the production of ethanol yields a higher net energy balance compared to corn — a 34% net gain for corn vs. a 100% gain for cellulosic biomass — based on a 1999 comparative study.[61] While corn's net energy balance (under optimistic assumptions concerning corn production

---

[58] For more information, see Biomass Research and Development Initiative, USDA/DOE, at [http://www.biomass.govtools.us/].

[59] more information on these energy bills, see CRS Report RL34136, *Biofuels Provisions in H.R. 3221 and H.R. 6: A Side-by-Side Comparison*, by Brent Yacobucci.

[60] For more information, see CRS Report RL34130 *Renewable Energy Policy in the 2007 Farm Bill* by Randy Schnepf.

[61] Argonne National Laboratory, Center for Transportation Research, *Effects of Fuel Ethanol Use on Fuel-Cycle Energy and Greenhouse Gas*, ANL/ESD-38, by M. Wang, C. Saricks, and D. Santini, January 1999, as referenced in DOE, DiPardo (2002).

and ethanol processing technology) was estimated at 67% by USDA in 2004, it is likely that cellulosic biomass's net energy balance would also have experienced parallel gains for the same reasons — improved crop yields and production practices, and improved processing technology. A recent review of research on ethanol's energy efficiency found that cellulose-based ethanol had a NEB ratio of 10, i.e., a 1,000% net gain.[62] As with corn-based ethanol, the NEB varies based on the production process used to grow, harvest, and process the feedstock.

Another factor that favors cellulosic ethanol's energy balance over corn-based ethanol relates to by-products. Corn-based ethanol's by-products are valued as animal feeds, whereas cellulosic ethanol's by-products are expected to serve directly as a processing fuel at the plant. This adaptation is expected to greatly improve both the economic efficiency and the net energy balance of cellulose ethanol over corn-based ethanol.

Long-Run Supply Issues. Cellulosic feedstock have an advantage over corn in that they grow well on marginal lands, whereas corn requires fertile cropland (as well as timely water and the addition of soil amendments). This greatly expands the potential area for growing cellulosic feedstock relative to corn. For example, in 2006 about 78 million acres were planted to corn, of which 75%, or about 59 million acres, were from the nine principal corn belts (IA, IL, IN, MN, MO, NE, OH, SD, WI). In contrast, that same year the United States had 243 million acres planted to the eight major field crops (corn, soybeans, wheat, cotton, barley, sorghum, oats, and rice), 433 million acres of total cropland (including forage crops and temporarily idled cropland), and 578 million acres of permanent pastureland, most of which is potentially viable for switchgrass production.[63]

A 2003 USDA study suggests that if 42 million acres of cropped, idle, pasture, and CRP acres were converted to switchgrass production, 188 million dry tons of switchgrass could be produced annually (at an implied yield of 4.5 metric tons per acre), resulting in the production of 16.7 billion gallons of ethanol or 10.9 billion GEG.[64] This would represent about 8% of U.S. gasoline use in 2005. Existing research plots have produced switchgrass yields of 15 dry tons per acre per year, suggesting tremendous long-run production potential. However, before any supply potential can be realized, research must first overcome the cellulosic conversion cost issue through technological developments.

In a 2005 study of U.S. biomass potential, USDA concluded that U.S. forest land and agricultural land had the potential to produce over 1.3 billion dry tons per year of biomass — 368 million dry tons from forest lands and 998 million dry tons from

---

[62] Alexander E. Farrel, Richard J. Pleven, Brian T. Turner, Andrew D. Jones, Michael O'Hare, and Daniel M. Kammon, "Ethanol Can Contribute to Energy and Environmental Goals," *Science*, vol. 311 (January 27, 2006), pp. 506-508.

[63] United Nations, Food and Agricultural Organization (FAO), FAOSTATS.

[64] USDA, Office of Energy Policy and New Uses (OEPNU), *The Economic Impacts of Bioenergy Crop Production on U.S. Agriculture*, AER 816, by Daniel De La Torre Ugarte et al., February 2003; available at [http://www.usda.gov/oce/reports/energy/index.htm].

agricultural lands — while still continuing to meet food, feed, and export demands.[65] According to the study, this volume of biomass would be more than ample to displace 30% or more of current U.S. petroleum consumption.

USDA's very optimistic assessment is tempered somewhat by a 2005 University of Minnesota study that uses the results from three major biofuels studies to estimate the potential supplies of biofuels from both corn-based ethanol and cellulosic-based ethanol from biomass crops and crop residue.[66] The analysis suggests that about 130.4 million tons of biomass could be produced directly from switchgrass with another 130.5 million tons from crop residue. If the biomass total of 260.9 million tons were converted to ethanol at a rate of 89.7 gallons per ton, it would produce 23.4 billion gallons of anhydrous ethanol. Adding 2% denaturant yields 23.9 billion gallons. Adding an additional 7 billion gallons of corn-based ethanol brings the total to 30.9 billion gallons or 20.7 billion GEG. This would represent about 22.7% of total U.S. gasoline consumption in 2005.

Methane from an Anaerobic Digester

An anaerobic digester is a device that promotes the decomposition of manure or "digestion" of the organics in manure by anaerobic bacteria (in the absence of oxygen) to simple organics while producing biogas as a waste product.[67] The principal components of biogas from this process are methane (60% to 70%), carbon dioxide (30% to 40%), and trace amounts of other gases. Methane is the major component of the natural gas used in many homes for cooking and heating, and is a significant fuel in electricity production. Biogas can also be used as a fuel in a hot water heater if hydrogen sulfide is first removed from the biogas supply. As a result, the generation and use of biogas can significantly reduce the cost of electricity and other farm fuels such as natural gas, propane, and fuel oil.

By early 2005, there were 100 digester systems in operation at commercial U.S. livestock farms, with an additional 94 planned for construction.[68] EPA estimates that anaerobic digester biogas systems are technically feasible at about 7,000 dairy and swine operations in the United States. The majority of existing systems are farm owned and operated using only livestock manure, and are found in the dairy production zones of California, Wisconsin, Pennsylvania, and New York. In 2005,

---

[65] USDA and DOE, Billion Ton Study (2005)

[66] Eidman, Vernon R. *Agriculture's Role in Energy Production: Current Levels and Future Prospects*, paper presented at a conference, "Energy from Agriculture: New Technologies, Innovative Programs and Success Stories," December 14-15, 2005, St. Louis, Missouri. The three studies used to generate the estimate are listed in the "For More Information" section as FAPRI (2005); Ugarte et al (2003); and Gallagher et al (2003).

[67] For more information on anaerobic digesters, see Appropriate Technology Transfer for Rural Areas (ATTRA), *Anaerobic Digestion of Animal Wastes: Factors to Consider*, by John Balsam, October 2002, at [http://www.attra.ncat.org/energy.html#Renewable]; or Iowa State University, Agricultural Marketing Resource Center, *Anaerobic Digesters*, at [http://www.agmrc.org/agmrc/commodity/biomass/].

[68] U.S. Environmental Protection Agency (EPA), *AgStar Digest*, Winter 2006; available at [http://www.epa.gov/agstar/].

they are estimated to have generated over 130 million kWh and to have reduced methane emissions by over 30,000 metric tons.

Anaerobic digestion system proposals have frequently received funding under the Renewable Energy Program (REP) of the 2002 farm bill (P.L. 107-171, Title IX, Section 9006). For example, in 2004 37 anaerobic digester proposals from 26 different states were awarded funding under the REP.[69] Also, the AgStar program — a voluntary cooperative effort by USDA, EPA, and DOE — encourages methane recovery at confined livestock operations that manage manure as liquid slurries.

Economic Efficiency. The primary benefits of anaerobic digestion are animal waste management, odor control, nutrient recycling, greenhouse gas reduction, and water quality protection. Except in very large systems, biogas production is a highly useful but secondary benefit. As a result, anaerobic digestion systems do not effectively compete with other renewable energy production systems on the basis of energy production alone. Instead, they compete with and are cost-competitive when compared to conventional waste management practices according to EPA.[70] Depending on the infrastructure design — generally some combination of storage pond, covered or aerated treatment lagoon, heated digester, and open storage tank — anaerobic digestion systems can range in investment cost from $200 to $500 per Animal Unit (i.e., per 1,000 pounds of live weight). In addition to the initial infrastructure investment, recurring costs include manure and effluent handling, and general maintenance. According to EPA, these systems can have financially attractive payback periods of three to seven years when energy gas uses are employed. On average, manure from a lactating 1,400-pound dairy cow can generate enough biogas to produce 550 Kilowatts per year.[71] A 200-head dairy herd could generate 500 to 600 Kilowatts per day. At 6¢ per kilowatt hour, this would represent potential energy cost savings of $6,000 to $10,000 per year.

The principal by-product of anaerobic digestion is the effluent (i.e., the digested manure). Because anaerobic digestion substantially reduces ammonia losses, the effluent is more nitrogen-rich than untreated manure, making it more valuable for subsequent field application. Also, digested manure is high in fiber, making it valuable as a high-quality potting soil ingredient or mulch. Other cost savings include lower total lagoon volume requirements for animal waste management systems (which reduces excavation costs and the land area requirement), and lower cover costs because of smaller lagoon surface areas.

Government Support. Federal assistance in the form of grants, loans, and loan guarantees is available under USDA's Renewable Energy Program (2002 farm bill,

[69] USDA, News Release No. 0386.04, September 15, 2004; *Veneman Announces $22.8 Million to Support Renewable Energy Initiatives in 26 States*, available at [http://www.usda.gov/Newsroom/0386.04.html]. For funding and program information on the Renewable Energy and Energy Efficiency Program, see [http://www.rurdev.usda.gov/rd/energy/].

[70] EPA, OAR, *Managing Manure with Biogas Recovery Systems*, EPA-430-F-02-004, winter 2002.

[71] ATTRA, *Anaerobic Digestion of Animal Wastes: Factors to Consider*, October 2002.

Title IX, Section 9006) and Rural Development Programs (Title VI, Sections 6013, 6017, and 6401). See the section below on public laws for more details.

Energy Efficiency. Because biogas is essentially a by-product of an animal waste management activity, and because the biogas produced by the system can be used to operate the system, the energy output from an anaerobic digestion system can be viewed as achieving even or positive energy balance. The principal energy input would be the fuel used to operate the manure handling equipment.

Long-Run Supply Issues. Anaerobic digesters are most feasible alongside large confined animal feeding operations (CAFOs). According to EPA, biogas production for generating cost effective electricity requires manure from more than 500 cows at a dairy operation or at least 2,000 head of swine at a pig feeding operation. As animal feeding operations steadily increase in size, the opportunity for anaerobic digestion systems will likewise increase. In addition, some digester systems may qualify for cost-share funds under USDA's Environmental Quality Incentives Program (EQIP).

Biodiesel

Biodiesel is an alternative diesel fuel that is produced from any animal fat or vegetable oil (such as soybean oil or recycled cooking oil). About 90% of U.S. biodiesel is made from soybean oil. As a result, U.S. soybean producers and the American Soybean Association (ASA) are strong advocates for greater government support for biodiesel production.

According to the National Biodiesel Board (NBB), biodiesel is nontoxic, biodegradable, and essentially free of sulfur and aromatics. In addition, it works in any diesel engine with few or no modifications and offers similar fuel economy, horsepower, and torque, but with superior lubricity and important emission improvements over petroleum diesel.[72] Biodiesel is increasingly being adopted by major fleets nationwide. The U.S. Postal Service, the U.S. military, and many state governments are directing their bus and truck fleets to incorporate biodiesel fuels as part of their fuel base.

U.S. biodiesel production has shown strong growth in recent years, increasing from under 1 million gallons in 1999 to an estimated 386 million gallons in 2006 (**Figure 5**). However, U.S. biodiesel production remains small relative to national diesel consumption levels. In 2005, estimated biodiesel production of 200 million gallons represented 0.4% of the 44.9 billion gallons of diesel fuel used nationally for vehicle transportation.[73] In addition to vehicle use, 18.5 billion gallons of diesel fuel were used for heating and power generation by residential, commercial, and industry, and by railroad and vessel traffic in 2005, bringing total U.S. diesel fuel use to nearly 63.1 billion gallons.

---

[72] For more information, visit the National Biodiesel Board at [http://www.biodiesel.org].

[73] Diesel consumption estimates are from DOE, IEA, *Annual Energy Outlook 2007.*

Figure 5. U.S. Biodiesel Production, 1998-2007

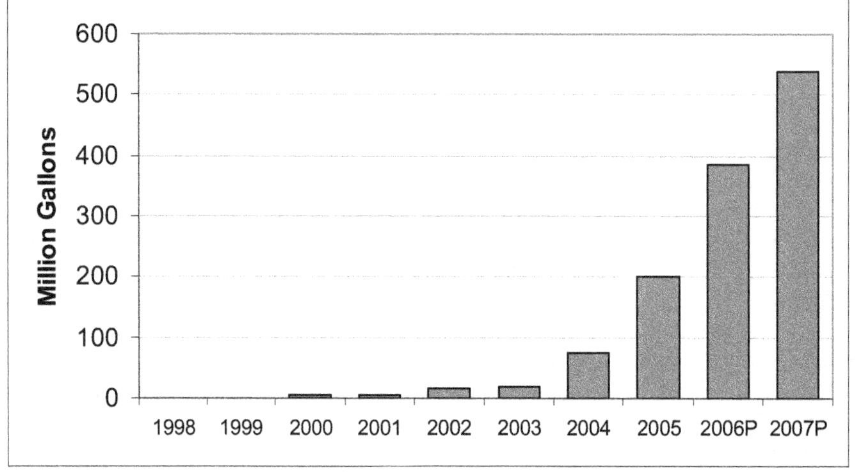

Source: 1998-2003: National Biodiesel Board; 2004-07 projected by FAPRI, March 2007.

According to the NBB, as of October 15, 2007, there were 165 companies in the United States with the potential to produce biodiesel commercially that were either in operation, under expansion, or under construction but scheduled to be completed within the next 18 months.[74] The NBB reported that the combined annual biodiesel production capacity (within the oleochemical industry) of these 165 plants once fully operational (including capacity under construction) would be an estimated 1.85 billion gallons per year. Because many of these plants also can produce other products such as cosmetics, estimated total capacity (and capacity for expansion) is far greater than actual biodiesel production.

Economic Efficiency. Despite the rapid expansion of biodiesel production capacity, there are two major economic impediments to future growth. First, biodiesel production is not profitable under the current set of market prices for vegetable oil feedstocks and the biodiesel produced. Second, the cost of producing biodiesel is generally more than the cost of producing its fossil fuel counterpart making the finished biodiesel uncompetitive with its fossil fuel counterpart in the marketplace. A 2004 DOE study suggests that, since the cost of the feedstock (whether vegetable oil or restaurant grease) is the largest single component of biodiesel production, the cost of producing biodiesel varies substantially with the choice of feedstock.[75] For example, in 2004/05 it cost $0.75 to produce a gallon of petroleum-based diesel, compared with about $2.86 to produce a gallon of biodiesel from soybean oil, and $1.59 from restaurant grease (all prices are quoted in 2002 dollars). However, wholesale soybean oil prices have doubled since 2002 — a pound of soybean oil sold

---

[74] A description of biodiesel production capacity with maps of existing and proposed plants is available at [http://www.biodiesel.org/resources/fuelfactsheets/default.shtm].

[75] Radich, Anthony. "Biodiesel Performance, Costs, and Use," Modeling and Analysis Papers, DOE/EIA, June 2004; available at [http://www.eia.doe.gov/oiaf/analysispaper/biodiesel/].

CRS-29

for $0.18 in U.S. wholesale markets in 2002 compared with the September 2007 average wholesale price of nearly $0.37.

The production cost differentials generally manifest themselves at the retail level as well. During July 2007, the retail price of B100 (100% biodiesel) averaged $3.27 per gallon, compared with $2.96 for conventional diesel fuel (**Table 2**).

Table 5. U.S. Diesel Fuel Use, 2005

	Total		Hypothetical scenario: 2% of total use[b]	
U.S. Diesel Use in 2004	**Million gallons[a]**	**%**	**Million gallons**	**Soybean oil equivalents: million pounds[a]**
Total Vehicle Use	44,887	71%	898	6,733
On-Road	38,053	60%	761	5,708
Off-Road	3,030	5%	61	455
Military	272	0%	5	41
Farm	3,532	6%	71	530
Total Non-vehicle Use	18,532	29%	365	2,736
All uses	63,129	100%	1263	9,469

Source: DOE, EIA, *U.S. Annual Adjusted Sales of Distillate Fuel Oil by End Use*.

a. Pounds are converted from gallons of oil using a 7.5 pounds-to-gallon conversion rate.
b. Hypothetical scenario included for comparison purposes only.

The prices of biodiesel feedstock, as well as petroleum-based diesel fuel, vary over time based on domestic and international supply and demand conditions. About 7.5 pounds of soybean oil are needed to produce a gallon of biodiesel. A comparison of the relative price relationship between soybean oil and petroleum diesel is indicative of the general economic viability of biodiesel production (**Figure 6**). As diesel fuel prices rise relative to biodiesel or biodiesel feedstock, and/or as biodiesel production costs fall through lower commodity prices or technological improvements in the production process, biodiesel becomes more economical. In addition, federal and state assistance helps to make biodiesel more competitive with diesel fuel.

Since late 2006, the soybean oil to diesel wholesale price comparison has turned against the use of soybean oil for biodiesel production — soybean oil prices have risen steadily (along with corn prices) above the 35¢ per pound ($2.63 per gallon) range, while diesel fuel have varied around $2 per gallon. Since early September 2007, the nearby CBOT futures contract for soybean oil has traded near 40¢ per pound ($3.00 per gallon), while more deferred contracts have been at or above 40¢. At such high soybean oil prices, biodiesel production is unprofitable even with the government subsidy of $1.00 per gallon.

Figure 6. Soybean Oil Versus Diesel Fuel Price, 2000-2007

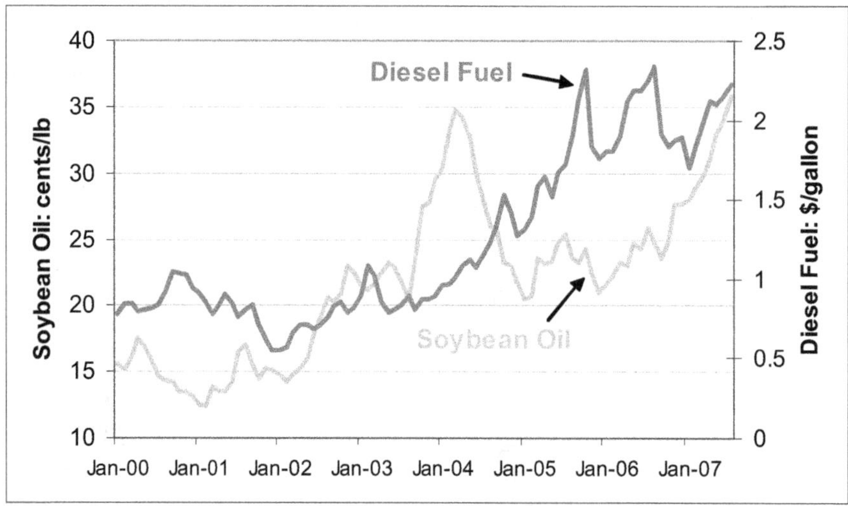

Source: No.2 diesel fuel national average wholesale price: DOE, EIA; soybean oil, Decatur, IL, USDA, FAS "Oilseed Circular."

The tremendous surge in soybean oil prices in particular, and vegetable oil prices in general that has occurred since 2001 is being driven in large part by rapid economic growth in China and India. As incomes of the lower economic strata grow in these countries, their demand for vegetable oil and other high quality food products has grown commensurately. The long-run outlook, as projected by both USDA and FAPRI, is for continued strong economic growth in China, India, and other developing countries over the next ten years. As a consequence, sustained strong vegetable oil prices are expected to curtail growth in the U.S. biodiesel sector under the current policy and market setting.

Government Support. The primary federal incentives for biodiesel production are somewhat similar to ethanol and include the following.[76]

- A production excise tax credit signed into law on October 22, 2004, as part of the American Jobs Creation Act of 2004 (Sec. 1344; P.L. 109-58). Under the biodiesel production tax credit, the subsidy amounts to \$1.00 for every gallon of agri-biodiesel (i.e., virgin vegetable oil and animal fat) that is used in blending with petroleum diesel. A 50¢ credit is available for every gallon of non-agri-biodiesel (i.e., recycled oils such as yellow grease). However, unlike the ethanol tax credit, which was extended through 2010, the biodiesel tax credit expires at the end of calendar year 2008.
- A small producer income tax credit (Sec. 1345; P.L. 109-58) of 10¢ per gallon for the first 15 million gallons of production for biodiesel

[76] See also section on "Public Laws That Support Agriculture-Based Energy Production and Use," below.

producers whose total output does not exceed 60 million gallons of biodiesel per year.

- Incentive payments (contingent on annual appropriations) on year-to-year production increases of renewable energy were previously available under USDA's Bioenergy Program (7 U.S.C. 8108); however, funding for this program expired at the end of FY2006.

Indirectly, other federal programs support biodiesel production by requiring federal agencies to give preference to biobased products in purchasing fuels and other supplies and by providing incentives for research on renewable fuels. Also, several states have their own incentives, regulations, and programs in support of renewable fuel research, production, and consumption that supplement or exceed federal incentives.

Energy Efficiency. Biodiesel appears to have a significantly better net energy balance than ethanol, according to a joint USDA-DOE 1998 study that found biodiesel to have an NEB of 3.2 — that is, 220% more energy was returned from a gallon of pure biodiesel than was used in its production.[77]

Long-Run Supply Issues. Both the ASA and the NBB are optimistic that the federal biodiesel tax incentive will provide the same boost to biodiesel production that ethanol has obtained from its federal tax incentive.[78] However, many commodity market analysts are skeptical of such claims. They contend that the biodiesel industry still faces several hurdles: the retail distribution network for biodiesel has yet to be established; the federal tax credit, which expires on December 31, 2008, does not provide sufficient time for the industry to develop; and potential domestic oil feedstock are relatively less abundant than ethanol feedstock, making the long-run outlook more uncertain.

In addition, biodiesel production confronts the same limited ability to substitute for petroleum imports and the same type of consumption tradeoffs as ethanol production. As an example consider a hypothetical scenario (as shown in **Table 5**) whereby a 2% usage requirement for vehicle diesel fuel were to be adopted. (This would replicate the European Union's goal of 2% of transportation fuels originating from biofuels by 2005, then growing to 5.75% by 2010). This hypothetical mandate would require about 898 million gallons of biodiesel (compared to estimated 2006 production of about 386 million gallons) or approximately 6.7 billion pounds of vegetable oil. During 2005/06, a total of 36.7 billion pounds of vegetable oils and animal fats were produced in the United States (**Table 6**); however, most of this production was committed to other food and industrial uses. Uncommitted biodiesel feedstock (as measured by the available stock levels on September 30, 2006) were about 4 billion pounds. Thus, after exhausting all available feedstock, an additional 2.7 billion pounds of oil would be needed to meet the hypothetical 2% biodiesel

---

[77] DOE, National Renewable Energy Laboratory (NREL), *An Overview of Biodiesel and Petroleum Diesel Life Cycles*, NREL/TP-580-24772, by John Sheehan et al., May 1998, available at [http://www.nrel.gov/docs/legosti/fy98/24772.pdf].

[78] For more information, see NBB, "Ground-Breaking Biodiesel Tax Incentive Passes," at [http://www.biodiesel.org/resources/pressreleases/gen/20041011_FSC_Passes_Senate.pdf].

blending requirement. This exceeds the 2 billion pounds of total vegetable oils exported by the United States in 2005/06, and is nearly double the 1.3 billion pounds of soybean oil exported that during same period.[79]

If U.S. soybean vegetable oil exports were to remain unchanged, the deficit biodiesel feedstock could be obtained either by reducing U.S. exports of whole soybeans by about 250 million bushels (then crushing them for their oil) or by expanding soybean production by approximately 6 million acres (assuming a yield of about 42 bushels of soybeans per acre).[80] Of course, any area expansion would likely come at the expense of some other crop such as corn, cotton, or wheat. Current high corn prices make such an area shift seem unlikely, at least in the near term. A further possibility is that U.S. oilseed producers could shift towards the production of higher-oil content crops such as canola or sunflower.

The bottom line is that a small increase in demand of fats and oils for biodiesel production could quickly exhaust available feedstock supplies and push vegetable oil prices significantly higher due to the low elasticity of demand for vegetable oils in food consumption.[81] Rising vegetable oil prices would reduce or eliminate biodiesel's competitive advantage vis-à-vis petroleum diesel, even with the federal tax credit, while increased oilseed crushing would begin to disturb feed markets.

As with ethanol production, increased soybean oil production (dedicated to biodiesel production) would generate substantial increases in animal feeds in the form of high-protein meals. When a bushel of soybeans is processed (or crushed), nearly 80% of the resultant output is in the form of soybean meal, while only about 18%-19% is output as soybean oil. Thus, for every 1 pound of soybean oil produced by crushing whole soybeans, over 4 pounds of soybean meal are also produced.

Crushing an additional 250 million bushels of soybeans for soybean oil would produce over 7.5 million short tons (s.t.) of soybean meal. In 2005/06, the United States produced 41.2 million s.t. of soybean meal. An additional 7.5 million s.t. of soybean meal (an increase of over 18%) entering U.S. feed markets would compete directly with the feed by-products of ethanol production (distillers dried grains, corn gluten feed, and corn gluten meal) with economic ramifications that have not yet been fully explored. Also similar to ethanol production, natural gas demand would likely rise with the increase in biodiesel processing.[82]

---

[79] U.S. export data is from USDA, FAS, PSD Online, February 9, 2007.

[80] Assuming 18% oil content per bushel of soybeans.

[81] ERS reported the U.S. own-price elasticity for "oils & fats" at -0.027 — i.e., a 10% increase in price would result in a 0.27% decline in consumption. In other words, demand declines only negligibly relative to a price rise. Such inelastic demand is associated with sharp price spikes in periods of supply shortfall. USDA, ERS, *International Evidence on Food Consumption Patterns*, Tech. Bulletin No. 1904, September 2003, p. 67.

[82] Assuming natural gas is the processing fuel, natural gas demand would increase due to two factors: (1) to produce the steam and process heat in oilseed crushing and (2) to produce methanol used in the conversion step. NREL, *An Overview of Biodiesel and Petroleum Diesel Life Cycles*, NREL/TP-580-24772, by John Sheehan et al., May 1998, p. 19.

CRS-33

Table 6. U.S. Potential Biodiesel Feedstock, 2005-2006

Oil type	Wholesale price[a] $/lb	Oil Production, 2005-2006		Ending Stocks: Sept. 30, 2006	
		Million pounds	Million gallons[b]	Million pounds	Million gallons[b]
Crops		**25827**	**3444**	**3711**	**495**
Soybean	**23**	20,393	2719	3020	403
Corn	28.4	2450	327	156	21
Cottonseed	28.9	950	127	101	14
Sunflowerseed	44.5	545	731	55	7
Canola	30.7	880	117	262	35
Peanut	50.6	219	298	60	8
Flaxseed/linseed	64.6	335	454	45	6
Safflower	72.2	56	7	11	1
Animal fat & other		**10885**	**1451**	**326**	**43**
Lard	21.1	795	106	13	2
Edible tallow	19.4	1780	237	22	3
Inedible tallow[c]	na	7110	948	238	32
Yellow grease[c]	11.6	1200	160	53	7
Total supply		**36,712**	**4895**	**4,037**	**538**

Source: USDA, ERS, *Oil Crops Yearbook*, OCS-2006, March 2006, Table 31. Rapeseed was calculated by multiplying oil production by a 40% conversion rate. The inedible tallow and yellow grease supplies come from Dept of Commerce, Bureau of Census, *Fats and Oils, Production, Consumption and Stocks, December 2006*; [http://www.census.gov/cir/www/311/m311k.html].

na = not available.

a. Average of monthly wholesale price quotes for vegetable oils are for 2005 calendar from USDA, FAS, *Oilseeds: World Markets and Trade*. Lard and edible tallow prices are for calendar 2005 from USDA, ERS, *Oil Crops Yearbook*, OCS-2006, March 2006, Tables 42 and 44. Yellow grease price is 1993-95 average from USDA, ERS, AER 770, Sept. 1998, p. 9.

b. Pounds are converted to gallons of oil using a 7.5 pounds-to-gallon conversion rate.

c. CRS annual projections for production and stocks based on Dept of Commerce Dec. 2006 monthly estimates from source cited above.

Wind Energy Systems

In 2006, electricity from wind energy systems accounted for about 0.1% of U.S. total energy consumption (**Table 1**). However, wind-generated electricity has been a much larger share of electricity used by the U.S. agriculture sector (28%), and of total direct energy used by U.S. agriculture (9%).[83] According to the American Wind Energy Association (AWEA), total installed wind energy production capacity has

[83] Data for agricultural use of wind-generated electricity is for 2003. For more information on energy consumption by U.S. agriculture, see CRS Report RL32677, *Energy Use in Agriculture: Background and Issues*, by Randy Schnepf.

expanded rapidly in the United States since the late 1990s, rising from 1,848 megawatts (MW) in 1998 to a reported 12,634 MW by June 30, 2007 (**Figure 7**).[84] According to the AWEA, on-line capacity is projected to expand to over 14,600 MW by the end of 2007. (See "Box: Primer on Measuring Electric Energy," later in this report, for a description of megawatts and other energy terminology.)

Figure 7. U.S. Installed Wind Energy Capacity, 1981-2007

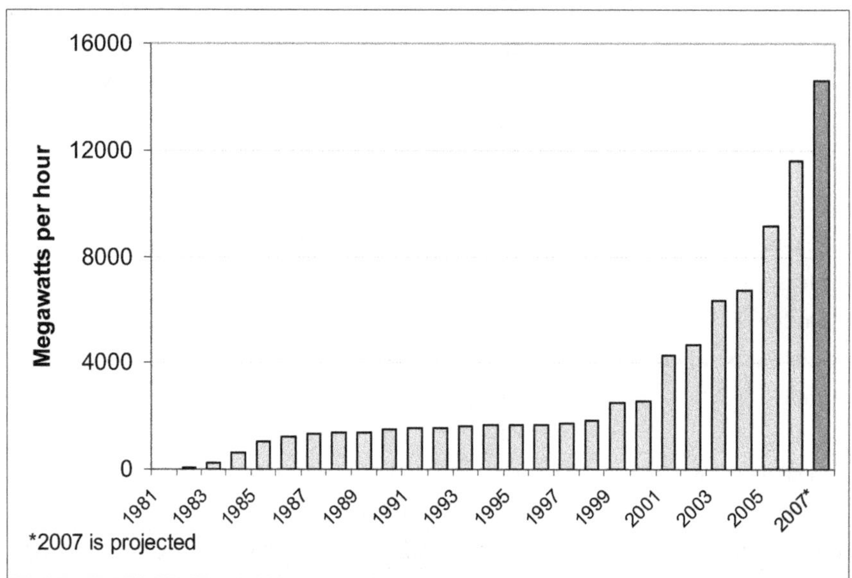

Source: American Wind Energy Association (AWEA).

About 7% of installed production capacity is in 10 predominantly midwestern and western states (see **Table 7**).

What Is Behind the Rapid Growth of Installed Capacity? Over the past 20 years, the cost of wind power has fallen approximately 90%, while rising natural gas prices have pushed up costs for gas-fired power plants, helping to improve wind energy's market competitiveness.[85] In addition, wind-generated electricity production and use is supported by several federal and state financial and tax incentives, loan and grant programs, and renewable portfolio standards.

---

[84] American Wind Energy Association (AWEA), at [http://www.awea.org/projects/].

[85] AWEA, *The Economics of Wind Energy*, March 2002.

CRS-35

Table 7. Installed Wind Energy Capacity by State, Ranked by Capacity as of December 31, 2006

	State	Current MW	Current Share	UC[a] or Planned MW	Total MW	Total Share
1	Texas	3,352	26.5%	1,246	4,599	23.7%
2	California	2,376	18.8%	565	2,941	15.2%
3	Iowa	967	7.7%	203	1,170	6.0%
4	Minnesota	897	7.1%	406	1,303	6.7%
5	Washington	818	6.5%	140	958	4.9%
6	Oklahoma	595	4.7%	95	689	3.6%
7	New Mexico	496	3.9%	-	496	2.6%
8	Oregon	438	3.5%	501	939	4.8%
9	New York	390	3.1%	282	672	3.5%
10	Colorado	366	2.9%	700	1,066	5.5%
11	Kansas	364	2.9%	-	364	1.9%
12	Illinois	305	2.4%	80	385	2.0%
13	Wyoming	288	2.3%	220	508	2.6%
14	Pennsylvania	179	1.4%	115	294	1.5%
15	North Dakota	178	1.4%	159	337	1.7%
16	Montana	146	1.2%	20	166	0.9%
	Other	478	3.8%	2,037	2,515	13.0%
	U.S. Total	12,634	100%	6,769	19,403	100%

Source: AWEA, [http://www.awea.org/projects/].
[a]UC = Under construction.

As of February 2007, renewable portfolio standards (RPSs) had been adopted by 21 states and the District of Columbia.[86] An RPS requires that utilities must derive a certain percentage of their overall electric generation from renewable energy sources such as wind power. Environmental and energy security concerns also have encouraged interest in clean, renewable energy sources such as wind power. Finally, rural incomes receive a boost from companies installing wind turbines in rural areas. sLandowners have typically received annual lease fees that range from $2,000 to $4000 per turbine per year for up to 20 years depending on factors such as the project size, the capacity of the turbines, and the amount of electricity produced.

Economic Efficiency. The per-unit cost of utility-scale wind energy is the sum of the various costs — capital, operations, and maintenance — divided by the

[86] AWEA, "State-Level Renewable Energy Portfolio Standards (RPS)," downloaded March 7, 2007, from [http://www.awea.org/legislative/pdf/State_RPS_Fact_Sheet_UPdated.pdf].

annual energy generation. Utility-scale wind power projects — those projects that generate at least 1 MW of electric power annually for sale to a local utility — account for over 90% of wind power generation in the United States.[87] For utility-scale sources of wind power, a number of turbines are usually built close together to form a wind farm.

In contrast with biofuel energy, wind power has no fuel costs. Instead, electricity production depends on the kinetic energy of wind (replenished through atmospheric processes). As a result, its operating costs are lower than costs for power generated from biofuels. However, the initial capital investment in equipment needed to set up a utility-scale wind energy system is substantially greater than for competing fossil or biofuels. Major infrastructure costs include the tower (30 meters or higher) and the turbine blades (generally constructed of fiberglass; up to 20 meters in length; and weighing several thousand pounds). Capital costs generally run about $1 million per MW of capacity, so a wind energy system of 10 1.5-MW turbines would cost about $15 million. Farmers generally find leasing their land for wind power projects easier than owning projects. Leasing is easier because energy companies can better address the costs, technical issues, tax advantages, and risks of wind projects. In 2004, less than 1% of wind power capacity installed nationwide was owned by farmers.[88]

While the financing costs of a wind energy project dominate its competitiveness in the energy marketplace, there are several other factors that also contribute to the economics of utility-scale wind energy production. These include:[89]

- the wind speed and frequency at the turbine location — the energy that can be tapped from the wind is proportional to the cube of the wind speed, so a slight increase in wind speed results in a large increase in electricity generation;
- improvements in turbine design and configuration — the taller the turbine and the larger the area swept by the blades, the more productive the turbine;
- economies of scale — larger systems operate more economically than smaller systems by spreading operations/maintenance costs over more kilowatt-hours;
- transmission and market access conditions (see below); and
- environmental and other policy constraints — for example, stricter environmental regulations placed on fossil fuel emissions enhance wind energy's economic competitiveness; or, alternately, greater protection of birds or bats,[90] especially threatened or endangered species, could reduce wind energy's economic competitiveness.

---

[87] GAO, *Wind Power*, GAO-04-756, September 2004, p. 66.

[88] Ibid., p. 6.

[89] AWEA, *The Economics of Wind Energy*, at [http://www.awea.org].

[90] Justin Blum, "Researchers Alarmed by Bat Deaths From Wind Turbines," *Washington Post*, by January 1, 2005.

A modern wind turbine can produce electricity for about 4.3¢ to 5.8¢ per kilowatt hour. In contrast to wind-generated electricity costs, modern natural-gas-fired power plants produce a kilowatt-hour of electricity for about 5.5¢ (including both fuel and capital costs) when natural gas prices are at $6 per million Btu's (or equivalently per 1,000 cu.ft.).[91]

Wellhead natural gas prices have shown considerable volatility since the late 1990s (**Figure 8**), but spiked sharply upward in September 2005 following Hurricane Katrina's damage to the Gulf Coast petroleum and natural gas importing and refining infrastructure. Prices have fallen back substantially from their November 2005 peak of $11.92 per 1,000 cu.ft., however, market conditions suggest that the steady price rise that has occurred since 2002 is unlikely to weaken anytime soon.[92] If natural gas prices continue to be substantially higher than average levels in the 1990s, wind power is likely to be competitive in parts of the country where good wind resources and transmission access can be coupled with the federal production tax credit.

Figure 8. Natural Gas Price, Wholesale, 1994-2007

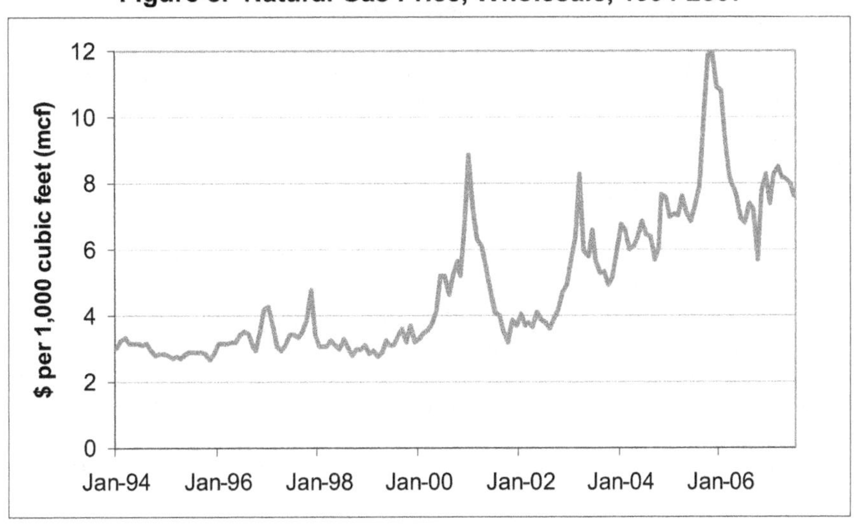

Source: DOE, EIA; monthly average wholesale (Industrial) price.

According to a wind energy consultant, wind turbine ownership offers substantially greater returns than crop farming.[93] For example, installing two 1.5 MW wind turbines producing 9 million kWh per year would generate about $325,000 (based on 3.5¢ per kWh and 16 miles per hour wind speed at 50 meters height). As an example, this compares with a 160-acre corn and soybean farm with average annual gross receipts of about $66,000 (80 acres of corn at 170 bushels/acre and $3.00 per

[91] Rebecca Smith, "Not Just Tilting Anymore," *Wall Street Journal*, October 14, 2004.

[92] For a discussion of natural gas market price factors, see CRS Report RL33714, *Natural Gas Markets in 2006*, by Robert Pirog.

[93] Tom Wind, Wind Utility Consulting, "Wind Power," presentation given on March 2, 2007, USDA Annual Outlook Forum 2007, Crystal City, VA.

bushel; and 80 acres of soybeans at 45 bushels at $7.00 per bushel). The consultant also suggests that carbon trading could potentially add another $100,000 per year within a decade.

Government Support. In addition to market factors, the rate of wind energy system development for electricity generation has been highly dependent on federal government support, particularly a production tax credit that provides a 1.8¢ credit for each kilowatt-hour of electricity produced by qualifying turbines built by the end of 2008 for a 10-year period.[94] The usefulness of the tax credit may be limited by a restriction under current U.S. tax law (IRC § 469) whereby individuals are not eligible to deduct losses incurred in businesses that they do not actively participate in. Legislation (H.R. 2007) was introduced in the 109th Congress to allow passive investors that provide capital for wind energy facilities and projects to be eligible for up to a $25,000 passive loss deduction in the Internal Revenue Code. The legislation was referred to the House Committee on Ways and Means but no further action was taken, and similar legislation has yet to be introduced in the 110th Congress. Currently, the $25,000 passive loss offset is only available for oil, gas, and real estate investments.

The inclusion of the federal tax credit reduces the cost of producing wind-generated electricity to 2.5¢ to 4¢ per kilowatt hour. In some cases the tax credit may be combined with a five-year accelerated depreciation schedule for wind turbines, as well as with grants, loans, and loan guarantees offered under several different programs.[95] To the extent that they offset a substantial portion (30% to 40%) of the price risk and initial financing charges, government incentives often provide the catalyst for stimulating new investments in rural wind energy systems.

Long-Run Supply Issues. Despite the advantages listed above, U.S. wind potential remains largely untapped, particularly in many of the states with the greatest wind potential, such as North and South Dakota (see **Figure 9**). Factors inhibiting growth in these states include lack of either (1) major population centers with large electric power demand needed to justify large investments in wind power, or (2) adequate transmission capacity to carry electricity produced from wind in sparsely populated rural areas to distant cities.

Areas considered most favorable for wind power have average annual wind speeds of about 16 miles per hour or more. The minimum wind velocity needed for electricity production by a wind turbine is 10 miles per hour.[96] The turbines operate

---

[94] The federal production tax credit was initially established as a 1.5¢ tax credit in 1992 dollars in the Energy Policy Act of 1992 (P.L. 102-146). The tax credit was extended through 2007 in the American Jobs Creation Act of 2004 (P.L. 108-357; Sec. 710), with an adjustment for annual inflation that raised it to its current value of 1.8¢ per kWh. The tax credit was further extended through 2008 by a provision in P.L. 109-432.

[95] A five-year depreciation schedule is allowed for renewable energy systems under the Economic Recovery Tax Act of 1981, as amended (P.L. 97-34; Stat. 230, codified as 26 U.S.C. § 168(e)(3)(B)(vi)).

[96] Tiffany, Douglas G. *Economic Analysis: Co-generation Using Wind and Biodiesel-*
(continued...)

at higher capacity with increasing wind speed until a "cut-out speed" is reached at about 50 miles per hour. At this speed the turbine is stopped and the blades are turned 90 degrees out of the wind and parked to prevent damage.

Figure 9. U.S. Areas with Highest Wind Potential

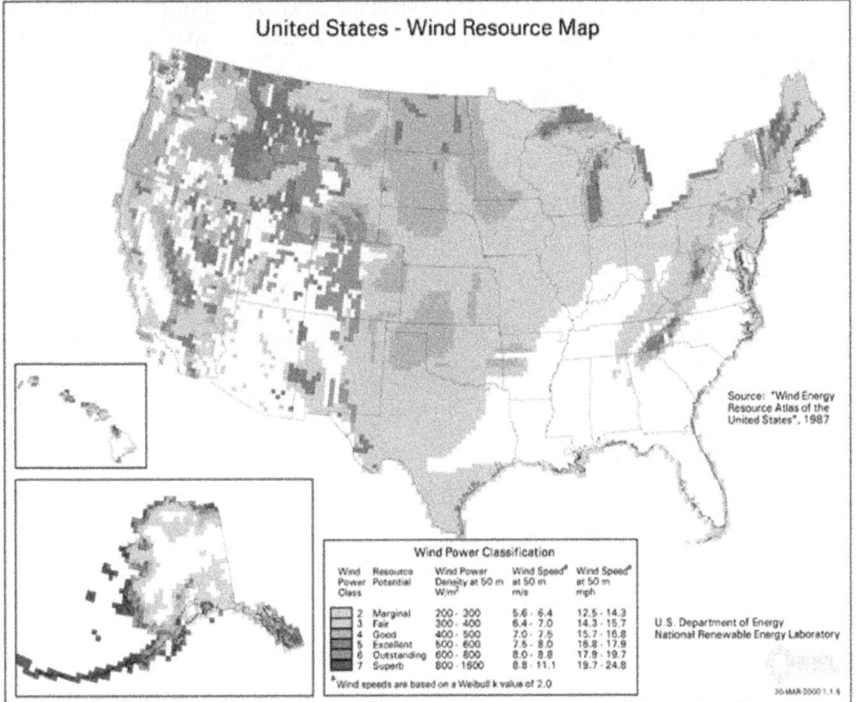

The DOE map of U.S. wind potential confirms that the most favorable areas tend to be located in sparsely populated regions, which may disfavor wind-generated electricity production for several reasons. First, transmission lines may be either inaccessible or of insufficient capacity to move surplus wind-generated electricity to distant demand sources. Second, transmission pricing mechanisms may disfavor moving electricity across long distances due to distance-based charges or according to the number of utility territories crossed. Third, high infrastructure costs for the initial hook-up to the power grid may discourage entry, although larger wind farms can benefit from economies of scale on the initial hook-up. Fourth, new entrants may see their access to the transmission power grid limited in favor of traditional customers during periods of heavy congestion. Finally, wind plant operators are often penalized for deviations in electricity delivery to a transmission line that result from the variability in available wind speed.

---

[96] (...continued)
Powered Generators, Staff Paper P05-10, Dept of Applied Econ., Univ. of Minn., October 2005.

Environmental Concerns. Three potential environmental issues — impacts on the visual landscape, bird and bat deaths, and noise issues — vary in importance based on local conditions. In some rural localities, the merits of wind energy appear to have split the environmental movement. For example, in the Kansas Flint Hills, local chapters of the Audubon Society and Nature Conservancy oppose installation of wind turbines, saying that they would befoul the landscape and harm wildlife; while Kansas Sierra Club leaders argue that exploiting wind power would help to reduce America's dependence on fossil fuels.

Box: Primer on Measuring Electric Energy

News stories covering electric generation topics often try to illustrate the worth of a megawatt (MW) in terms of how many homes a particular amount of generation could serve. However, substantial variation may appear in implied household usage rates. So what really is a MW and how many homes can one MW of generation really serve?

Basics. A watt (W) is the basic unit used to measure electric power. Watts measure instantaneous power. In contrast, a watt-hour (Wh) measures the total amount of energy consumed in an hour. For example, a 100 W light bulb is rated to consume 100 W of power when turned on. If a 100 W bulb were on for 4 hours it would consume 400 Wh of energy. A kilowatt (kW) equals 1,000 W and a megawatt (MW) equals 1,000 kW or 1 million W. Electricity production and consumption are measured in kilowatt-hours (kWh), while generating capacity is measured in kilowatts or megawatts. If a power plant that has 1 MW of capacity operated nonstop (i.e, 100%) during all 8,760 hours in the year, it would produce 8,760,000 kWh.

More realistically, a 100 MW rated wind farm is capable of producing 100 MW during peak winds, but will produce much less than its rated amount when winds are light. As a result of these varying wind speeds, over the course of a year a wind farm may only average 30 MW of power production. On average, wind power turbines typically operate the equivalent of less than 40% of the peak (full load) hours in the year due to the intermittency of the wind. Wind turbines are "on-line" — actually generating electricity — only when wind speeds are sufficiently strong (i.e., at least 9 to 10 miles per hour).

Average MW per Household. In its 2004 analysis of the U.S. wind industry, the Government Accountability Office (GAO) assumed that an average U.S. household consumed about 10,000 kWh per year (GAO, *Renewable Energy: Wind Power's Contribution to Electric Power Generation and Impact on Farms and Rural Communities*, GAO-04-756, Sept. 2004). However, the amount of electricity consumed by a typical residential household varies dramatically by region of the country. According to 2001 Energy Information Administration (EIA) data, New England residential homes consumed the least amount of electricity, averaging 653 kWh of load in a month, while the East South Central region, which includes states such as Georgia and Alabama and Tennessee, consumed nearly double that amount at 1,193 kWh per household. The large regional disparity in electric consumption is driven by many factors including the heavier use of air conditioning in the South. As a result, a 1 MW generator in the Northeast would be capable of serving about twice as many households as the same generator located in the South because households in the Northeast consume half the amount of electricity as those in the South.

So how many homes can a wind turbine rated at 1 MW really serve? In the United States, a wind turbine with a peak generating capacity of 1 MW, rated at 30% annual capacity, placed on a tower situated on a farm, ranch, or other rural land, can generate about 2.6 million kilowatt-hours [=(1MW)*(30%)*(8.76 kWh)] in a year which is enough electricity to serve the needs of 184 (East South Central) to 354 (New England) average U.S. households depending on which region of the country you live in.

Source: Bob Bellemare, UtiliPoint International Inc., Issue Alert, June 24, 2003; available at [http://www.utilipoint.com/issuealert/article.asp?id=1728].

Public Laws That Support Agriculture-Based Energy Production and Use

This section provides a brief overview of the major pieces of legislation that support agriculture-based renewable energy production. It is noteworthy that many of the federal programs that currently support renewable energy production in general, and agriculture-based energy production in particular, are outside the purview of USDA and have legislative origins outside of the farm bill.

Federal support is provided in the form of excise and income tax credits; loans, grants, and loan guarantees; research, development, and demonstration assistance; educational program assistance; procurement preferences; user mandates, and a tariff on imported ethanol from countries outside of the Caribbean Basin Initiative.[97] In addition, this section briefly reviews major proposals by the Administration and bills introduced by the 110th Congress that relate to agriculture-based renewable energy.

Tariff on Imported Ethanol

A most-favored-nation tariff of 54 cents per gallon is imposed on most imported ethanol. The tariff is intended to offset the 51-cents-per-gallon production tax credit available for every gallon of ethanol blended in gasoline. Exceptions to the tariff are ethanol imports from the Caribbean region and Central America under the Caribbean Basin Initiative (CBI). The CBI — which is designed to promote development and stability in the Caribbean region and Central America — allows the imports of most products, including ethanol, duty-free. In many cases, the tariff presents a significant barrier to imports as it negates lower production costs in other countries. For example, by some estimates, Brazilian production costs are 40% to 50% lower than in the United States.[98] The tariff, which was set to expire October 1, 2007, was extended through 2008 by a provision in P.L. 109-432.

Clean Air Act Amendments of 1990 (CAAA; P.L. 101-549)

The Reformulated Gasoline and Oxygenated Fuels programs of the CAAA have provided substantial stimuli to the use of ethanol.[99] In addition, the CAAA requires the Environmental Protection Agency (EPA) to identify and regulate air emissions from all significant sources, including on- and off-road vehicles, urban buses, marine engines, stationary equipment, recreational vehicles, and small engines used for lawn and garden equipment. All of these sources are candidates for biofuel use.

---

[97] For more information on federal incentives for biofuel production, see CRS Report RL33572, *Biofuels Incentives: A Summary of Federal Programs*, by Brent D. Yacobucci.

[98] For more information, see CRS Report RS21930,*Ethanol Imports and the Caribbean Basin Initiative*, by Brent D. Yacobucci.

[99] CRS Report RL33290, *Fuel Ethanol: Background and Public Policy Issues*, by Brent D. Yacobucci.

Energy Policy Act of 1992 (EPACT; P.L. 102-486)

Energy security provisions of EPACT favor expanded production of renewable fuels. Provisions related to agriculture-based energy production included:

- EPACT's alternative-fuel motor fleet program implemented by DOE requires federal, state, and alternative fuel providers to increase purchases of alternative-fueled vehicles. Under this program, DOE has designated neat (100%) biodiesel as an environmentally positive or "clean" alternative fuel.[100]
- A 1.5¢ per kilowatt/hour production tax credit (PTC) for wind energy was established. The PTC is applied to electricity produced during a wind plant's first ten years of operation.

Biomass Research and Development Act of 2000 (Biomass Act; Title III, P.L. 106-224)

The Biomass Act (Title III of the Agricultural Risk Protection Act of 2000 [P.L. 106-224]) contains several provisions to further research and development in the area of biomass-based renewable fuel production.

- (Sec. 304) The Secretaries of Agriculture and Energy shall cooperate with respect to, and coordinate, policies and procedures that promote research and development leading to the production of biobased fuels and products.
- (Sec. 305) A Biomass Research and Development Board is established to coordinate programs within and among departments and agencies of the Federal Government for the purpose of promoting the use of biofuels and products.
- (Sec. 306) A Biomass Research and Development Technical Advisory Committee is established to advise, facilitate, evaluate, and perform strategic planning on activities related to research, development, and use of biobased fuels and products.
- (Sec. 307) A Biomass Research and Development Initiative (BRDI) is established under which competitively awarded grants, contracts, and financial assistance are provided to eligible entities undertaking research on, and development and demonstration of, biobased fuels and products.[101]
- (Sec. 309) The Secretaries of Agriculture and Energy are obliged to submit an annual joint report to Congress accounting for the nature and use of any funding made available under this initiative.[102]

---

[100] NBB, "Biodiesel Emissions," at [http://www.biodiesel.org/pdf_files/fuelfactsheets/emissions.pdf].

[101] The official website for the Biomass Research and Development Initiative may be found at [http://www.brdisolutions.com/].

[102] This report is available at [http://www.brdisolutions.com/].

- (Sec. 310) To undertake these activities, Commodity Credit Corporation (CCC) funds of $49 million per year were authorized for FY2002-FY2005.

Biomass-related program funding levels were expanded through FY2007 by Section 9008 of the 2002 farm bill (P.L. 107-171) which also made available (until expended) new funding of $5 million in FY2002 and $14 million in each of FY2003-FY2007; however, FY2006 funding was reduced to $12 million (P.L. 109-97; Title VII, Sec. 759). Subsequently, Title II of the Healthy Forest Restoration Act of 2003 (P.L. 108-148) raised the annual authorization from $49 million to $54 million. Finally Sections 942-948 of the Energy Policy Act of 2005 (P.L. 109-58) raised the annual authorization from $54 million to $200 million starting in FY2006, and extended it through FY2015. In addition to new funding, many of the original biomass-related provisions were expanded and new provisions were added by these same laws as described below.

Energy Provisions in the 2002 Farm Bill (P.L. 107-171)[103]

In the 2002 farm bill, three separate titles — Title IX: Energy, Title II: Conservation, and Title VI: Rural Development — each contain programs that encourage the research, production, and use of renewable fuels such as ethanol, biodiesel, and wind energy systems.

Federal Procurement of Biobased Products (Title IX, Section 9002). Federal agencies are required to purchase biobased products under certain conditions. A voluntary biobased labeling program is included. Legislation provides funding of $1 million annually through the USDA's Commodity Credit Corporation (CCC) for FY2002-FY2007 for testing biobased products. USDA published final rules in the *Federal Register* (vol. 70, no. 1, pp. 41-50, January 3, 2005). The regulations define what a biobased product is under the statue, identify biobased product categories, and specify the criteria for qualifying those products for preferred procurement.

Biorefinery Development Grants (Title IX, Section 9003). Federal grants are provided to ethanol and biodiesel producers who construct or expand their production capacity. Funding for this program was authorized in the 2002 farm bill, but no funding was appropriated. Through FY2006, no funding had yet been proposed; therefore, no implementation regulations have been developed.

Biodiesel Fuel Education Program (Title IX, Section 9004). Administered by USDA's Cooperative State Research, Education, and Extension Service, competitively awarded grants are made to nonprofit organizations that educate governmental and private entities operating vehicle fleets, and educate the public about the benefits of biodiesel fuel use. Final implementation rules were published in the *Federal Register* (vol. 68, no. 189, September 30, 2003). Legislation provides funding of $1 million annually through the CCC for FY2003-FY2007 to fund

[103] USDA, *2002 Farm Bill*, "Title IX — Energy," online information available at [http://www.ers.usda.gov/Features/Farmbill/titles/titleIXenergy.htm]. For more information, see CRS Report RL31271, *Energy Provisions of the Farm Bill: Comparison of the New Law with Previous Law and House and Senate Bills*, by Brent D. Yacobucci.

the program. As of January 2006, only two awardees — the National Biodiesel Board and the University of Idaho — had been selected.[104]

Energy Audit and Renewable Energy Development Program (Title IX, Section 9005). This program is intended to assist producers in identifying their on-farm potential for energy efficiency and renewable energy use. Funding for this program was authorized in the 2002 farm bill, but through FY2006 no funding has been appropriated. As a result, no implementation regulations have been developed.

Renewable Energy Systems and Energy Efficiency Improvements (Renewable Energy Program) (Title IX; Section 9006). Administered by USDA's Rural Development Agency, this program authorizes loans, loan guarantees, and grants to farmers, ranchers, and rural small businesses to purchase renewable energy systems and make energy efficiency improvements.[105] Grant funds may be used to pay up to 25% of the project costs. Combined grants and loans or loan guarantees may fund up to 50% of the project cost. Eligible projects include those that derive energy from wind, solar, biomass, or geothermal sources. Projects using energy from those sources to produce hydrogen from biomass or water are also eligible. Legislation provides that $23 million will be available annually through the CCC for FY2003-FY2007 for this program. Unspent money lapses at the end of each year. Final implementation rules, including program guidelines for receiving and reviewing future loan and loan guarantee applications, were published in the *Federal Register* (vol. 708, no. 136, July 18, 2005).

Prior to each fiscal year, USDA publishes a Notice of Funds Availability (NOFA) in the *Federal Register* inviting applications for the Renewable Energy Program, most recently on February 22, 2006, when the availability of $22.8 million (half as competitive grants, and half for guaranteed loans) was announced. Not all applications are accepted. On February 22, 2006, USDA announced that $11.8 million in grants for FY2006 and $176.5 million in loan guarantees were available for renewable energy and energy efficient projects.[106] USDA estimates that loans and loan guarantees are more effective than grants in assisting renewable energy projects, because program funds would be needed only for the credit subsidy costs (i.e., government payments made minus loan repayments to the government).[107]

Hydrogen and Fuel Cell Technologies (Title IX, Section 9007). Legislation requires that USDA and DOE cooperate on research into farm and rural applications for hydrogen fuel and fuel cell technologies under a memorandum of understanding. No new budget authority is provided.

---

[104] These awardees were selected in August 2003; more information is available at [http://www.biodiesel.org/usda/].

[105] For more information on this program, see [http://www.rurdev.usda.gov/rbs/farmbill/index.html].

[106] USDA News Release 0051.06, February 22, 2006.

[107] USDA News Release 0261.05, July 15, 2005. For more information on the broader potential of loan guarantees see, GAO, *Wind Power*, GAO-04-756, September 2004, pp. 54-55.

CRS-46

Biomass Research and Development (Title IX; Section 9008).[108] This provision extends an existing program — created under the Biomass Research and Development Act (BRDA) of 2000 — that provides competitive funding for research and development projects on biofuels and bio-based chemicals and products, administered jointly by the Secretaries of Agriculture and Energy. Under the BRDA, $49 million per year was authorized for FY2002-FY2005. Section 9008 extended the $49 million in budget authority through FY2007, and added new funding levels of $5 million in FY2002 and $14 million for FY2003-FY2007 — unspent funds may be carried forward, making the additional funding total $75 million for FY2002-FY2007. (The $49 million in annual funding for FY2002-FY2007 was raised to $54 million for that same period by P.L. 108-148, then raised to $200 million per year for FY2006-FY2015 by Sec. 941 of P.L. 109-58; see below). In November 2006, USDA and DOE jointly announced the selection of 17 projects to receive total funding of approximately $17.5 million from the agencies under the BRDI. Cost-sharing by private sector partners increases the total value to over $27 million.[109]

Cooperative Research and Development — Carbon Sequestration (Title IX; Section 9009). This provision amends the Agricultural Risk Protection Act of 2000 (P.L. 106-224, Sec. 221) to extend through FY2011 the one-time authorization of $15 million of the Carbon Cycle Research Program, which provides grants to land-grant universities for carbon cycle research with on-farm applications.

Bioenergy Program (Title IX; Section 9010). This is an existing program (7 C.F.R. 1424) in which the Secretary makes payments from the CCC to eligible bioenergy producers — ethanol and biodiesel — based on any year-to-year increase in the quantity of bioenergy that they produce (fiscal year basis). The goal is to encourage greater purchases of eligible commodities used in the production of bioenergy (e.g., corn for ethanol or soybean oil for biodiesel). The Bioenergy Program was initiated on August 12, 1999, by Executive Order 13134. On October 31, 2000, then-Secretary of Agriculture Glickman announced that, pursuant to the executive order, $300 million of discretionary CCC funds ($150 million in both FY2001 and FY2002) would be made available to encourage expanded production of biofuels. The 2002 farm bill extended the program and its funding by providing that $150 million would be available annually through the CCC for FY2003-FY2006. The final rule for the Bioenergy Program was published in the *Federal Register* (vol. 68, no. 88, May 7, 2003).

The FY2003 appropriations act limited spending for the Bioenergy Program funding for FY2003 to 77% ($115.5 million) of the $150 million; however, the full $150 million was eventually spent. In FY2004, no limitations were imposed. However, a $50 million reduction from the $150 million was contained in the FY2005 appropriations act, followed by a $90 million reduction in the FY2006 appropriations act. Funding authority for this program ended after FY2006.

---

108 For more information, see the joint USDA-DOE website at [http://www.biomass.govtools.us/].

109 Ibid.

CRS-47

Renewable Energy on Conservation Reserve Program (CRP) Lands (Title II; Section 2101). This provision amends Section 3832 of the Farm Security Act of 1985 (1985 farm bill) to allow the use of CRP lands for biomass (16 USC 3832(a)(7)(A)) and wind energy generation (16 USC 3832(a)(7)(B)) harvesting for energy production.

Rural Development Loan and Grant Eligibility Expanded to More Renewables (Title VI). Section 6013 — Loans and Loan Guarantees for Renewable Energy Systems — amends Section 310B of the Consolidated Farm and Rural Development Act (CFRDA) (7 U.S.C. 1932(a)(3)) to allow loans for wind energy systems and anaerobic digesters. Section 6017(g)(A) — Business and Industry Direct and Guaranteed Loans — amends Section 310B of CFRDA (7 U.S.C. 1932) to expand eligibility to include farmer and rancher equity ownership in wind power projects. Limits range from $25 million to $40 million per project. Section 6401(a)(2) — Value-Added Agricultural Product Market Development Grants — amends Section 231 of CFRDA (7 U.S.C. 1621 note; P.L.106-224) to expand eligibility to include farm- or ranch-based renewable energy systems. Competitive grants are available to assist producers with feasibility studies, business plans, marketing strategies, and start-up capital. The maximum grant amount is $500,000 per project.

Additional support for renewable energy projects is available in the form of various loans and grants from USDA's Rural Development Agency under other programs such as the Small Business Innovation Research (SBIR) grants and Value-Added Producer Grants (VAPG).[110] In keeping with a trend started in 2003, USDA is giving priority consideration to grant applications that dedicate at least 51% of the project costs to biomass energy. Most recently, on January 9, 2006, Agriculture Secretary Johanns announced the availability of $19 million in grants in support of the development of renewable energy projects and value-added agricultural business ventures.[111]

The Healthy Forest Restoration Act of 2003 (P.L. 108-148)

Title II of P.L. 108-148 amended the Biomass Act of 2000 by expanding the use of grants, contracts, and assistance for biomass to include a broader range of forest management activities. In addition, Sec. 201(b) increased the annual amount of discretionary funding available under the Biomass Act for FY2002-FY2007 from $49 million to $54 million (7 USC 8101 note). Section 202 granted authority to the Secretary of Agriculture to establish a program to accelerate adoption of biomass-related technologies through community-based marketing and demonstration activities, and to establish small-scale businesses to use biomass materials. It also authorized $5 million annually to be appropriated for each of FY2004-FY2008 for such activities. Finally, Sec. 203 established a biomass utilization grant program to provide funds to offset the costs incurred in purchasing biomass materials for

---

[110] For more information see [http://www.rurdev.usda.gov/rd/energy/].

[111] USDA News Release 0002.06, January 9, 2006.

qualifying facilities. Funding of $5 million annually was authorized to be appropriated for each of FY2004-FY2008 for this biomass utilization grant program.

The American Jobs Creation Act of 2004 (P.L. 108-357)

The American Jobs Creation Act — signed into law on October 22, 2004 — contains two provisions (Sections 301 and 701) that provide tax exemptions for three agri-based renewable fuels: ethanol, biodiesel, and wind energy.

Federal Fuel Tax Exemption for Ethanol (Section 301). This provision provides for an extension and replaces the previous federal ethanol tax incentive (26 U.S.C. 40). The tax credit is revised to allow for blenders of gasohol to receive a federal tax exemption of $0.51 per gallon for every gallon of pure ethanol. Under this volumetric orientation, the blending level is no longer relevant to the calculation of the tax credit. Instead, the total volume of ethanol used is the basis for calculating the tax.[112] The tax credit for alcohol fuels was extended through December 31, 2010.

Federal Fuel Tax Exemption for Biodiesel (Section 301). This provision provides for the first ever federal biodiesel tax incentive — a federal excise tax and income tax credit of $1.00 for every gallon of agri-biodiesel (i.e., virgin vegetable oil and animal fat) that is used in blending with petroleum diesel; and a 50¢ credit for every gallon of non-agri-biodiesel (i.e., recycled oils such as yellow grease). The tax credits for biodiesel fuels were extended through December 31, 2006 (extended through 2008 by P.L. 109-58; see below).

Federal Production Tax Exemption for Wind Energy Systems (Section 710). This provision renews a federal production tax credit (PTC) that expired on December 31, 2003. The renewed tax credit provides a 1.5¢ credit (adjusted annually for inflation) for a 10-year period for each kilowatt-hour of electricity produced by qualifying turbines that are built by the end of 2005 (extended through 2007 by P.L. 109-58; see below). The inflation-adjusted PTC stood at 1.8¢ per kWh as of December 2003.

Energy Policy Act of 2005 (EPACT; P.L. 109-58)

The Energy Policy Act of 2005 — signed into law on August 8, 2005 — contains several provision related to agriculture-based renewable energy production including the following.[113]

National Renewable Fuels Standard (RFS) (Sec. 1501). Requires that 4.0 billion gallons of renewable fuel be used domestically in 2006, increasing to 7.5 billion gallons by 2012.

---

[112] For more information, see the American Coalition for Ethanol, *Volumetric Ethanol Excise Tax Credit (VEETC)* at [http://www.ethanol.org/veetc.html]

[113] For more information, see CRS Report RL32204, *Omnibus Energy Legislation: Comparison of Non-Tax Provisions in the H.R. 6 Conference Report and S. 2095*, by Mark Holt and Carol Glover, coordinators.

Minimum Quantity of Ethanol from Cellulosic Biomass (Sec. 1501). For calendar 2013 and each year thereafter, the RFS volume shall contain a minimum of 250 million gallons derived from cellulosic biomass.

Special Consideration for Cellulosic Biomass or Waste Derived Ethanol (Sec. 1501). For purposes of the RFS, each gallon of cellulosic biomass ethanol or waste derived ethanol shall be counted as the equivalent of 2.5 gallons of renewable fuel.

Small Ethanol Producer Credit Adjusted (Sec. 1347). The definition of a small ethanol producer was extended from 30 million gallons per year to 60 million gallons per year. Qualifying producers are eligible for an additional tax credit of 10¢ per gallon on the first 15 million gallons of production.

Biodiesel Tax Credit Extension Through 2008 (Sec. 1344). Extends the $1.00 per gallon biodiesel tax credit through 2008.

Small Biodiesel Producer Credit Established (Sec. 1345). Agri-biodiesel producers with a productive capacity not in excess of 60 million gallons are eligible for an additional tax credit of 10¢ per gallon on the first 15 million gallons of production.

Funding Support for Research, Development, and Demonstration of Alternate Biofuel Processes. Several alternate forms of assistance including (Sec. 1512) grants for conversion assistance of cellulosic biomass, waste-derived ethanol, and approved renewable fuels; (Sec. 1514) establish a demonstration program for advanced biofuel technologies; (Sec. 1515) extend biodiesel feedstock sources to include animal and municipal waste; and (Sec. 1516) provide loan guarantees for demonstration projects for ethanol derived from sugarcane, bagasse, and other sugarcane byproducts.

Wind PTC Extension Through 2007 (Sec. 1301). Provides a two-year extension through December 31, 2007, for the production tax credit for wind; maintains the PTC inflation adjustment factor of current law; and (Sec. 1302) extends the PTC to agricultural cooperatives.

Agricultural Biomass Research and Development Programs (Sec. 942-948). This section of EPACT provides several amendments to the BRDA as follows. Section 941 updates BRDA to intensify focus on achieving the scientific breakthroughs (particularly with respect to cellulosic biomass) required for expanded deployment of biobased fuels, products, and power, including:

- increased emphasis on feedstock production and delivery, including technologies for harvest, handling and transport of crop residues;
- research and demonstration (R&D) of opportunities for synergy with existing biofuels production, such as use of dried distillers grains (DDGs) as a bridge feedstock;
- support for development of new and innovative biobased products made from corn, soybeans, wheat, sunflower, and other raw agricultural commodities;

CRS-50

- ensuring a balanced and focused R&D approach by distributing funding by technical area (20% to feedstock production; 45% to overcoming biomass recalcitrance; 30% to product diversification; and 5% to strategic guidance), and within each technical area by value category (15% to applied fundamentals; 35% to innovation; and 50% to demonstration); and
- increasing annual program authorization from the current $54 million to $200 million for 10 years — FY2006-FY2015.

Section 942 expands the production incentives for cellulosic biofuels by directing the Secretary of Energy to establish a program of production incentives to deliver the first billion gallons of annual cellulosic biofuels production by 2015. Funds are allocated for proposed projects through set payments on a per gallon basis for the first 100 million gallons of annual production, followed by a reverse auction competitive solicitation process to secure low-cost cellulosic biofuels production contracts. Production incentives are awarded to the lowest bidders, with not more than 25% of the funds committed for each auction awarded to a single bid. Awards may not exceed $100 million in any year, nor $1 billion over the lifetime of the program. The first auction shall take place within one year of the first year of annual production of 100 million gallons of cellulosic biofuels, with subsequent auctions each year thereafter until annual cellulosic biofuels production reaches 1 billion gallons. Funding of $250 million, until expended, is authorized to carry out this section subject to appropriations.

Section 943 expands the Biobased Procurement Program authorized under Section 9002 of the 2002 farm bill by applying the provision to federal government contractors. Currently the program requires only federal agencies to give preference to biobased products for procurement exceeding $10,000 when suitable biobased products are available at reasonable cost. Section 943 also directs the Architect of the Capitol, the Sergeant at Arms of the Senate, and the Chief Administrative Officer of the House of Representatives to comply with the Biobased Procurement Program for procurement of the United States Capitol Complex. Furthermore, it directs the Architect of the Capitol to establish within the Capitol Complex a program of public education regarding its use of biobased products.

Sections 944-946 establish USDA grants programs to assist small biobased businesses with marketing and certification of biobased products (Sec. 944; funding of $1 million is authorized for FY2006, and such sums as necessary thereafter); to assist regional bioeconomy development associations and Land Grant institutions in supporting and promoting the growth of regional bioeconomies (Sec. 945; funding of $1 million is authorized for FY2006, and such sums as necessary thereafter); and for demonstrations by farmer-owned enterprises of innovations in pre-processing of feedstocks and multiple crop harvesting techniques, such as one-pass harvesting, to add value and lower the investment cost of feedstock processing at the biorefinery (Sec. 946; annual funding of $5 million is authorized for FY2006-FY2010).

Section 947 establishes a USDA program of education and outreach consisting of (1) training and technical assistance for feedstock producers to promote producer ownership and investment in processing facilities; and (2) public education and outreach to familiarize consumers with biobased fuels and products. Annual funding of $1 million is authorized for FY2006-FY2010. Finally, Section 948 requires a report on the economic potential of biobased products through the year 2025 as well as the economic potential by product area (within one year of enactment or by August 8, 2006), and analysis of economic indicators of the biobased economy (within two years of enactment or by August 8, 2007).

Tax Relief and Health Care Act of 2006 (P.L. 109-432)

The Tax Relief and Health Care Act of 2006 — signed into law on December 20, 2006 — contains two major provisions related to agriculture-based renewable energy production.

Extension of Production Tax Credit. The production tax credit available for electricity produced from certain renewable resources including wind energy (referred to earlier in P.L. 109-58; Section 1301) was extended by one year through December 31, 2008.

Extension of Ethanol Import Tariff. The 54¢ per gallon most-favored-nation tariff on most imported ethanol (referred to in the earlier section of this report "Tariff on Imported Ethanol") was extended through December 31, 2008.

Agriculture-Related Energy Bills in 110th Congress

As of October 15, 2007, numerous bills have been introduced in the 110th Congress that seek to enhance or extend current provisions in existing law that support agriculture-based energy production and use. For a listing of related legislation in the 110th Congress, see CRS Report RL33831, *Energy Efficiency and Renewable Energy Legislation in the 110th Congress* by Fred Sissine.

House-passed New Farm Bill — H.R. 2419. On July 27, 2007, the House approved a new farm bill — the Farm, Nutrition, and Bioenergy Act of 2007 (H.R. 2419) — which includes an energy title (Title IX). H.R. 2419, as amended and passed by the House, expands and extends several provisions from the energy title of the enacted 2002 farm bill with substantial increases in funding and a heightened focus on developing cellulosic ethanol production. A key departure from current farm-bill related energy provisions is that most new funding would be directed away from corn-starch-based ethanol production and towards either cellulosic-based biofuels production or to new as-yet-undeveloped technologies with some type of agricultural linkage. A funding complication relating to "pay-go" budget restrictions on new energy funding arose during the House Agriculture Committee's (HAC's) bill preparation because the Congressional Budget Office's March baseline showed no funding in the farm bill for a new energy title. However, the House Ways and Means Committee resolved this issue for the HAC by finding $2.4 billion in revenue offsets from outside of the agriculture budget.

The Senate Agriculture Committee (SAC) is expected to mark up its version of a 2007 farm bill in late-October. The Senate is also expected to adhere to "pay-go" budget restrictions which may, in and of themselves, lead to different funding choices than made by the House. For more information, see CRS Report RL34130, *Renewable Energy Policy in the 2007 Farm Bill*, by Randy Schnepf.

In addition to 2007 farm bill developments, both the House and Senate have passed different versions of new energy bills that contain many biofuel provisions similar to those contained in Title IX of the 2002 farm bill.[114] The Senate approved its version of an energy bill, H.R. 6 (the Renewable Fuels, Consumer Protection, and Energy Efficiency Act of 2007) on June 21, 2007, while the House approved its own energy bill, H.R. 3221 (the New Direction for Energy Independence, National Security, and Consumer Protection Act of 2007) on August 4, 2007. In particular, Title V of H.R. 3221 contains provisions similar or identical to provisions passed in Title IX of H.R. 2419. For more information on these energy bills, see CRS Report RL34136, *Biofuels Provisions in H.R. 3221 and H.R. 6: A Side-by-Side Comparison*, by Brent Yacobucci.

Also, on October 15, 2007, the House approved H.Con.Res. 25 expressing the sense of Congress that it is the goal of the United States that, not later than January 1, 2025, the agricultural, forestry, and working land of the United States should provide from renewable resources not less than 25 percent of the total energy consumed in the United States and continue to produce safe, abundant, and affordable food, feed, and fiber. A similar bill, S.Con.Res. 3, has been introduced in the Senate and referred to the Agriculture Committee.

In addition, several bills will likely be introduced that seek to provide incentives for the production and use of alternative fuel vehicles. See CRS Report RL33564, *Alternative Fuels and Advanced Technology Vehicles: Issues in Congress*, by Brent D. Yacobucci, for a listing of proposed legislation on alternative fuel vehicles. See CRS Report RS22351, *Tax Incentives for Alternative Fuel and Advanced Technology Vehicles*, by Brent D. Yacobucci, for a description of existing alternative-fuel vehicle tax incentives.

State Laws and Programs

Several state laws and programs influence the economics of renewable energy production and use by providing incentives for research, production, and consumption of renewable fuels such as biofuels and wind energy systems.[115] In addition, demand for agriculture-based renewable energy is being driven, in part, by state Renewable Portfolio Standards (RPS) that require utilities to obtain set percentages of their electricity from renewable sources by certain target dates. The amounts and deadlines vary, but as of January 2006, 34 states had laws instituting RPSs requiring, at a minimum, that state vehicle fleets procure certain volumes or percentages of renewable fuels. In several states, the RPS applied state-wide on all motor vehicles; for example see Minnesota Statutes Section 239.77 which requires that all diesel fuel

---

[114] For more information on these energy bills, see CRS Report RL34136, *Biofuels Provisions in H.R. 3221 and H.R. 6: A Side-by-Side Comparison*, by Brent Yacobucci.

[115] For more information on state and federal programs, see *State and Federal Incentives and Laws*, at the DOE's Alternative Fuels Data Center, [http://www.eere.energy.gov/afdc/laws/incen_laws.html].

sold or offered for sale in the state for use in internal combustion engines must contain at least 2% biodiesel fuel by volume. This mandate was to take effect by June 30, 2005, provided certain market conditions were met.[116]

Administration Proposals

State of the Union (SOU) 2006. In his 2006 SOU, President Bush introduced the notion of "switchgrass" as a potential energy source and announced the "Advanced Energy Initiative," which included a goal of making cellulosic ethanol cost competitive with corn-based ethanol by 2012.[117]

State of the Union (SOU) 2007. In his 2007 SOU, President Bush announced his "20 in 10" plan, which calls for reducing U.S. gasoline consumption by 20% in 10 years (i.e., by 2017).[118] The President proposed two major approaches for achieving his "20 in 10" goal.

- Increasing the supply of renewable and alternative fuels by setting a mandatory fuels standard to require 35 billion gallons of renewable and alternative fuels in 2017. This would be nearly five times the 2012 RFS target of 7.5 billions now in law. If accomplished, the Administration estimates that this would displace 15% of projected annual gasoline use in 2017.

- Reforming and modernizing corporate average fuel economy (CAFÉ) standards for cars and extending the current light truck rule. According to the Administration, this will reduce projected annual gasoline use by up to 8.5 billion gallons by 2017, a further 5% reduction that, in combination with increasing the supply of renewable and alternative fuels, will bring the total reduction in projected annual gasoline use to 20%.

According to the Administration, the President's FY2008 budget will request $2.7 billion for the advanced energy initiative — an increase of 26% above the 2007 request and 53% above 2006. President Bush has called for increased federal investment in hydrogen fuel technology research, as well as increased investment in advanced batteries for hybrids and plug-in hybrids, biodiesel fuels, and new methods of producing ethanol and other biofuels.

USDA's New Farm Bill Proposal (January 2007). On January 31, 2007, Agriculture Secretary Mike Johanns announced USDA's 2007 farm bill proposal. The

---

[116] For more information on Minnesota vehicle fuel acquisition requirements, visit [http://www.eere.energy.gov/afdc/laws/incen_laws.html].

[117] More information on the Advanced Energy Initiative from the 2006 SOU is available at [http://www.whitehouse.gov/stateoftheunion/2006/energy/index.html].

[118] More information see the White House "Fact Sheet: Strengthening America's Energy Security and Improving the Environment," January 24, 2007; available at [http://www.whitehouse.gov/news/releases/2007/01/20070124-5.html].

proposal includes $1.6 billion in new funding over ten years for energy innovation, including bio-energy research, energy efficiency grants, and $2 billion in loans for cellulosic ethanol plants.[119]

For More Information

Renewable Energy

DOE, Energy Information Agency (EIA), [http://www.eia.doe.gov/].

DOE, National Renewable Energy Laboratory (NREL), *Renewable Energy*, [http://www.nrel.gov/].

USDA, Oak Ridge National Laboratory, Energy Efficiency and Renewable Energy Program, *Renewable Energy*, [http://www.ornl.gov/sci/eere/renewables/index.htm].

USDA, Office of the Chief Economist, Office of Energy Policy and New Uses (OEPNU), [http://www.usda.gov/oce/energy/index.htm].

The Sustainable Energy Coalition, [http://www.sustainableenergy.org/].

Eidman, Vernon R. "Agriculture as a Producer of Energy," presentation at USDA conference *Agriculture as a Producer and Consumer of Energy*, June 24, 2004.

Biofuels

American Coalition for Ethanol, [http://www.ethanol.org/].

Bio, *Achieving Sustainable Production of Agricultural Biomass for Biorefinery Feedstock*, ©2006 Biotechnology Industry Organization; available at [http://www.bio.org/ind/biofuel/SustainableBiomassReport.pdf].

CRS Report RL33290, *Fuel Ethanol: Background and Public Policy Issues*, by Brent D. Yacobucci.

CRS Report RL33564, *Alternative Fuels and Advanced Technology Vehicles: Issues in Congress*, by Brent D. Yacobucci.

CRS Report RL33572, *Biofuels Incentives: A Summary of Federal Programs*, by Brent D. Yacobucci.

Distillery and Fuel Ethanol Worldwide Network, [http://www.distill.com/].

---

[119] Title-by-Title details of USDA's 2007 Farm Bill proposal are available at [http://www.usda.gov/wps/portal/usdafarmbill?navtype=SU&navid=FARM_BILL_FORUMS].

DOE, Energy Efficiency and Renewable Energy (EERE), *Alternative Fuels Data Center*, [http://www.eere.energy.gov/afdc/].

Eidman, Vernon R. *Agriculture's Role in Energy Production: Current Levels and Future Prospects*, paper presented at a conference, "Energy from Agriculture: New Technologies, Innovative Programs and Success Stories," December 14-15, 2005, St. Louis, Missouri; available at [http://www.farmfoundation.org/projects/documents/EIDMANpaperrevisedTOPOST12-19-05.pdf].

Environmental Protection Agency (EPA), Fuels and Fuel Additives, *Alternative Fuels*, [http://www.epa.gov/otaq/consumer/fuels/altfuels/altfuels.htm].

Hill, Jason, Erik Nelson, David Tilman, Stephen Polasky, and Douglas Tiffany. "Environmental, economic, and energetic costs and benefits of biodiesel and ethanol biofuels," *Proceedings of the National Academy of Sciences*, Vol. 103, No. 30, July 25, 2006, 11206-11210.

Koplow, Doug. *Biofuels — At What Cost? Government Support for Ethanol and Biodiesel in the United States*, Global Subsidies Initiative of the International Institute for Sustainable Development, Geneva, Switzerland, October 2006; available at [http://www.globalsubsidies.org].

National Biodiesel Board (NBB), [http://www.biodiesel.org/].

Renewable Fuels Association (RFA), [http://www.ethanolrfa.org/].

USDA, *The Economic Feasibility of Ethanol Production from Sugar in the United States*, Office of Energy Policy and New Uses (OEPNU), Office of the Chief Economist (OCE), USDA, and Louisiana State University (LSU), July 2006; available at [http://www.usda.gov/oce/reports/energy/EthanolSugarFeasibilityReport3.pdf].

USDA/Dept. Of Energy (DOE). *Biomass as Feedstock for a Bioenergy and Bioproducts Industry: The Technical Feasibility of a Billion-Ton Annual Supply*, April 2005; available at [http://feedstockreview.ornl.gov/pdf/billion_ton_vision.pdf].

Economic Benefits of Biofuel Production

De La Torre Ugarte, D., M. Walsh, H. Shapouri, and S. Slinsky. *The Economic Impacts of Bioenergy Crop Production on U.S. Agriculture*, AER 816, USDA, Office of the Chief Economist (OCE), Office of Energy Policy and New Uses (OEPNU), February 2003; available at [http://www.usda.gov/oce/reports/energy/AER816Bi.pdf].

Food and Agricultural Policy Research Institute (FAPRI), *Implications of Increased Ethanol Production for U.S. Agriculture*, FAPRI-UMC Report #10-05, August 22, 2005; available at [http://www.fapri.missouri.edu/].

FAPRI, *Returns to Biofuels Production*, FAPRI-UMC Report #06-07, March 2007.

FAPRI, *Economic Impacts of Not Extending Biofuels Subsidies*, FAPRI-UMC Report #17-07, May 2007.

CRS-56

FAPRI, *Impacts of a 15 Billion Gallon Biofuels Use Mandate*, FAPRI-UMC Report #22-07, June 2007.

Gallagher, P. Otto, H. Shapouri, J. Price, G. Schamel, M. Dikeman, and H. Brubacker, *The Effects of Expanding Ethanol Markets on Ethanol Production, Feed Markets, and the Iowa Economy*, Staff Paper #342, Dept. of Economics, Iowa St. Univ., June 30, 2001.

Gallagher, P., E. Wailes, M. Dikeman, J. Fritz, W. Gauther, and H. Shapouri, *Biomass From Crop Residues: Cost and Supply Estimates,* AER 819, USDA, OCE, OEPNU, March 2003; available at [http://www.usda.gov/oce/reports/energy/ AER819.pdf].

McElroy, Michael B. "Chapter 12: Ethanol from Biomass: Can it Substitute for Gasoline?" Draft from book in progress available at [http://www-as.harvard.edu/people/faculty/mbm/Ethanol_chapter1.pdf]

Radich, Anthony. "Biodiesel Performance, Costs, and Use," Modeling and Analysis Papers, DOE/EIA, June 2004; available at [http://www.eia.doe.gov/oiaf/analysispaper/biodiesel/].

Shapouri, Hosein, James Duffield, Andrew McAloon, Michael Wang. "The 2001 Net Energy Balance of Corn-ethanol." Paper presented at the Corn Utilization and Technology Conference, June 7-9, 2004, Indianapolis, IN.

Shapouri, Hosein; James A. Duffield, and Michael Wang. *The Energy Balance of Corn Ethanol: An Update*. USDA, Office of the Chief Economist, Office of Energy Policy and New Uses. Agricultural Economic Report (AER) No. 813, July 2002; available at [http://www.usda.gov/oce/reports/energy/index.htm].

Swenson, Dave S. *Input-Outrageous: The Economic Impacts of Modern Biofuels Production*, Iowa State University webpapers, June 2006; available at [http://www.econ.iastate.edu/research/webpapers/paper_12644.pdf].

Tiffany, Douglas G. and Vernon R. Eidman. *Factors Associated with Success of Fuel Ethanol Producers*, Dept of Applied Economics, Univ. of Minnesota, Staff Paper P03-7, August 2003.

Urbanchuk, J. M. *The Contribution of the Ethanol Industry to the American Economy in 2004*, March 12, 2004, available at [http://www.ncga.com/ethanol/pdfs/EthanolEconomicContributionREV.pdf]

Urbanchuk, J. M. and J. Kapell, *Ethanol and the Local Community*, June 20, 2002, available at [http://www.ncga.com/ethanol/pdfs/EthanolLocalCommunity.pdf].

Urbanchuk, J. M. *An Economic Analysis of Legislation for a Renewable Fuels Requirement for Highway Motor Fuels*, November 7, 2001.

USDA. *An Analysis of the Effects of an Expansion in Biofuel Demand on U.S. Agriculture*, OCE/ERS, May 2007.

CRS-57

Anaerobic Digestion Systems

The Agricultural Marketing Research Center, *Bio-Mass/Forages*, at [http://www.agmrc.org/agmrc/commodity/biomass/].

Consequences of Expanded Agriculture-Based Biofuel Production

Avery, Dennis. "Biofuels, Food, or Wildlife? The Massive Land Costs of U.S. Ethanol," No. 5, Competitive Enterprise Institute, September 21, 2006.

Babcock, Bruce, and D.A. Hennessy, "Getting More Corn Acres From the Corn Belt" *Iowa Ag Review*, Vol. 12, No. 4, Fall 2006, pp. 6-7

Doering, Otto C. "U.S. Ethanol Policy: Is It the Best Alternative?" *Current Agriculture, Food & Resource Issues*, No. 5, 2004, pp. 204-211.

Doornbosch, Richard, and Ronald Steenblik. *Biofuels: is the Cure Worse that the Disease?*, SG/SD/RT(2007)3, paper presented at OECD Roundtable on Sustainable Development, Paris, September 11-12, 2007.

Elobeid, Amani, S. Tokgoz, D.J. Hayes, B.A. Babcock, and C.E. Hart. "The Long-Run Impact of Corn-Based Ethanol on the Grain, Oilseed, and Livestock Sectors: A Preliminary Assessment," CARD Briefing Paper 06-BP 49, November 2006.

Hart, Chad E. "Feeding the Ethanol Boom: Where Will the Corn Come From?" *Iowa Ag Review*, Vol. 12, No. 4, Fall 2006, pp. 2-3

Kohlmeyer, Bob. "The Other Side of Ethanol's Bonanza," *Ag Perspectives* (World Perspectives, Inc.), December 14, 2004.

McElroy, Michael B. "Chapter 12: Ethanol from Biomass: Can it Substitute for Gasoline?" Draft from book in progress available at [http://www-as.harvard.edu/people/faculty/mbm/Ethanol_chapter1.pdf]

National Academies of Science. *Water Implications of Biofuels Production in the United States*, National Research Council, ISBN: 0-309-11360-1, 2007; available at [http://www.nap.edu/catalog/12039.html].

Taylor, Richard D., J.W. Mattson, J. Andino, and W.W. Koo. *Ethanol's Impact on the U.S. Corn Industry*, Agribusiness & Applied Economics Report No. 580, Center for Agricultural Policy and Trade Statistics, North Dakota St. Univ., March 2006.

Tokgaz, Simla, and Amani Elobeid. "An Analysis of the Link between Ethanol, Energy, and Crop Markets," Working Paper 06-EP 435, CARD, November 2006.

Wisner, R., and P. Baumel, "Ethanol, Exports, and Livestock: Will There be Enough Corn to Supply Future Needs?," *Feedstuffs*, no. 30, vol. 76, July 26, 2004.

CRS-58

Wind Energy Systems

American Wind Energy Association (AWEA), [http://www.awea.org/].

DOE, Wind Energy Program, [http://www1.eere.energy.gov/windandhydro/].

The Utility Wind Interest Group, [http://www.uwig.org].

Tiffany, Douglas G. *Co-Generation Using Wind and Biodiesel-Powered Generators*, Dept of Applied Econ., Univ. of Minnesota, Staff Paper P05-10, October 2005.

Biofuels Incentives: A Summary of Federal Programs

Brent D. Yacobucci
Specialist in Energy and Environmental Policy

January 27, 2010

Congressional Research Service
7-5700
www.crs.gov
R40110

CRS Report for Congress
Prepared for Members and Committees of Congress

Summary

With recent high energy prices, the passage of major energy legislation in 2005 (P.L. 109-58) and 2007 (P.L. 110-140), and the passage of a new Farm Bill in 2008 (P.L. 110-246) there is ongoing congressional interest in promoting alternatives to petroleum fuels. Biofuels—transportation fuels produced from plants and other organic materials—are of particular interest.

Ethanol and biodiesel, the two most widely used biofuels, receive significant government support under federal law in the form of mandated fuel use, tax incentives, loan and grant programs, and certain regulatory requirements. The 22 programs and provisions listed in this report have been established over the past three decades, and are administered by five separate agencies and departments: Environmental Protection Agency, U.S. Department of Agriculture, Department of Energy, Internal Revenue Service, and Customs and Border Protection. These programs target a variety of beneficiaries, including farmers and rural small businesses, biofuel producers, petroleum suppliers, and fuel marketers. Arguably, the most significant federal programs for biofuels have been tax credits for the production or sale of ethanol and biodiesel. However, with the establishment of the renewable fuel standard (RFS) under P.L. 109-58, Congress has mandated biofuels use; P.L. 110-140 significantly expanded that mandate. In the long term, the mandate may prove even more significant than tax incentives in promoting the use of these fuels.

The 2008 Farm Bill—The Food, Conservation, and Energy Act of 2008—amended or established various biofuels incentives, including lowering the value of the ethanol excise tax credit, establishing a tax credit for cellulosic biofuel production, extending import duties on fuel ethanol, and establishing several new grant and loan programs.

Some key biofuels incentives have expired or are set to expire (e.g., tax credits for biodiesel and renewable diesel), and there is congressional interest in extending these credits.

This report outlines federal programs that provide direct or indirect incentives for biofuels. For each program described, the report provides details including administering agency, authorizing statute(s), annual funding, and expiration date. The **Appendix** provides summary information in a table format.

Contents

Tables

Appendixes

Contacts

Introduction

With recent high energy prices, the passage of the Energy Policy Act of 2005 (P.L. 109-58) and the Energy Independence and Security Act of 2007 (P.L. 110-140), and the passage of the 2008 Farm Bill (P.L. 110-246), there is ongoing congressional interest in promoting greater use of alternatives to petroleum fuels. Biofuels—transportation fuels produced from plants and other organic materials—are of particular interest. Ethanol and biodiesel, the two most widely used biofuels, receive significant federal support in the form of tax incentives, loan and grant programs, and regulatory programs. The 2008 Farm Bill also modified existing incentives—including ethanol tax credits and import duties—and established a new tax credit for cellulosic biofuels. The Farm Bill also authorized new biofuels loan and grant programs, some of which were funded in the FY2010 appropriations cycle.

This report outlines 22 current, expired, or pending federal programs that provide direct or indirect incentives for biofuels. The programs are grouped below by administering agency. The incentives for biofuels are summarized in the **Appendix**. This information is compiled from authorizing statutes, committee reports, and Administration budget request documents.

Environmental Protection Agency (EPA)—Renewable Fuel Standard

- Administered by: EPA
- Established: 2005 by the Energy Policy Act of 2005, §1501 (P.L. 109-58); expanded by the Energy Independence and Security Act of 2007, §202 (P.L. 110-140)
- Scheduled termination: None
- Description: The Energy Policy Act of 2005 established a renewable fuel standard (RFS) for automotive fuels. The RFS was expanded by the Energy Independence and Security Act of 2007. The RFS requires the blending of renewable fuels (including ethanol and biodiesel) in transportation fuel. In 2008, fuel suppliers were required to blend 9.0 billion gallons of renewable fuel into gasoline; this requirement increases annually to 36 billion gallons in 2022. The expanded RFS also specifically mandates the use of "advanced biofuels"—fuels produced from non-corn feedstocks and with 50% lower lifecycle greenhouse gas emissions than petroleum fuel—starting in 2009. Of the 36 billion gallons required in 2022, at least 21 billion gallons must be advanced biofuel. There are also specific quotas for cellulosic biofuels and for biomass-based diesel fuel. On May 1, 2007, EPA issued a final rule on the original RFS program detailing compliance standards for fuel suppliers, as well as a system to trade renewable fuel credits between suppliers. On May 26, 2009, EPA proposed rules for the expanded program (RFS2), including lifecycle analysis methods necessary to categorize fuels as advanced biofuels, and new rules for credit verification and trading. While this program is not a direct subsidy for the construction of biofuels plants, the guaranteed market created by the renewable fuel standard is expected to stimulate growth of the biofuels industry.

Internal Revenue Service (IRS)

Various tax credits and other incentives are available for the production, blending, and/or sale of biofuels and biofuel blends. Tax credits vary by the type of fuel and the size of the producer. Before the enactment of the Energy Improvement and Extension Act of 2008 (P.L. 110-343, Division B), some of the credits allowed taxpayers to blend biofuels produced outside the United States with conventional fuels, export the blended fuel, and claim the tax credit. Section 203 of P.L. 110-343 effectively eliminated this so-called "splash-and-dash" practice by requiring that any fuel eligible for the credit must be produced and/or used within the United States.

Volumetric Ethanol Excise Tax Credit

- Administered by: Internal Revenue Service
- Established: 2005 by the American Jobs Creation Act of 2004, §301 (P.L. 108-357); modified by the Food, Conservation, and Energy Act of 2008, §15331 (P.L. 110-246); further amended by the Energy Improvement and Extension Act of 2008 (P.L. 110-343, Division B), §203
- Scheduled termination: December 31, 2010
- Description: Gasoline suppliers who blend ethanol with gasoline are eligible for a tax credit of 45 cents per gallon of ethanol.
- Qualified applicant: Blenders of gasohol (i.e., gasoline suppliers and marketers)
- For more information: IRS Publication 510, Chapter 2: Fuel Tax Credits and Refunds http://www.irs.gov/publications/p510/ch02.html

Small Ethanol Producer Credit

- Administered by: Internal Revenue Service
- Established:1990 by the Omnibus Budget Reconciliation Act of 1990, §11502 (P.L. 101-508); extended by the American Jobs Creation Act of 2004, §301 (P.L. 108-357); expanded by the Energy Policy Act of 2005, §1347 (P.L. 109-58); amended by the Energy Improvement and Extension Act of 2008 (P.L. 110-343, Division B), §203
- Scheduled termination: December 31, 2010
- Description: The small ethanol producer credit is valued at 10 cents per gallon of ethanol produced. The credit may be claimed on the first 15 million gallons of ethanol produced by a small producer in a given year.
- Qualified applicant: Any ethanol producer with production capacity below 60 million gallons per year
- For more information: IRS Publication 510, Chapter 2: Fuel Tax Credits and Refunds http://www.irs.gov/publications/p510/ch02.html

Biodiesel Tax Credit

- Administered by: Internal Revenue Service
- Established: 2005 by the American Jobs Creation Act of 2004, §302 (P.L. 108-357); extended by the Energy Policy Act of 2005, §1344 (P.L. 109-58); amended by the Energy Improvement and Extension Act of 2008 (P.L. 110-343, Division B), §202-203
- Scheduled termination: December 31, 2009—Expired
- Description: Biodiesel producers (or producers of diesel/biodiesel blends) can claim a per-gallon tax credit. The credit is valued at $1.00 per gallon. Before amendment by P.L. 110-343, the credit was valued at $1.00 per gallon of "agri-biodiesel" (biodiesel produced from virgin agricultural products such as soybean oil or animal fats), or 50 cents per gallon of biodiesel produced from previously used agricultural products (e.g., recycled fryer grease).
- Qualified applicant: Biodiesel producers and blenders
- For more information: IRS Publication 510, Chapter 2: Fuel Tax Credits and Refunds http://www.irs.gov/publications/p510/ch02.html

Small Agri-Biodiesel Producer Credit

- Administered by: Internal Revenue Service
- Established: 2005 by the Energy Policy Act of 2005, §1345 (P.L. 109-58); amended by the Energy Improvement and Extension Act of 2008 (P.L. 110-343, Division B), §202-203
- Scheduled termination: December 31, 2009—Expired
- Description: The small agri-biodiesel producer credit is valued at 10 cents per gallon of "agri-biodiesel" (see Biodiesel Tax Credit, above) produced. The credit may be claimed on the first 15 million gallons of ethanol produced by a small producer in a given year.
- Qualified applicant: Any agri-biodiesel producer with production capacity below 60 million gallons per year
- For more information: IRS Publication 510, Chapter 2: Fuel Tax Credits and Refunds http://www.irs.gov/publications/p510/ch02.html

Renewable Diesel Tax Credit

- Administered by: Internal Revenue Service
- Established: 2005 by the Energy Policy Act of 2005, §1346 (P.L. 109-58); amended by the Energy Improvement and Extension Act of 2008 (P.L. 110-343, Division B), §202-203
- Scheduled termination: December 31, 2009—Expired
- Description: Producers of biomass-based diesel fuel (or producers of diesel/renewable biodiesel blends) can claim $1.00 per gallon tax credit.

Renewable diesel is similar to biodiesel, but it is produced through different processes and thus is ineligible for the (above) biodiesel credits.

- Qualified applicant: Renewable diesel producers and blenders
- For more information: IRS Publication 510, Chapter 2: Fuel Tax Credits and Refunds http://www.irs.gov/publications/p510/ch02.html

Credit for Production of Cellulosic Biofuel

- Administered by: Internal Revenue Service
- Established: January 1, 2009, by the Food, Conservation, and Energy Act of 2008, §15321 (P.L. 110-246)
- Scheduled termination: December 31, 2012
- Description: Producers of cellulosic biofuel can claim $1.01 per gallon tax credit. For producers of cellulosic ethanol, the value of the credit is reduced by the amount of the volumetric ethanol excise tax credit and the small ethanol producer credit (see above)—currently, the value is 46 cents per gallon. The credit applies to fuel produced after December 31, 2008.
- Qualified applicant: Cellulosic biofuel producers
- Note: The credit for cellulosic ethanol varies with other ethanol credits such that the total *combined* value of all credits is $1.01 per gallon. As the volumetric ethanol excise tax credit and/or the small ethanol producer credits decrease, the per-gallon credit for cellulosic ethanol production increases by the same amount.

Special Depreciation Allowance for Cellulosic Biofuel Plant Property

- Administered by: Internal Revenue Service
- Established: 2006 by the Tax Relief and Health Care Act of 2006, §209 (P.L. 109-432); amended by the Energy Improvement and Extension Act of 2008 (P.L. 110-343, Division B), §201
- Scheduled termination: December 31, 2012
- Description: A taxpayer may take a depreciation deduction of 50% of the adjusted basis of a new cellulosic biofuel plant in the year it is put in service. Any portion of the cost financed through tax-exempt bonds is exempted from the depreciation allowance. Before amendment by P.L. 110-343, the accelerated depreciation applied only to cellulosic ethanol plants that break down cellulose through enzymatic processes—the amended provision applies to all cellulosic biofuel plants.
- Qualified applicant: Any cellulosic ethanol plant acquired after December 20, 2006, and placed in service before January 1, 2013. Any plant that had a binding contract for acquisition before December 20, 2006, does not qualify.

- For more information: See Senate Finance Committee, Summary of House-Senate Agreement on Tax, Trade, Health, and Other Provisions, December 7, 2006.

Alternative Fuel Station Credit

- Administered by: Internal Revenue Service
- Established: 2005 by the Energy Policy Act of 2005 §1342 (P.L. 109-58); extended by the Energy Improvement and Extension Act of 2008, §207 (P.L. 110-343, Division B); expanded by the American Recovery and Reinvestment Act, §1123 (P.L. 111-5)
- Scheduled termination: December 31, 2012
- Description: A taxpayer may take a 50% credit for the installation of alternative fuel infrastructure, up to $50,000, including E85 (85% ethanol and 15% gasoline) infrastructure. Residential installations qualify for a $2,000 credit (biofuels pumps are not generally installed in residential applications)
- Qualified applicant: Individual or business that installs alternative fuel infrastructure

Department of Agriculture (USDA)

Biorefinery Assistance

- Administered by: Rural Business-Cooperative Service (RBS)
- Annual funding: $74 million in mandatory spending for FY2009, $245 million for FY2010; authorization of and additional $150 million annually for FY2009-FY2012
- FY2010 Appropriation: $245 million to guarantee $691 million in loans
- Established: 2008 by the Food, Conservation, and Energy Act of 2008, §9001 (P.L. 110-246)
- Scheduled termination: End of FY2012
- Description: Grants to biorefineries that use renewable biomass to reduce or eliminate fossil fuel use.
- Qualified applicant: Biorefineries in existence at the date of enactment.
- For more information: See RBS website—http://www.rurdev.usda.gov/rbs/busp/baplg9003.htm

Repowering Assistance

- Administered by: RBS

- Annual funding: $35 million in mandatory funding for FY2009, to remain available until expended, plus $15 million authorized annually for FY2009 through FY2012
- FY2010 Appropriation: None ($35 million in FY2009)
- Established: 2008 by the Food, Conservation, and Energy Act of 2008, §9001 (P.L. 110-246)
- Scheduled termination: End of FY2012
- Description: Grants to biorefineries that use renewable biomass to reduce or eliminate fossil fuel use. RBS issued a Notice of Funding Availability June 12, 2009—http://www.rurdev.usda.gov/rbs/busp/9004%20FR%20NOFA.pdf.
- Qualified applicant: Biorefineries in existence at the date of enactment.
- For more information: See RBS website—http://www.rurdev.usda.gov/rbs/busp/RepoweringAssistance.htm

Bioenergy Program for Advanced Biofuels

- Administered by: RBS
- Annual funding: Mandatory funding of $55 million for FY2009, $55 million for FY2010, $85 million for FY2011, and $105 million for FY2012, plus $25 million authorized annually for FY2009-FY2012
- FY2010 Appropriation: $55 million ($30 million in FY2009)
- Established: 2008 by the Food, Conservation, and Energy Act of 2008, §9001 (P.L. 110-246)
- Scheduled termination: End of FY2012
- Description: Provides payments to producers to support and expand production of advanced biofuels. RBS issued a Notice of Contract Proposal June 12, 2009—http://www.rurdev.usda.gov/rbs/busp/NOCP%20FR%209005.pdf
- Qualified applicant: Producer of advanced biofuels
- For more information: See RBS website—http://www.rurdev.usda.gov/rbs/busp/9005Biofuels.htm

Feedstock Flexibility Program for Producers of Biofuels (Sugar)

- Administered by: Commodity Credit Corporation (CCC)
- Annual funding: Such sums as necessary are authorized to be appropriated—no appropriation to date
- FY2010 Appropriation: None
- Established: 2008 by the Food, Conservation, and Energy Act of 2008, §9001 (P.L. 110-246)
- Scheduled termination: None

- Description: Authorizes the use of CCC funds to purchase surplus sugar, to ensure the sugar program operates at no-net-cost, to be resold as a biomass feedstock to produce bioenergy.
- Qualified applicant: Producer of biofuels using eligible sugar as a feedstock

Biomass Crop Assistance Program (BCAP)

- Administered by: Farm Service Agency (FSA)
- Annual funding: Commodity payment—depends on the number of applications
- Established: 2008 by the Food, Conservation, and Energy Act of 2008, §9001 (P.L. 110-246)
- Scheduled termination: End of FY2012
- Description: Dollar-for-dollar matching payments for collection, harvesting, storage, and transportation (CHST) of biomass to qualified biofuel production facilities (as well as bioenergy or biobased products), up to $45 per ton
- Qualified applicant: Person who delivers eligible biomass to a qualified facility
- For more information: See FSA website—http://www.fsa.usda.gov/FSA/webapp?area=home&subject=ener&topic=bcap

Rural Energy for America Program (REAP)

- Administered by: RBS
- Annual funding: $255 million total in mandatory spending from FY2009-FY2012; an additional authorization of $25 million annually for FY2009-FY2012
- FY2010 Appropriation: $99.4 million in discretionary and mandatory spending to cover $19.7 million in grants and $388 million in loans ($60 million in FY2009 to cover $300 million in loans and $30 million in grants)
- Established: 2008 by the Food, Conservation, and Energy Act of 2008, §9001 (P.L. 110-246)
- Scheduled termination: End of FY2012
- Description: This program replaced the Renewable Energy Systems and Energy Efficiency Improvements program in the 2002 Farm Bill. The program provides grants and loans for a variety of rural energy projects, including efficiency improvements and renewable energy projects. Although REAP is not exclusively aimed at biofuels projects, the program could be a significant source of loan funds for such projects.

Biomass Research and Development

- Administered by: National Institute of Food and Agriculture (NIFA)

- Annual funding: $118 million total mandatory spending for FY2009-FY2012; up to $35 million additional discretionary funding annually
- FY2010 Appropriation: $28 million ($20 million in FY2009)
- Established: FY2001 by the Biomass Research and Development Act of 2000, §307 (P.L. 106-224); program extended and mandatory appropriations provided by the Farm Security and Rural Investment Act of 2002, §9008 (P.L. 107-171); program extended and funding authorization expanded by the Energy Policy Act of 2005, §941 (P.L. 109-58); significantly modified by the Food, Conservation and Energy Act of 2008, §9008 (P.L. 110-246)
- Scheduled termination: End of FY2012
- Description: Grants are provided for biomass research, development, and demonstration projects. Eligible projects include ethanol and biodiesel demonstration plants.
- Qualified applicant: Wide range of possible applicants
- For more information: http://www.brdisolutions.com/default.aspx

Other USDA Programs

The following programs within the Rural Business Cooperative Service could possibly be used to assist biofuels producers indirectly:

- Business and Industry (B&I) Guaranteed Loans
- Rural Business Enterprise Grants (RBEG)
- Value-Added Grants
- Rural Economic Development Loan and Grant Programs

Department of Energy (DOE)

Biorefinery Project Grants

- Administered by: Office of Energy Efficiency and Renewable Energy
- Annual funding: Approximately $200 million appropriated annually for the biomass program—not all of this funding will go toward biorefinery project grants
- FY2010 Appropriation: $220 million for entire biomass program
- Established: FY2001 through funding authorized in various statutes
- Scheduled termination: None
- Description: This program provides funds for cooperative biomass research and development for the production of fuels, electric power, chemicals, and other products.

- Qualified applicant: Varies from year to year, depending on program goals in a given year
- For more information: http://www.eere.energy.gov/biomass/

Loan Guarantees for Ethanol and Commercial Byproducts from Cellulose, Municipal Solid Waste, and Sugar Cane

- Administered by: DOE
- Annual funding: Not specified
- FY2010 Appropriation: $0
- Established: 2005 by the Energy Policy Act of 2005, §§1510, 1511, and 1516 (P.L. 109-58)
- Scheduled termination: Varies, depending on specific program
- Description: The Energy Policy Act of 2005 authorizes several programs to provide loan guarantees for the construction of facilities that produce ethanol and other commercial products from cellulosic material, municipal solid waste, or sugar cane.
- Qualified applicant: Private lending institutions, to guarantee loans for the construction of biofuels plants

DOE Loan Guarantee Program

- Administered by: DOE
- Annual funding: $5.4 million for administrative expenses in FY2008; authority for $51 billion in loan guarantees for energy projects in FY2008, including $10 billion for renewable energy and energy efficiency; $6 billion additional FY2009 appropriation to cover $49 billion in loans to all projects (not just biofuels)
- FY2010 Appropriation: $43 million for administrative expenses—to be offset by loan application fees
- Established: 2005 by the Energy Policy Act of 2005, Title XVII (P.L. 109-58)
- Scheduled termination: Not specified
- Description: Title XVII of the Energy Policy Act of 2005 authorizes DOE to provide loan guarantees for energy projects that reduce air pollutant and greenhouse gas emissions, including biofuels projects.
- Qualified applicant: Private lending institutions, to guarantee loans for clean energy projects.
- For more information: http://www.lgprogram.energy.gov/

Cellulosic Ethanol Reserve Auction

- Administered by: DOE

- Annual funding: $1 billion total authorized for all fiscal years; not more than $100 million may be paid in any given year
- FY2010 Appropriation: None; $5 million in FY2008 for administrative expenses
- Established: 2005 by the Energy Policy Act of 2005, §942 (P.L. 109-58)
- Scheduled termination: Not specified
- Description: Section 942 of the Energy Policy Act of 2005 authorizes DOE to provide per-gallon incentive payments for cellulosic biofuels until annual U.S. production reaches 1 billion gallons or 2015, whichever is earlier. DOE finalized regulations on October 15, 2009. http://www.epa.gov/fedrgstr/EPA-IMPACT/2009/October/Day-15/i24778.htm.
- Qualified applicant: Any U.S. cellulosic biofuel production facility that meets applicable requirements.

U.S. Customs and Border Protection—Import Duty for Fuel Ethanol

- Administered by: U.S. Customs and Border Protection
- Annual funding: N/A
- Established: 1980 by the Omnibus Reconciliation Act of 1980 (P.L. 96-499); amended by the Tax Reform Act of 1986, §423 (P.L. 99-514) extended by the Tax Relief and Health Care Act of 2006, §302 (P.L. 109-432); further extended by the Food, Conservation, and Energy Act of 2008, §15333 (P.L. 110-246)
- Scheduled termination: December 31, 2010
- Description: A 2.5% ad valorem tariff and a most-favored-nation duty of $0.54 per gallon of ethanol (for fuel use) applies to imports into the United States from most countries; most ethanol from Caribbean Basin Initiative (CBI) countries may be imported duty-free.
- Covered Entities: Fuel ethanol importers
- For more information: CRS Report RS21930, *Ethanol Imports and the Caribbean Basin Initiative (CBI)*, by Brent D. Yacobucci; Senate Finance Committee, Summary of House-Senate Agreement on Tax, Trade, Health, and Other Provisions, December 7, 2006.

Department of Transportation—Manufacturing Incentive for Flexible Fuel Vehicles

- Administered by: National Highway Traffic Safety Administration
- Annual funding: N/A

- Established: 1975 by the Energy Policy and Conservation Act of 1975 (P.L. 94-163); amended by various statutes, most recently the Energy Independence and Security Act of 2007, §109 (P.L. 110-140)
- Scheduled termination: After model year 2019
- Description: Automakers are required to meet Corporate Average Fuel Economy (CAFE) standards for their passenger cars and light trucks. Manufacturers may gain credits for the sale of alternative fuel vehicles, including ethanol/gasoline flexible fuel vehicles (FFVs). However, the credits are limited—the maximum fuel economy increase allowed through the use of these credits is 1.2 miles per gallon through model year (MY) 2014. The credits are phased out after MY2014 and are completely eliminated after MY2019.

Appendix. Summary of Federal Incentives Promoting Biofuels

Table A-1. Federal Biofuels Incentives by Agency

Administering Agency	Program	Description	Original Authorizing Legislation	FY2010 Appropriation	Expiration Date
Environmental Protection Agency	Renewable Fuel Standard	Mandated use of renewable fuel in gasoline: 4.0 billion gallons in 2006, increasing to 7.5 billion gallons in 2012	P.L. 109-58 §1501	N/A	None
Internal Revenue Service	Volumetric Ethanol Excise Tax Credit	Gasoline suppliers who blend ethanol with gasoline are eligible for a tax credit of 45 cents per gallon of ethanol	P.L. 108-357 §301	N/A	End of 2010
	Small Ethanol Producer Credit	An ethanol producer with less than 60 million gallons per year in production capacity may claim a credit of 10 cents per gallon on the first 15 million gallons produced in a year	P.L. 101-508	N/A	End of 2010
	Biodiesel Tax Credit	Producers of biodiesel or diesel/biodiesel blends may claim a tax credit of $1.00 per gallon of biodiesel.	P.L. 108-357	N/A	End of 2009
	Small Agri-Biodiesel Producer Credit	An agri-biodiesel (produced from virgin agricultural products) producer with less than 60 million gallons per year in production capacity may claim a credit of 10 cents per gallon on the first 15 million gallons produced in a year	P.L. 109-58	N/A	End of 2009
	Renewable Diesel Tax Credit	Producers of renewable diesel (similar to biodiesel, but produced through a different process) may claim a tax credit of $1.00 per gallon of renewable diesel	P.L. 109-58	N/A	End of 2009

Administering Agency	Program	Description	Original Authorizing Legislation	FY2010 Appropriation	Expiration Date
	Credit for Production of Cellulosic Biofuel	Producers of cellulosic biofuel may claim a tax credit of $1.01 per gallon. For cellulosic ethanol producers, the value of the production tax credit is reduced by the value of the volumetric ethanol excise tax credit and the small ethanol producer credit—the credit was currently valued at 46 cents per gallon. The credit applies to fuel produced after December 31, 2008.	P.L. 110-246	N/A	End of 2012
	Special Depreciation Allowance for Cellulosic Biofuel Plant Property	Plants producing cellulosic biofuels may take a 50% depreciation allowance in the first year of operation, subject to certain restrictions	P.L. 109-432	N/A	End of 2012
	Alternative Fueling Station Credit	A credit of up to $50,000 is available for the installation of alternative fuel pumps, including E85 (85% ethanol and 15% gasoline)	P.L. 109-58 §1342	N/A	End of 2010
Department of Agriculture	Biorefinery Assistance	Loan guarantees and grants for the construction and retrofitting of biorefineries to produce advanced biofuels	P.L. 110-246 §9001	$245 million to guarantee $691 million in loans	End of FY2012
	Repowering Assistance	Grants to biorefineries that use renewable biomass to reduce or eliminate fossil fuel use	P.L. 110-246 §9001	$35 million in FY2009	End of FY2012
	Bioenergy Program for Advanced Biofuels	Provides payments to producers to support and expand production of advanced biofuels	P.L. 110-246 §9001	$55 million	End of FY2012
	Feedstock Flexibility Program for Producers of Biofuels (Sugar)	Authorizes the use of CCC funds to purchase surplus sugar, to be resold as a biomass feedstock to produce bioenergy	P.L. 110-246 §9001	No appropriation to date	None
	Biomass Crop Assistance Program (BCAP)	Provides financial assistance for biomass crop establishment costs and annual payments for biomass production; also provides payments to assist with costs for biomass collection, harvest, storage, and transportation	P.L. 110-246 §9001	Dollar-for dollar commodity payment	End of FY2012

Administering Agency	Program	Description	Original Authorizing Legislation	FY2010 Appropriation	Expiration Date
	Rural Energy for America Program (REAP)	Loan guarantees and grants for a wide range of rural energy projects, including biofuels.	P.L. 110-246 §9001	$99.4 million to cover $19.7 million in grants and $388 million in loan guarantees	End of FY2012
	Biomass Research and Development	Grants for biomass research, development, and demonstration projects	P.L. 106-224	$28 million	End of FY2015
Department of Energy	Biorefinery Project Grants	Funds cooperative R&D on biomass for fuels, power, chemicals, and other products	Various statutes	$220 million total for biomass program	None
	Loan Guarantees for Ethanol and Commercial Byproducts from Various Feedstocks	Several programs of loan guarantees to construct facilities that produce ethanol and other commercial products from cellulosic material, municipal solid waste, and/or sugarcane	P.L. 109-58 §§1510, 1511, and 1516	No appropriation to date	Varies
	DOE Loan Guarantee Program	Loan guarantees for energy projects that reduce air pollutant and greenhouse gas emissions, including biofuels projects	P.L. 109-58 Title XVII	$43 million FY2010 appropriation to be offset by loan fees Approximately $100 billion in loan authority from FY2008 and FY2009 appropriations; $10 billion in loan authority for renewable energy and energy efficiency	None
	Cellulosic Ethanol Reserve Auction	Authorizes DOE to provide per-gallon payments to cellulosic biofuel producers	P.L. 109-58 §942	No FY2010 appropriation $5 million in FY2008 for administrative expenses	August 8, 2015
U.S. Customs and Border Protection	Import Duty for Fuel Ethanol	All imported ethanol is subject to a 2.5% ad valorem tariff; fuel ethanol is also subject to a most-favored-nation added duty of 54 cents per gallon (with some exceptions)	P.L. 96-499	N/A	End of 2010

Administering Agency	Program	Description	Original Authorizing Legislation	FY2010 Appropriation	Expiration Date
Department of Transportation	Flexible Fuel Vehicle Production Incentive	Automakers subject to Corporate Average Fuel Economy (CAFE) standards may accrue credits under that program for the production and sale of alternative fuel vehicles, including ethanol/gasoline flexible fuel vehicles (FFVs)	P.L. 94-163	N/A	Incentive expires after model year 2019

Source: CRS.

Author Contact Information

Brent D. Yacobucci
Specialist in Energy and Environmental Policy
byacobucci@crs.loc.gov, 7-9662

Order Code RL33290

CRS Report for Congress

Fuel Ethanol: Background and Public Policy Issues

Updated April 24, 2008

Brent D. Yacobucci
Specialist in Energy and Environmental Policy
Resources, Science, and Industry Division

Prepared for Members and Committees of Congress

Fuel Ethanol: Background and Public Policy Issues

Summary

Ethanol plays a key role in policy discussions about energy, agriculture, taxes, and the environment. In the United States it is mostly made from corn; in other countries it is often made from cane sugar. Fuel ethanol is generally blended in gasoline to reduce emissions, increase octane, and extend gasoline stocks. Recent high oil and gasoline prices have led to increased interest in alternatives to petroleum fuels for transportation. Further, concerns over climate change have raised interest in developing fuels with lower fuel-cycle greenhouse-gas emissions. Supporters of ethanol argue that its use can lead to lower emissions of toxic and ozone-forming pollutants, and greenhouse gases, especially if higher-level blends are used. They further argue that ethanol use displaces petroleum imports, thus promoting energy security. Ethanol's detractors argue that various federal and state policies supporting ethanol distort the market and amount to corporate welfare for corn growers and ethanol producers. Further, they argue that the energy and chemical inputs needed to turn corn into ethanol actually increase emissions and energy consumption, although most recent studies have found modest energy and emissions benefits from ethanol use relative to gasoline, depending on how the ethanol is produced.

The market for fuel ethanol is heavily dependent on federal incentives and regulations. Ethanol production is encouraged by a federal tax credit of 51 cents per gallon. This incentive allows ethanol — which has historically been more expensive than conventional gasoline — to compete with gasoline and other blending components. In addition to the above tax credit, small ethanol producers qualify for an additional production credit. It has been argued that the fuel ethanol industry could scarcely survive without these incentives.

In addition to the above tax incentives, the Energy Policy Act of 2005 (P.L. 109-58) established a renewable fuel standard (RFS). This RFS was expanded by the Energy Independence and Security Act of 2007 (P.L. 110-140), and requires the use of 9.0 billion gallons of renewable fuels in 2008, increasing each year to 36 billion gallons in 2022. Much of this requirement will likely be met with ethanol. In addition, the bill requires that an increasing share of the mandate be met with "advanced biofuels" — biofuels produced from feedstocks other than corn starch. Potential "advanced biofuels" include domestic ethanol from cellulosic material (such as perennial grasses and municipal solid waste), ethanol from sugar cane, and diesel fuel substitutes produced from a variety of feedstocks. The United States consumed approximately 6.8 billion gallons of ethanol in 2007, mostly from corn. A sigificant supply of cellulosic ethanol is likely several years off. Some analysts believe the RFS could have serious effects on motor fuel suppliers, leading to higher fuel prices.

Other issues of congressional interest include support for purer blends of ethanol as an alternative to gasoline (as opposed to a gasoline blending component), promotion of ethanol vehicles and infrastructure, and imports of ethanol from foreign countries. This report supersedes CRS Report RL30369, *Fuel Ethanol: Background and Public Policy Issues* (out of print but available from the author).

Contents

List of Tables

Fuel Ethanol: Background and Public Policy Issues

Introduction

The promotion of alternatives to petroleum, including fuel ethanol, has been an ongoing goal of U.S. energy policy. This promotion has led to the establishment of significant federal policies beneficial to the ethanol industry, including tax incentives, import tariffs, and mandates for ethanol use. The costs and benefits of ethanol — and the policies that support it — have been questioned. Areas of concern include whether ethanol yields more or less energy than the fossil fuel inputs needed to produce it; whether ethanol decreases reliance on petroleum in the transportation sector; whether its use increases or decreases greenhouse gas emissions; and whether various federal policies should be maintained.

This report provides background and discussion of policy issues relating to U.S. ethanol production, especially ethanol made from corn. It discusses U.S. fuel ethanol consumption both as a gasoline blending component and as an alternative to gasoline. The report discusses various costs and benefits of ethanol, including fuel costs, pollutant emissions, and energy consumption. It also outlines key areas of congressional debate on policies beneficial to the ethanol industry.

Ethanol Basics

Fuel ethanol (ethyl alcohol) is made by fermenting and distilling simple sugars. It is the same compound found in alcoholic beverages. The biggest use of fuel ethanol in the United States is as an additive in gasoline. It serves as an oxygenate, to prevent air pollution from carbon monoxide and ozone; as an octane booster, to prevent early ignition, or "engine knock"; and as an extender of gasoline stocks. In purer forms, it can also be used as an alternative to gasoline in automobiles specially designed for its use. It is produced and consumed mostly in the Midwest, where corn — the main feedstock for domestic ethanol production — is grown.

The initial stimulus for ethanol production in the mid-1970s was the drive to develop alternative and renewable supplies of energy in response to the oil embargoes of 1973 and 1979. Since the 1970s, production of fuel ethanol has been encouraged through federal tax incentives for ethanol-blended gasoline. The use of fuel ethanol was further stimulated by the Clean Air Act Amendments of 1990, which required the use of oxygenated or reformulated gasoline (RFG). The Energy Policy Act of 2005 (P.L. 109-58) established a renewable fuels standard (RFS), which mandates the use of ethanol and other transportation renewable fuels.

CRS-2

Approximately 99% of fuel ethanol consumed in the United States is "gasohol"[1] or "E10" (blends of gasoline with up to 10% ethanol). About 1% is consumed as "E85" (85% ethanol and 15% gasoline), and alternative to gasoline.[2]

Fuel ethanol is usually produced in the United States from the distillation of fermented simple sugars (e.g. glucose) derived primarily from corn, but also from wheat, potatoes, or other vegetables.[3] However, ethanol can also be produced from cellulosic material such as switchgrass, rice straw, and sugar cane waste (known as bagasse). The alcohol in fuel ethanol is identical chemically to ethanol used for other purposes such as distilled spirit beverages and industrial products.[4]

Ethanol and the Agricultural Economy[5]

Corn constitutes about 95% of the feedstock for ethanol production in the United States. The other 5% is largely grain sorghum, along with some barley, wheat, cheese whey and potatoes. Corn is used because it is a relatively low cost source of starch that can be relatively easily converted to simple sugars, and then fermented and distilled. The U.S. Department of Agriculture (USDA) estimates that about 3.2 billion bushels of corn will be used to produce about 6 billion gallons of fuel ethanol during the 2007/2008 corn marketing year (September 2007 through August 2008).[6] This is roughly 25% of the projected 13 billion bushels of total corn utilization for all purposes.[7] However, it should be noted that ethanol production capacity is expanding rapidly, and corn demand for ethanol production may exceed USDA's projection.[8]

In the absence of the ethanol market, lower corn prices probably would stimulate increased corn utilization in other markets, but sales revenue would not be as high. The lower prices and sales revenue would likely result in higher federal spending on corn subsidy payments to farmers, as long as corn prices were to stay below the price triggering federal loan deficiency subsidies.

---

[1] Technically, gasohol is any blend of ethanol and gasoline, but the term most often refers to the 10% blend.

[2] U.S. Department of Energy (DOE), Energy Information Administration (EIA), *Alternatives to Traditional Transportation Fuels 2005*, updated November 2007.

[3] In some other countries, most notably Brazil, ethanol is produced from cane sugar.

[4] Industrial uses include perfumes, aftershaves, and cleansers.

[5] For a more detailed discussion of ethanol's role in agriculture, see CRS Report RL32712, *Agriculture-Based Renewable Energy Production*, by Randy Schnepf.

[6] One bushel of corn generates approximately 2.7 gallons of ethanol.

[7] Utilization data are used, rather than production, due to the existence of carryover stocks. Corn utilization data address the total amount of corn used within a given period.

[8] As of March 2008, the Renewable Fuels Association reported U.S. production capacity at 8.3 billion gallons, with an additional 5.1 billion gallons of capacity under construction (including expansions of existing plants). See [http://www.ethanolrfa.org/industry/locations/] (updated March 17, 2008).

CRS-3

Table 1. Corn Utilization, 2007-2008 Forecast

	Quantity (million bushels)	Share of total use
Livestock feed & residual	5,950	45.9%
Food, seed & industrial:	4,555	35.2%
— Fuel alcohol	**3,200**	**24.7%**
— High fructose corn syrup	500	3.9%
— Glucose & dextrose	235	1.8%
— Starch	270	2.1%
— Cereals & other products	193	1.5%
— Beverage alcohol	135	1.0%
— Seed	22	0.2%
Exports	2,450	18.9%
Total Use	**12,955**	**100.0%**
Total Production	**13,074**	

Source: Basic data are from USDA, Economic Research Service, *Feed Outlook*, March 13, 2008.

Note: Annual use can exceed production through the use of stocks carried over from previous years.

Ethanol Refining and Production

According to the Renewable Fuels Association,[9] about 80% of the corn used for ethanol is processed by "dry" milling plants (which use a grinding process) and the other 20% is processed by "wet" milling plants (which use a chemical extraction process). The basic steps of both processes are similar. First, the corn is processed, with various enzymes added to separate fermentable sugars from other components such as protein and fiber; some of these other components are used to make coproducts, such as animal feed. Next, yeast is added to the mixture for fermentation to make alcohol. The alcohol is then distilled to fuel-grade ethanol that is 85%-95% pure. Then the ethanol is partially dehydrated to remove excess water. Finally, for fuel and industrial purposes the ethanol is denatured with a small amount of a displeasing or noxious chemical to make it unfit for human consumption.[10] In the United States, the denaturant for fuel ethanol is gasoline.

Ethanol is produced largely in the Midwest corn belt, with roughly 70% of the national output occurring in five states: Iowa, Nebraska, Illinois, Minnesota and South Dakota. Because it is generally less expensive to produce ethanol close to the

---

[9] [http://www.ethanolrfa.org/].

[10] Renewable Fuels Association, *Ethanol Industry Outlook 2002, Growing Homeland Energy Security.*

feedstock supply, it is not surprising that the top corn-producing states in the U.S. are also the main ethanol producers. This geographic concentration is an obstacle to the use of ethanol on the East and West Coasts. Most ethanol use is in the metropolitan centers of the Midwest, where it is produced. When ethanol is used in other regions, shipping costs tend to be high, since ethanol-blended gasoline cannot travel through petroleum pipelines, but must be transported by truck, rail, or barge. However, due to Clean Air Act requirements,[11] concerns over other fuel additives, and the establishment of a renewable fuels standard, ethanol use on the East and West Coasts is growing steadily. For example, in 1999 California and New York accounted for 5% of U.S. ethanol consumption, increasing to 22% in 2003, and 33% in 2004.[12]

The potential for expanding production geographically is one motivation behind research on cellulosic ethanol. If regions could locate production facilities closer to the point of consumption, the costs of using ethanol could be lessened. Furthermore, if regions could produce fuel ethanol from local crops, there could be an increase in regional agricultural income.

Table 2. Top 10 Ethanol Producers by Capacity, March 2008

(existing production capacity — million gallons per year)

POET	1,208
Archer Daniels Midland (ADM)	1,070
VeraSun Energy Corporation	560
U.S. BioEnergy Corp.	420
Hawkeye Renewables	225
Aventine Renewable Energy	207
Abenoga Bioenergy Corp.	198
White Energy	148
Renew Energy	130
Cargill	120
All others	4,024
Total	8,310

Source: Renewable Fuels Association, U.S. Fuel Ethanol Industry Plants and Production Capacity, March 17, 2008.

---

[11] P.L. 109-58 amended the Clean Air Act to eliminate the reformulated gasoline oxygenate standard, one of the key federal policies promoting the use of ethanol. However, the act also established a renewable fuels standard, effectively mandating the use of ethanol. (See "Renewable Fuels Standard" below.)

[12] U.S. Department of Transportation, Federal Highway Administration, *Highway Statistics Series*, 1999, 2003, and 2004.

CRS-5

Historically, ethanol production was concentrated among a few large producers. However, that concentration has declined over the past several years. **Table 2** shows that currently, the top five companies account for approximately 42% of production capacity, and the top ten companies account for approximately 48% of production capacity. Critics of the ethanol industry in general — and specifically of the ethanol tax incentives — have argued that the tax incentives for ethanol production equate to "corporate welfare" for a few large producers.[13] However, the share of production capacity controlled by the largest producers has been dropping as more producers have entered the market.

Section 1501(a)(2) of the Energy Policy Act of 2005 required the Federal Trade Commission (FTC) to study whether there is sufficient competition in the U.S. ethanol industry. The FTC concluded that "the level of concentration in ethanol production would not justify a presumption that a single firm, or a small group of firms, could wield sufficient market power to set prices or coordinate on prices or output."[14] Further, they concluded that the level of concentration has been decreasing in recent years.

Overall, at the beginning of 2008, domestic ethanol production capacity was approximately 8 billion gallons per year, and is expected to grow to 13 billion gallons per year, counting existing plants and plants under construction.[15] Under various federal and state laws and incentives, consumption has increased from 1.8 billion gallons per year in 2001 to 6.8 billion gallons per year in 2007. Domestic production capacity will continue increasing to meet the growing demand, including increased demand resulting from implementation of the renewable fuels standard established by the Energy Policy Act of 2005.

Fuel is not the only output of an ethanol facility, however. Coproducts play an important role in the profitability of a plant. In addition to the primary ethanol output, the corn wet milling process generates corn gluten feed, corn gluten meal, and corn oil, and dry milling process creates distillers grains. Corn oil is used as a vegetable oil and is priced higher than soybean oil; the other coproducts are used as livestock feed. In 2004, U.S. ethanol mills produced 7.3 million metric tons of distillers grains, 2.4 million metric tons of corn gluten feed, 0.4 million metric tons of corn gluten meal, and 560 million pounds of corn oil.[16]

Revenue from the ethanol byproducts help offset the cost of corn used in ethanol production. The net cost of corn relative to the price of ethanol and the difference between ethanol and wholesale gasoline prices are the major economic determinants

---

[13] Erin M. Hymel, The Heritage Foundation, *Ethanol Producers Get a Handout from Consumers*, October 16, 2002.

[14] Federal Trade Commission, *2006 Report on Ethanol Market Concentration*, December 1, 2006. p. 2.

[15] Renewable Fuels Association, *U.S. Fuel Ethanol Industry Plants and Production Capacity*, March 2008.

[16] Renewable Fuels Association, *Ethanol Industry Outlook 2005, Homegrown for the Homeland*, February 2005.

of the level of ethanol production. Higher the corn prices lead to lower profits for ethanol producers; higher gasoline prices lead to higher profits. Recently, high corn prices have cut into corn ethanol producers' profits.

Fuel Consumption

Approximately 7 billion gallons of ethanol fuel were consumed in the United States in 2007, mainly blended into E10 gasohol (a blend of 10% ethanol and 90% gasoline). This figure represents only 5% of the approximately 140 billion gallons of gasoline consumption in the same year.[17] Under the renewable fuels standard, motor fuel will be required to contain 36 billion gallons of renewable fuel annually by 2022. It is expected that much of this requirement will be met with ethanol.

Ethanol consumption in 2007 accounted for approximately 4% of combined gasoline and diesel fuel consumption.[18] Because of its physical properties, ethanol can be more easily substituted for — or blended into — gasoline, which powers most passenger cars and light trucks. However, heavy-duty vehicles are generally diesel-fueled. For this reason, research is ongoing into ethanol-diesel blends.

A key barrier to wider use of fuel ethanol is its cost relative to gasoline. Even with tax incentives for ethanol use (see the section on Economic Effects), the fuel is often more expensive than gasoline per gallon.[19] Further, since fuel ethanol has a somewhat lower energy content per gallon, more fuel is required to travel the same distance. This energy loss leads to a 2%-3% decrease in miles-per-gallon vehicle fuel economy with 10% gasohol. This is due to the fact that there is simply less energy in one gallon of ethanol than in one gallon of gasoline, as opposed to any detrimental effect on the efficiency of the engine.[20]

However, ethanol's chemical properties make it very useful for some applications, especially as an additive in gasoline. The oxygenate requirement of the Clean Air Act Reformulated Gasoline (RFG) program provided a major boost to the use of ethanol.[21] Oxygenates are used to promote more complete combustion of gasoline, which reduces carbon monoxide (CO) emissions and may reduce volatile organic compound (VOC) emissions.[22] In addition, oxygenates can replace other chemicals in gasoline, such as benzene, a toxic air pollutant. Conversely, the higher

---

[17] DOE, EIA, *Alternatives to Traditional Transportation Fuels 2005*, Table C1.

[18] Ibid.

[19] However, gasoline prices have been high recently, making ethanol more attractive as a blending component.

[20] In fact, there is some evidence that the combustion efficiency of an engine improves with the use of ethanol relative to gasoline. In this way, a greater percentage of energy in the fuel is transferred to the wheels. However, this improved efficiency does not completely negate the fact that there is less energy in a gallon of ethanol than in a gallon of gasoline.

[21] Section 211, Subsection k; 42 U.S.C. 7545.

[22] CO, VOCs and nitrogen oxides are the main precursors to ground-level ozone.

volatility of ethanol-blended gasoline can in some cases lead to higher VOC emissions (see "Air Quality," below).

The two most common oxygenates are ethanol and methyl tertiary butyl ether (MTBE). Until recently, MTBE, made primarily from natural gas or petroleum products, was preferred to ethanol in most regions because it was generally much less expensive, easier to transport and distribute, and available in greater supply. Because of different distribution systems and gasoline blending processes, substituting one oxygenate for another can lead to significant transitional costs, in addition to the cost differential between the two additives.

Despite the cost differential, there are several possible advantages of using ethanol over MTBE. Since ethanol is produced from agricultural products, it has the potential to be a sustainable fuel, while MTBE is produced from fossil fuels, either natural gas or petroleum. In addition, ethanol is readily biodegradable, eliminating some of the potential concerns about groundwater contamination that have surrounded MTBE (see the section on MTBE). However, there is concern that ethanol use can increase the risk of groundwater contamination by benzene and other toxic compounds.[23]

Both ethanol and MTBE also can be blended into otherwise non-oxygenated gasoline to raise the octane rating of the fuel. High-performance engines and older engines often require higher octane fuel to prevent early ignition, or "engine knock." Other chemical additives may be used for the same purpose, but some of these alternatives are highly toxic, and some are regulated as pollutants under the Clean Air Act.[24] Furthermore, since these other additives do not contain oxygen, their use may not lead to the same emissions reductions as oxygenated gasoline.

E85 Consumption

In purer forms, such as E85, ethanol can also be used as an alternative to gasoline in vehicles specifically designed to use it. Currently, this use represents only approximately 1% of ethanol consumption in the United States. To promote the development of E85 and other alternative fuels, Congress has enacted various legislative requirements and incentives. The Energy Policy Act of 1992 requires the federal government and state governments, along with businesses in the alternative

---

[23] Gasoline contains many different chemical compounds, including toxic substances such as benzene. In the case of a leaking gasoline storage tank, various compounds within the gasoline, based on their physical properties, will travel different distances through the ground. The concern with ethanol is that there is very limited evidence that plumes of benzene and other toxic substances travel farther if ethanol is blended into gasoline. However, this property has not been firmly established, as it has not been studied in depth. Susan E. Powers, David Rice, Brendan Dooher, and Pedro J. J. Alvarez, "Will Ethanol-Blended Gasoline Affect Groundwater Quality?," *Environmental Science and Technology*, January 1, 2001, p. 24A.

[24] Lead was commonly used as an octane enhancer until it was phased-out through the mid-1980s (lead in gasoline was completely banned in 1995), due to the fact that it disables emissions control devices, and because it is toxic to humans.

fuel industry, to purchase alternative-fueled vehicles.[25] In addition, under the Clean Air Act Amendments of 1990, municipal fleets can use alternative fuel vehicles as one way to mitigate air quality problems. Both E85 and E95 (95% ethanol with 5% gasoline) are currently considered alternative fuels by the Department of Energy.[26] The small amount of gasoline added to the alcohol helps prevent corrosion of engine parts and aids ignition in cold weather.

Table 3. Estimated U.S. Consumption of Fuel Ethanol, Gasoline, and Diesel

(million gasoline-equivalent gallons)

	1996	1998	2000	2002	2004
E85	1	2	7	10	22
E95	3	0[a]	0	0	0
Ethanol in Gasohol (E10)	660	890	1,110	1,120	2,052[b]
Gasoline[c]	117,800	122,850	125,720	130,740	136,370
Diesel	30,100	33,670	36,990	38,310	40,740

Source: Department of Energy, Alternatives to Traditional Transportation Fuels 1999.

a. A major drop in E95 consumption occurred between 1997 and 1998 because the number of E95-fueled vehicles in operation dropped from 347 to 14, due to the elimination of an ethanol-fueled municipal bus fleet in California. This fleet was eliminated due to higher fuel and maintenance costs. DOE currently reports that no E95 vehicles were in operation in 2004.

b. An estimated 3.4 billion gallons of ethanol were consumed in 2004. However, due to ethanol's lower energy content, the number of equivalent gallons is lower.

c. Gasoline consumption includes ethanol in gasohol.

Approximately 22 million gasoline-equivalent gallons (GEG)[27] of E85 were consumed in 2004, mostly in Midwestern states.[28] (See **Table 3**.) A key reason for the relatively low consumption of E85 is that there are relatively few vehicles that operate on E85. In 2006 the National Ethanol Vehicle Coalition estimated that there were approximately six million E85-capable vehicles on U.S. roads,[29] as compared

---

[25] P.L. 102-486. For example, of the light-duty vehicles purchased by a federal agency in a given year, 75% must be alternative fuel vehicles.

[26] More diluted blends of ethanol, such as E10, are considered to be "extenders" of gasoline, as opposed to alternatives.

[27] Since different fuels produce different amounts of energy per gallon when consumed, the unit of a gasoline-equivalent gallon (GEG) is used to compare total energy consumption. It takes roughly 1.4 gallons of E85 to equal the energy content in one gallon of gasoline.

[28] DOE, EIA, *Alternatives to Traditional Transportation Fuels.*

[29] National Ethanol Vehicle Coalition, *Frequently Asked Questions*, accessed February 3, 2006 [http://www.e85fuel.com/e85101/faq.php].

CRS-9

to approximately 230 million gasoline- and diesel-fueled vehicles.[30] Most E85-capable vehicles are "flexible fuel vehicles" or FFVs. An FFV can operate on any mixture of gasoline and 0% to 85% ethanol. A large majority of FFVs on U.S. roads are fueled exclusively on gasoline. In 2004, approximately 146,000 flexible fuel vehicles (FFVs) were actually fueled by E85.[31] Proponents of E85 and FFVs argue that even though few FFVs are operated on E85, the large number of these vehicles already on the road means that incentives to expand E85 infrastructure are more likely to be successful.

One obstacle to the use of alternative fuel vehicles is that they generally have a higher purchase price than conventional vehicles, although this margin has decreased in recent years with newer technology. Another obstacle is that, as stated above, fuel ethanol is often more expensive than gasoline or diesel fuel. In addition, there are very few fueling sites for E85, especially outside of the Midwest. As of February 2006, there were 556 fuel stations with E85, as compared to roughly 120,000 gasoline stations across the country. Further, 362 (65%) of these stations were located in the five highest ethanol-producing states: Minnesota, Illinois, Iowa, South Dakota, and Nebraska. In February 2006, there were only 60 stations in 10 states along the east and west coasts, where population — and thus fuel demand — is higher. However, E85 capacity is expanding rapidly, and the number of E85 stations nearly tripled (to 1,365) between February 2006 and March 2008, and the number along the coasts had increased to 146 stations in 13 states (although roughly half of all stations are still in the top five ethanol-producing states).

Development of Cellulosic Feedstocks

A key barrier to ethanol's expanded role in U.S. fuel consumption is its price differential with gasoline. Since a major part of the total production cost is the cost of feedstock, reducing feedstock costs could lead to lower wholesale ethanol costs. For this reason, there is a great deal of interest in producing ethanol from cellulosic feedstocks. Cellulosic materials include low-value waste products such as recycled paper and rice hulls, or dedicated fuel crops, such as switchgrass[32] and fast-growing trees. A dedicated fuel crop would be grown and harvested solely for the purpose of fuel production.

However, as the name indicates, cellulosic feedstocks are high in cellulose. Cellulose forms a majority of plant matter, but it is generally fibrous and cannot be

---

[30] Federal Highway Administration, *Highway Statistics 2003*, November 2004.

[31] DOE, EIA, *Alternatives to Traditional Transportation Fuels*. In 1997, some manufacturers began making flexible E85/gasoline fueling capability standard on some models. However, some owners may not be aware of their vehicles' flexible fuel capability.

[32] Switchgrass is a tall, fast-growing perennial grass native to the North American tallgrass prairie. It is of key interest because it readily grows with limited fertilizer use in marginal growing areas. Further, its cultivation can improve soil quality.

directly fermented.[33] It must first be broken down into simpler molecules, which is currently expensive. A 2000 study by USDA and the National Renewable Energy Laboratory (NREL) estimated a 70% increase in production costs with large-scale ethanol production from cellulosic biomass compared with ethanol produced from corn.[34] Therefore, federal research has focused on both reducing the process costs for cellulosic ethanol and improving the availability of cellulosic feedstocks. The Natural Resources Defense Council estimates that with mature technology, advanced ethanol production facilities could produce significant amounts of fuel at $0.59 to $0.91 per gallon (before taxes) by 2012, a price that is competitive with Energy Information Administration (EIA) projections for gasoline prices in 2012.[35]

Other potential benefits from the development of cellulosic ethanol include lower greenhouse gas and air pollutant emissions and a higher energy balance[36] than corn-based ethanol.[37] Further, expanding the feedstocks for ethanol production could allow areas outside of the Midwest to produce ethanol with local feedstocks.

In his 2006 State of the Union Address, President Bush announced an expansion of biofuels research at the Department of Energy.[38] A stated goal in the speech is to make cellulosic ethanol "practical and competitive within six years," with a potential goal of reducing Middle East oil imports by 75% by 2025.[39] This goal would require an increase in ethanol consumption to as much as 60 billion gallons, from 4.9 billion gallons in 2004.[40] As part of the FY2007 DOE budget request, the Administration sought an increase of 65% above FY2006 funding for "Biomass and Biorefinery Systems R&D," which includes research into cellulosic ethanol.[41] In his 2007 State of the Union Address, President Bush further defined a goal of increasing the use of

---

[33] Lee R. Lynd, Dartmouth College, *Cellulosic Ethanol Fact Sheet*, June 13, 2003. For the National Commission on Energy Policy Forum: The Future of Biomass and Transportation Fuels.

[34] Andrew McAloon, Frank Taylor, and Winnie Yee (USDA), and Kelly Ibsen and Robert Wolley (NREL), *Determining the Cost of Producing Ethanol from Corn Starch and Lignocellulosic Feedstocks*, October 2000.

[35] Nathanael Greene, Natural Resources Defense Council, *Growing Energy - How Biofuels Can Help End America's Oil Dependence*, December 2004, Table 18.

[36] The ratio of the energy needed to produce a fuel to that fuel's energy output. For more details, see section below on "Energy Balance."

[37] Alexander E. Farrell, Richard J. Plevin, Brian T. Turner, Andrew D. Jones, Michael O'Hare, and Daniel M. Kammen, "Ethanol Can Contribute to Energy and Environmental Goals," *Science*, January 27, 2006, pp. 506-508.

[38] President George W. Bush, *State of the Union Address*, January 31, 2006, [http://www.whitehouse.gov/news/releases/2006/01/20060131-10.html].

[39] Ibid.

[40] Peter Rhode, "Bush Biofuel Goal Likely Means Speeding Current Plans By Decades," *New Fuels and Vehicles.com*, February 3, 2006.

[41] The FY2006 appropriation was $91 million; the FY2007 request is $150 million. DOE, *FY2007 Congressional Budget Request*, February 2006, vol. 3, p. 141.

renewable and alternative fuels to 35 billion gallons by 2017.[42] This would mean a roughly seven-fold increase from 2006 levels. Such an increase would most likely be infeasible using corn and other grains as feedstocks. Therefore, the President's goal will likely require significant breakthroughs in technology to convert cellulose into motor fuels.

As stated above, the Energy Independence and Security Act of 2007 (P.L.110-140) expanded the renewable fuel standard (RFS). Further, starting in 2016, and increasing share of the RFS must come from "advanced biofuels," such as cellulosic ethanol, ethanol from sugar cane, and biodiesel. Further, of the advanced biofuel mandate (which reaches 21 billion gallons in 2022), there is a specific carve-out for cellulosic biofuels (reaching 16 billion gallons in 2022).

Economic Effects

Ethanol's relatively high price is a major constraint on its use as an alternative fuel and as a gasoline additive. As a result, ethanol has not been competitive with gasoline except with incentives. Wholesale ethanol prices, excluding incentives from the federal government and state governments, are significantly higher than wholesale gasoline prices. With federal and state incentives, however, the effective price of ethanol is reduced. Furthermore, gasoline prices have risen recently, making ethanol more attractive as both a blending component and as an alternative fuel.

Before 2004, the primary federal incentive supporting the ethanol industry was a 5.2 cents per gallon exemption that blenders of gasohol (E10) received from the 18.4¢ federal excise tax on motor fuels. Because the exemption applied to blended fuel, of which ethanol comprises only 10%, the exemption provided for an effective subsidy of 52 cents per gallon of pure ethanol. The 108th Congress replaced this exemption with an income tax credit of 51 cents per gallon of pure ethanol used in blending (P.L. 108-357).[43] **Table 4** shows that ethanol and gasoline prices are competitive on a per gallon basis when the ethanol tax credit is factored in. However, the energy content of a gallon of ethanol is about one third lower than a gallon of gasoline. As Table 4 shows, on an equivalent energy basis, ethanol can be significantly more expensive than gasoline, even with the tax credit.

The comparative cost figures in **Table 4** are for ethanol as a blending component in gasoline. However, the use of E85 in flexible fuel vehicles has been associated with improved combustion efficiency. The National Ethanol Vehicle Coalition estimates that FFVs run on E85 experience a 5% to15% decrease in miles-per-gallon fuel economy,[44] as opposed to the 29% drop in Btu content per gallon. Therefore, on a per-mile basis, E85's cost premium is likely in the middle of these above estimates.

---

[42] President George W. Bush, *State of the Union Address*, January 23, 2007, [http://www.whitehouse.gov/news/releases/2007/01/20070123-2.html].

[43] 26 U.S.C. 40.

[44] National Ethanol Vehicle Coalition, *op. cit.*

Table 4. Wholesale Price of Pure Ethanol Relative to Gasoline

(August 2006 to January 2008)

	Relative price by volume	Relative price on an equivalent energy basis[c]
Ethanol Wholesale Price[a]	150 to 250 cents/gallon	227 to 379 cents/equivalent gallon
Alcohol Fuel Tax Incentive	51 cents/gallon	77 cents/equivalent gallon
Effective Price of Ethanol	99 to 199 cents/gallon	150 to 302 cents/equivalent gallon
Gasoline Wholesale Price[b]	135 to225 cents/gallon	135 to 225 cents/gallon
Wholesale Price Difference[d]	(-)101 to (+)39 cents/gallon	(-)50 to (+)142 cents/gallon

Source: CRS analysis of Chicago Board of Trade, *Ethanol Derivatives, Updated through January 2008*, February 13, 2008; "US Wholesale Posted Prices," *Platt's Oilgram Price Report*. August 1, 2006, through January 31, 2007.

a. This is the average Chicago daily terminal price for pure ("neat") ethanol.
b. This is the average Chicago price for regular gasoline.
c. A gallon of gasoline contains 115,000 British thermal units (Btu) of energy, while a gallon of ethanol contains 76,000 Btu. Therefore it takes roughly 1.51 gallons of pure ethanol to equal the Btu content of one gallon of gasoline.
d. The wholesale price difference is computed on a daily basis.

Many proponents and opponents agree that the ethanol industry might not survive without tax incentives. An economic analysis conducted in 1998 by the Food and Agriculture Policy Research Institute, concurrent with the congressional debate over extension of the excise tax exemption, concluded that elimination of the exemption would cause annual ethanol production from corn to decline roughly 80% from 1998 levels.[45]

The tax incentives for ethanol are criticized by some as "corporate welfare,"[46] encouraging the inefficient use of agricultural and other resources and depriving the government of needed revenues.[47] In 1997, the General Accounting Office estimated that the excise tax exemption reduced Highway Trust Fund by $7.5 to $11 billion over the 22 years from FY1979 to FY2000.[48]

---

[45] Food and Agriculture Policy Research Institute, *Effects on Agriculture of Elimination of the Excise Tax Exemption for Fuel Ethanol*, Working Paper 01-97, April 8, 1997.

[46] Erin Hymel, *op. cit.*

[47] U.S. General Accounting Office (GAO), *Effects of the Alcohol Fuels Tax Incentives*, March 1997.

[48] Jim Wells, GAO, *Petroleum and EthanolFuels: Tax Incentives and Related GAO Work*,
(continued...)

Proponents of the tax incentive argue that ethanol leads to better air quality and reduced greenhouse gas emissions, and that substantial benefits flow to the agriculture sector due to the increased demand for corn to produce ethanol. Furthermore, they argue that the increased market for ethanol reduces oil imports and strengthens the U.S. trade balance.

Air Quality

One often-cited benefit of ethanol use is improvement in air quality. The Clean Air Act Amendments of 1990 (P.L. 101-549) created the Reformulated Gasoline (RFG) program, which was a major impetus to the development of the U.S. ethanol industry. The Energy Policy Act of 2005 (P.L. 109-58) made significant changes to that program that directly affect U.S. markets for gasoline and ethanol.

Before the Energy Policy Act of 2005. Through 2005, ethanol was primarily used in gasoline to meet a minimum oxygenate requirement for RFG.[49] RFG is used to reduce vehicle emissions in areas that are in severe or extreme nonattainment of National Ambient Air Quality Standards (NAAQS) for ground-level ozone.[50] Ten metropolitan areas, including New York, Los Angeles, Chicago, Philadelphia, and Houston, are covered by this requirement, and many other areas with less severe ozone problems have opted into the program, as well.[51] In these areas, RFG is used year-round.

EPA states that RFG has led to significant improvements in air quality, including a 17% reduction in volatile organic compound (VOC) emissions from vehicles, and a 30% reduction in emissions of toxic air pollutants.[52] Furthermore, according to EPA, "ambient monitoring data from the first year (1995) of the RFG program also showed strong signs that RFG is working. For example, detection of benzene (one of the air toxics controlled by RFG, and a known human carcinogen) declined dramatically, with a median reduction of 38% from the previous year."[53]

---

[48] (...continued)
September 25, 2000.

[49] Clean Air Act, Section 211, Subsection k; 42 U.S.C. 7545.

[50] Ground-level ozone is an air pollutant that causes smog, adversely affects health, and injures plants. It should not be confused with stratospheric ozone, which is a natural layer some 6 to 20 miles above the earth and provides a degree of protection from harmful radiation.

[51] Under new ozone standards recently promulgated by EPA, the number of RFG areas will likely increase.

[52] The RFG program defines "toxic air pollutants" as benzene, 1,3-butadiene, polycyclic organic matter, acetaldehyde, and formaldehyde.

[53] Margo T. Oge, Director, Office of Mobile Sources, U.S. EPA, *Testimony Before the Subcommittee on Energy and Environment of the Committee on Science, U.S. House of Representatives*, September 14, 1999.

However, the benefits of oxygenates in RFG have been questioned. Although oxygenates lead to lower emissions of carbon monoxide (CO), in some cases they may lead to higher emissions of nitrogen oxides (NO_X) and VOCs. Since all three contribute to the formation of ozone, the National Research Council concluded that while RFG certainly leads to improved air quality, the oxygenate requirement in RFG may have little overall impact on ozone formation.[54] In fact, in some areas, the use of low-level blends of ethanol (10% or less) may actually lead to increased ozone formation due to atmospheric conditions in that specific area.[55] Some argue that the main benefit of oxygenates is that they displace other, more dangerous compounds found in gasoline such as benzene. Furthermore, high gasoline prices have also raised questions about the cost-effectiveness of the RFG program.

Evidence that the most widely used oxygenate, methyl tertiary butyl ether (MTBE), contaminates groundwater led to a push by some to eliminate the oxygen requirement in RFG. MTBE has been identified as an animal carcinogen, and there is concern that it is a possible human carcinogen. In California, New York, and Connecticut, MTBE was banned as of January 2004, and several states have followed suit.

Some refiners claimed that the environmental goals of the RFG program could be achieved through cleaner, although potentially more costly, gasoline that does not contain any oxygenates.[56] These claims added to the push to remove the oxygenate requirement and allow refiners to produce RFG in the most cost-effective manner, whether or not that includes the use of oxygenates. However, since oxygenates also displace other harmful chemicals in gasoline, some environmental groups were concerned that eliminating the oxygenate requirements would compromise air quality gains resulting from the current standards. This potential for "backsliding" is a result of the fact that the current performance of RFG is substantially better than the Clean Air Act requires. If the oxygenate standard were eliminated, environmental groups feared that refiners would only meet the requirements of the law, as opposed to maintaining the current overcompliance. The amendments to the RFG program in P.L. 109-58 require refiners to blend gasoline in a way that maintains the toxic emissions reductions achieved in 2001 and 2002.[57]

Following the Energy Policy Act of 2005. P.L. 109-58 made substantial changes to the RFG program. Section 1504(a) eliminated the RFG oxygenate standard as of May 2006, and required EPA to revise its regulations on the RFG

---

[54] National Research Council, *Ozone-Forming Potential of Reformulated Gasoline*, May, 1999.

[55] Wisconsin Department of Natural Resources, Bureau of Air Management, *Ozone Air Quality Effects of a 10% Ethanol Blended Gasoline in Wisconsin*, September 6, 2005.

[56] Al Jessel, Senior Fuels Regulatory Specialist of Chevron Products Company, *Testimony Before the House Science Committee Subcommittee on Energy and Environment*, September 30, 1999.

[57] P.L. 109-58, Section 1504(b).

program to allow the sale of non-oxygenated RFG. This revision is effective May 6, 2006 in most areas of the country.[58]

E85 and Air Quality. The air quality benefits from purer forms of ethanol can be substantial. Compared to gasoline, use of E85 can result in a significant reduction in ozone-forming vehicle emissions in urban areas.[59] And while the use of ethanol also leads to increased emissions of acetaldehyde, a toxic air pollutant, as defined by the Clean Air Act, these emissions can be controlled through the use of advanced catalytic converters.[60] However, as stated above, purer forms of ethanol have not been widely used.

Energy Consumption and Greenhouse Gas Emissions

Energy Balance. A frequent argument for the use of ethanol as a motor fuel is that it reduces U.S. reliance on oil imports, making the U.S. less vulnerable to a fuel embargo of the sort that occurred in the 1970s. To analyze the net energy consumption of ethanol, the entire fuel cycle must be considered. The fuel cycle consists of all inputs and processes involved in the development, delivery and final use of the fuel. For corn-based ethanol, these inputs include the energy needed to produce fertilizers, operate farm equipment, transport corn, convert corn to ethanol, and distribute the final product. According to a fuel-cycle study by Argonne National Laboratory, with current technology the use of corn-based E10 leads to a 3% reduction in fossil energy use per vehicle mile relative to gasoline, while use of E85 leads to roughly a 40% reduction in fossil energy use.[61]

However, other studies question the Argonne study, suggesting that the amount of energy needed to produce ethanol is roughly equal to the amount of energy obtained from its combustion. Since large amounts of fossil fuels are used to make fertilizer for corn production and to run ethanol plants, ethanol use could lead to little or no net reduction in fossil energy use. Nevertheless, a recent meta-study of

---

[58] Environmental Protection Agency, "Regulation of Fuels and Fuel Additives: Removal of Reformulated Gasoline Oxygen Content Requirement and Revision of Commingling Prohibition to Address Non-Oxygenated Reformulated Gasoline, Direct Final Rule," *71 Federal Register 8973*, February 22, 2006.

[59] It should be noted that the overall fuel-cycle ozone-forming emissions from corn-based E85 are roughly equivalent to those from gasoline. However, some of the emissions attributable to E85 are in rural areas where corn is grown and the ethanol is produced — areas where ozone formation is potentially less of a concern. Norman Brinkman and Trudy Weber (General Motors Corporation), Michael Wang (Argonne National Laboratory), and Thomas Darlingon (Air Improvement Resource, Inc.), *Well-to-Wheels Analysis of Advanced Fuel/Vehicle Systems — A North American Study of Energy Use, Greenhouse Gas Emissions, and Criteria Pollutant Emissions*, May 2005.

[60] California Energy Commission, *Ethanol-Powered Vehicles.*

[61] M. Wang, C. Saricks, and D. Santini, "Effects of Fuel Ethanol on Fuel-Cycle Energy and Greenhouse Gas Emissions," Argonne National Laboratory, January 1999.

research on ethanol's energy balance and greenhouse gas emissions found that most studies give corn-based ethanol a slight positive energy balance.[62] However, because most of the energy used to produce ethanol comes from natural gas or electricity, most studies conclude that overall *petroleum* dependence (as opposed to *energy* dependence) can be significantly diminished through expanded use of ethanol.

Despite the fact that ethanol displaces gasoline, the benefits to energy security from corn-based ethanol are not certain. As stated above, fuel ethanol only accounts for approximately 2.5% of gasoline consumption in the United States by volume. In terms of energy content, ethanol accounts for approximately 1.5%. This small market share led the Government Accountability Office (formerly the General Accounting Office) to conclude that the ethanol tax incentive has done little to promote energy security.[63] Further, as long as ethanol remains dependent on the U.S. corn supply, any threats to this supply (e.g. drought), or increases in corn prices, would negatively affect the supply and/or cost of ethanol. In fact, that happened when high corn prices caused by strong export demand in 1995 contributed to an 18% decline in ethanol production between 1995 and 1996.

Cellulosic Ethanol Energy Balance. Because cellulosic feedstocks require far less fertilizer for their production, the energy balance and other benefits of cellulosic ethanol could be significant. The Argonne study concluded that with advances in technology, the use of cellulose-based E10 could reduce fossil energy consumption per mile by 8%, while cellulose-based E85 could reduce fossil energy consumption by roughly 70%.[64]

Greenhouse Gas Emissions. Directly related to fossil energy consumption is the question of greenhouse gas emissions. Proponents of ethanol argue that over the entire fuel cycle it has the potential to reduce greenhouse gas emissions from automobiles relative to gasoline, therefore reducing the risk of possible global warming.

Fuel Cycle Analysis. Because ethanol contains carbon,[65] combustion of the fuel necessarily results in emissions of carbon dioxide (CO_2), the primary greenhouse gas. Further, greenhouse gases are emitted through the production and use of nitrogen-based fertilizers, as well as the operation of farm equipment and vehicles to transport feedstocks and finished products. However, since photosynthesis (the process by which plants convert light into chemical energy) requires absorption of CO_2, the growth cycle of the feedstock crop can serve — to some extent — as a "sink" to absorb some fuel-cycle greenhouse emissions.

According to the Argonne study, overall fuel-cycle greenhouse gas emissions from corn-based E10 (measured in grams per mile) are approximately 1% lower than

---

[62] Farrell, et al., p. 506.

[63] U.S. General Accounting Office, *Effects of the Alcohol Fuels Tax Incentives*, March 1997.

[64] Wang, et al., table 7.

[65] The chemical formula for ethanol is C_2H_5OH.

CRS-17

from gasoline, while emissions are approximately 20% lower with E85.[66] Other studies that conclude higher fuel-cycle energy consumption for ethanol production also conclude higher greenhouse gas emissions for the fuel. The meta-study on energy consumption and greenhouse gas emissions concluded that pure ethanol results in 13% lower greenhouse gas emissions, with approximately a 10% reduction using E85.[67]

Fuel Cycle Emissons from Cellulosic Ethanol. Because of the limited use of fertilizers, fossil energy consumption — and thus greenhouse gas emissions — is significantly reduced with ethanol production from cellulosic feedstocks. The Argonne study concludes that with advances in technology, cellulosic E10 could reduce greenhouse gas emissions by 7% to 10% relative to gasoline, while cellulosic E85 could reduce greenhouse gas emissions by 67% to 89%.[68] The meta-study of energy consumption and greenhouse gas emissions found a similar potential for greenhouse gas reductions.[69]

Lifecycle Analysis. A key criticism of fuel-cycle analyses is that they generally do not take changes in land use into account. For example, if a previously uncultivated piece of land is tilled to plant biofuel crops, some of the carbon stored in the field could be released. In that case, the overall GHG benefit of biofuels could be compromised. One study estimates that taking land use into account (a lifecycle analysis as opposed to a fuel-cycle analysis), the GHG reduction from corn ethanol is less than 3% per mile relative to gasoline,[70] while cellulosic biofuels have a life-cycle reduction of 50%.[71] Other recent studies indicate even smaller GHG reductions.

This is of key interest because under the RFS, as amended by P.L. 110-140, to qualify under the mandate, all fuels from new biofuel refineries must achieve at least a 20% reduction in *lifecycle* greenhouse gas emissions. Further, to qualify as advanced biofuels, they must achieve a 50% reduction in lifecycle emissions. EPA is tasked with developing regulations to rate fuels on their lifecycle emissions, and determining which fuels qualify under the new standard.

---

[66] Wang, et al., table 7.

[67] Farrell, et al., p. 506.

[68] Wang, et al., table 7.

[69] Farrell, et al., p. 507.

[70] Mark A. Delucchi, *Draft Report: Life cycle Analyses of Biofuels*. 2006.

[71] While a 50% life-cycle reduction is still significant, it is far less than the 90% reduction suggested by fuel-cycle analyses.

CRS-18

Policy Concerns and Congressional Activity

Recent congressional interest in ethanol fuels has mainly focused on six policies and issues: (1) the renewable fuel standard; (2) "boutique" fuels; (3) the alcohol fuel tax incentives; (4) ethanol imports through Caribbean Basin Initiative (CBI) countries; (5) fuel economy credits for dual fuel vehicles; and (6) the role of biofuels in the upcoming Farm Bill. In the 109th and 110th Congresses, several of these issues were debated during consideration of the Energy Policy Act of 2005 (P.L. 109-58) and the Energy Independence and Security Act of 2007 (P.L. 110-140).

Renewable Fuel Standard (RFS)

The renewable fuels standard requires motor fuel to contain a minimum amount of fuel produced from renewable sources such as biomass, solar, or wind energy. Proposals to establish an RFS gained traction as part of the discussion over comprehensive energy policy. Supporters argued that demand for ethanol creates jobs, and that there are major environmental and energy security benefits to using renewable fuels. However, opponents argued that any renewable fuel standard would only exacerbate a situation of artificial demand for ethanol created by tax incentives and fuel quality standards. Any requirement above the existing level for ethanol would require the construction and/or expansion of ethanol plants, and likely would lead to increased fuel prices and further instability in an already tight fuel supply chain. Further, they argued that a renewable fuel standard would lead to increased corn prices caused by higher demand.

On August 8, 2005, President Bush signed the Energy Policy Act of 2005 (P.L. 109-58). Section 1501 required the use of at least 4.0 billion gallons of renewable fuel in 2006, increasing to 7.5 billion gallons in 2012 (see **Table 5**). Through 2007 the requirement was largely met using ethanol, although other fuels such as biodiesel played a limited role.[72] The law directed EPA to establish a credit trading system to provide flexibility to fuel producers. Further, under the RFS, ethanol produced from cellulosic feedstocks was granted extra credit: a gallon of cellulosic ethanol counted as 2.5 gallons of renewable fuel under the RFS. Also, P.L. 109-58 required that 250 million gallons of cellulosic ethanol be blended in gasoline annually starting in 2013.[73] The Energy Independence and Security Act of 2007 (P.L. 110-140), signed by President Bush on December 19, 2007, significantly expanded the RFS, requiring the use of 9.0 billion gallons of renewable fuel in 2008, increasing to 36 billion gallons in 2022. Further, P.L. 110-140 requires an increasing amount of the mandate be met with "advanced biofuels" — biofuels produced from feedstocks other than corn starch (and with 50% lower lifecycle greenhouse gas emissions than petroleum fuels). Within the advanced biofuel mandate, there are specific carve-outs for cellulosic biofuels and bio-based diesel substitutes (e.g., biodiesel).

---

[72] Biodiesel is a synthetic diesel fuel made from oils such as soybean oil. For more information, see CRS Report RL30758, *Alternative Transportation Fuels and Vehicles: Energy, Environment, and Development Issues*, by Brent D. Yacobucci.

[73] Currently, world production of cellulosic ethanol is limited. No plants currently exist in the United States, although some small plants are in the planning phase.

Table 5. Expanded Renewable Fuel Standard Requirements Under P.L. 110-140

Year	Previous RFS (billion gallons)	Expanded RFS (billion gallons)	Advanced Biofuel Mandate (billion gallons)[a]	Cellulosic Biofuel Mandate (billion gallons)[b]	Biomass-Based Diesel Fuel (billion gallons)[b]
2006	4.0				
2007	4.7				
2008	5.4	**9.0**			
2009	6.1	**11.1**	0.6		0.5
2010	6.8	**12.95**	0.95	0.1	0.65
2011	7.4	**13.95**	1.35	0.25	0.8
2012	7.5	**15.2**	2.0	0.5	1.0
2013	7.6 (est.)	**16.55**	2.75	1.0	1.0
2014	7.7 (est.)	**18.15**	3.75	1.75	1.0
2015	7.8 (est.)	**20.5**	5.5	3.0	1.0
2016	7.9 (est.)	**22.25**	7.25	4.25	1.0
2017	8.1 (est.)	**24.0**	9.0	5.5	1.0
2018	8.2 (est.)	**26.0**	11.0	7.0	1.0
2019	8.3 (est.)	**28.0**	13.0	8.5	1.0
2020	8.4 (est.)	**30.0**	15.0	10.5	1.0
2021	8.5 (est.)	**33.0**	18.0	13.5	1.0
2022	8.6 (est.)	**36.0**	21.0	16.0	1.0

a. The advanced biofuel (i.e. non-corn-starch ethanol) mandate is a subset of the expanded RFS. The difference between the expanded RFS mandate and the advanced biofuel mandate — 15 billion gallons in 2015 onward) is effectively a cap on corn ethanol.

b. The cellulosic biofuel and biomass-based diesel fuel mandates are subsets of the advanced biofuel mandate.

Ethanol producers are rapidly expanding capacity in order to meet the increased demand created by the RFS. Between January 2005 and January 2008, U.S. ethanol production capacity expanded from 3.6 billion gallons per year to 7.2 billion gallons per year.

EPA is required to establish a system for suppliers to generate and trade credits earned for exceeding the standard in a given year. Credits can then be purchased by other suppliers to meet their quotas. On May 1, 2007, EPA released a final rulemaking for 2007 and beyond. Included in the rule were provisions for credit trading, as well as provisions for generating credits from the sale of biodiesel and other fuels.[74] Because of the changes in the RFS from P.L. 110-140, EPA will need to publish new rules to reflect those changes. Perhaps most importantly, EPA will need to develop rules for determining the lifecycle greenhouse gas emissions (see "Greenhouse Gas Emissions," above). Fuels from new biorefineries must achieve at least a 20% lifecycle greenhouse gas reduction relative to petroleum fuels, and advanced biofuels must achieve at least a 50% reduction.

The effects of such an increase in ethanol production could be significant, especially if that ethanol comes from corn.[75] These effects include increased corn demand and higher corn prices, leading to higher costs for food (especially in places where corn is a significant part of the local diet) and higher animal feed prices (and higher meat prices). Expanded ethanol use could also strain an already tight ethanol distribution system that is dependent on rail cars for transport, since ethanol may not be transported by pipeline in the United States. Other concerns include the potential for increased water use for corn cultivation and the increased use of chemical fertilizers and pesticides.[76]

"Boutique" Fuels[77]

As a result of the federal reformulated gasoline requirements, as well as related state and local environmental requirements, gasoline suppliers may face several different standards for gasoline quality in different parts of one state or in adjacent states. These different standards sometimes require a supplier to provide several different fuel formulations in a region.[78] These different formulations are sometimes

---

[74] Environmental Protection Agency, "Regulation of Fuels and Fuel Additives: Renewable Fuel Standard Program; Final Rule," *72 Federal Register* 23899-23948, May 1, 2007.

[75] For more information, see CRS Report RL33928, *Ethanol and Biofuels: Agriculture, Infrastructure, and Market Constraints Related to Expanded Production*, by Brent Yacobucci and Randy Schnepf.

[76] For more information on some of the potential concerns from an expanded RFS, see CRS Report RL34265, *Selected Issues Related to an Expansion of the Renewable Fuel Standard (RFS)*, by Brent D. Yacobucci and Tom Capehart.

[77] For more information on boutique fuels, see CRS Report RL31361, *"Boutique Fuels" and Reformulated Gasoline: Harmonization of Fuel Standards*, by Brent D. Yacobucci.

[78] These various formulations should not be confused with gasoline "grades" — "regular," "mid-grade," and "premium" octane level fuels — which are not required by federal law but (continued...)

referred to as "boutique" fuels.[79] Because of varying requirements, if there is a disruption to the supply of fuel in one area, refiners producing fuel for other nearby areas may not be able to supply fuel quickly enough to meet the increased demand.

EPA conducted a study on the effects of harmonizing standards and released a staff white paper in October 2001.[80] EPA modeled several scenarios, some with limited changes to the existing system, others with drastic changes. In its preliminary analysis, EPA concluded that some minor changes could be made that might mitigate supply disruptions without significantly increasing costs or adversely affecting vehicle emissions. However, all of the changes modeled in EPA's study would require amendments to various provisions in the Clean Air Act.

Congressional interest has centered on the question of whether the various standards could be harmonized to reduce the number of gasoline formulations. Section 1504(c) of P.L. 109-58 consolidates two summertime RFG formulations into one fuel, eliminating one class of fuel. Further, P.L. 109-58 prohibits the number of state fuel blends from exceeding the number as of September 1, 2004. However, many of the larger systemic issues were not addressed.

Alcohol Fuel Tax Incentives[81]

As stated above, the ethanol tax incentives are controversial. The incentives allow fuel ethanol to compete with other additives, since the wholesale price of ethanol is so high. Proponents of ethanol argue that the incentives lower dependence on foreign imports, promote air quality, and benefit farmers.[82]

Opponents argue that the tax incentives support an industry that could not exist on its own. Despite objections from opponents, Congress in 1998 extended the motor fuels excise tax exemption through 2007, but at slightly lower rates.[83] To eliminate concerns over Highway Trust Fund revenue losses, the 108th Congress replaced the excise tax exemption with an income tax credit, effectively transferring

---

[78] (...continued)
are desired by consumers and required in some engine designs.

[79] EPA, Office of Transportation and Air Quality, *Staff White Paper: Study of Unique Gasoline Fuel Blends ("Boutique Fuels"), Effects on Fuel Supply and Distribution and Potential Improvements*, October 2001.

[80] Harmonization refers to an attempt to aggregate fuels with similar requirements under a single requirement, thus limiting the number of possible formulations.

[81] For more information, see CRS Report RL32979, *Alcohol Fuels Tax Incentives*, by Salvatore Lazzari.

[82] U.S. General Accounting Office (GAO), *Effects of the Alcohol Fuels Tax Incentives*, March 1997.

[83] P.L. 105-178.

the effects of the incentive from the Highway Trust Fund to the general treasury, and extending the incentive through 2010.[84]

Ethanol Imports

There is growing concern over ethanol imports among some stakeholders. Because of lower production costs and/or government incentives, ethanol prices in Brazil and other countries can be significantly lower than in the United States. To offset the U.S. tax incentives that all ethanol (imported or domestic) receives, most imports are subject to a relatively small 2.5% ad valorem tariff, but more significantly an added duty of $0.54 per gallon. This duty effectively negates the tax incentives for covered imports, and has been a significant barrier to ethanol imports.

However, under certain conditions imports of ethanol from Caribbean Basin Initiative (CBI) countries are granted duty-free status.[85] This is true even if the ethanol was actually produced in a non-CBI country. In this scenario the ethanol is dehydrated in a CBI country, then shipped to the United States.[86] This avenue for avoiding the duty by imported ethanol has been criticized by some stakeholders, including some Members of Congress.

On December 20, 2006, President Bush signed the Tax Relief and Health Care Act of 2006 (P.L. 109-432). Among other provisions, the act extended the duty on imported ethanol through December 31, 2008.

Fuel Economy Credits for Dual Fuel Vehicles

The Energy Policy and Conservation Act (EPCA) of 1975[87] requires Corporate Average Fuel Economy (CAFE) standards for motor vehicles.[88] Under EPCA, the average fuel economy of all vehicles of a given class that a manufacturer sells in a model year must be equal to or greater than the standard for that class. These standards were first enacted in response to the desire to reduce petroleum consumption and promote energy security after the Arab oil embargo. The model year 2007 standard for passenger cars is 27.5 miles per gallon (mpg), while the standard for light trucks is 22.2 mpg.

---

[84] P.L. 108-357.

[85] The CBI countries include Costa Rica, Jamaica, and El Salvador, which represent a significant percentage of U.S. fuel ethanol imports. For more information on ethanol imports from CBI countries, see CRS Report RS21930, *Ethanol Imports and the Caribbean Basin Initiative*, by Brent D. Yacobucci.

[86] Dehydration is the final step in the ethanol production process. Excess water is removed from the ethanol to make it usable as motor fuel. For more information, see section above on "Ethanol Refining and Production."

[87] P.L. 94-163.

[88] For more information on CAFE standards, see CRS Report RL33413, *Automobile and Light Truck Fuel Economy: The CAFE Standards*, by Brent D. Yacobucci and Robert Bamberger.

EPCA (and subsequent amendments to it) provides manufacturing incentives for alternative fuel vehicles, including ethanol vehicles.[89] For each alternative fuel vehicle a manufacturer produces, the manufacturer generates credits toward meeting the CAFE standards. These credits can be used to increase the manufacturer's average fuel economy. Credits apply to both dedicated and dual fuel vehicles. Dual fuel vehicles can be operated on both a conventional fuel (gasoline or diesel) and an alternative fuel, usually ethanol. Opponents have raised concerns that while manufacturers are receiving credits for production of these dual fuel vehicles, they are generally operated solely on gasoline, because of the cost and unavailability of alternative fuels. This claim is supported by the fact that EIA estimates that only about 2% of flexible fuel vehicles are currently operated on E85. Supporters of the credits counter that the incentives are necessary for the production of alternative fuel vehicles, and that as the number of vehicles increases, the infrastructure for alternative fuels will grow. However, the success of this strategy has been limited to date.

The credits were set to expire at the end of the 2004 model year. However, in 2004 the Department of Transportation (DOT) issued a final rule extending the credits through model year 2008.[90] Section 772 of P.L. 109-58 extended the credits through model year 2010, and extended DOT's authority (to continue the credits) through 2014.

The 2008 Farm Bill

It is expected that the 110th Congress will reauthorize existing farm programs and promote new programs as part of a new Farm Bill. Most of the provisions of the most recent Farm Bill — The Farm Security and Rural Investment Act of 2002 (P.L. 107-171) — expired in 2007. On July 27, 2007, the House approved a new farm bill, H.R. 2419. Title IX would expand and extend several provisions from the 2002 farm bill's energy title, with substantial increases in funding and a heightened focus on developing cellulosic energy production, and a move away from corn-starch-based ethanol. The Senate passed its version on December 14, 2007. The Senate version would expand 2002 Farm Bill programs, create new tax incentives for cellulosic ethanol, and require studies on expanded biofuel infrastructure. As of March 2008, a conference on the House and Senate bills was pending.[91]

---

[89] 49 U.S.C. 32905.

[90] 60 *Federal Register* 7689, February 19, 2004.

[91] For more information on biofuel provisions in the Farm Bill, see CRS Report RL34239, *Biofuels in the 2007 Energy and Farm Bills: A Side-by-Side Comparison*, by Brent D. Yacobucci, Randy Schnepf, and Tom Capehart.

Conclusion

Although the use of fuel ethanol has been limited to date (only about 3% to 5% of gasoline consumption), it has the potential to significantly displace *petroleum* demand. However, the overall benefits in terms of *energy* consumption and greenhouse gases are limited, especially in the case of corn-based ethanol. With only a slight net energy benefit from the use of corn-based ethanol, transportation energy demand is essentially transferred from one fossil fuel (petroleum) to another (natural gas and/or coal). There may be strategic benefits from this transfer, especially if the replacement fuel comes from domestic sources or from foreign sources in more stable areas. However, the benefits in terms of greenhouse gas emissions reductions is limited.

Cellulosic feedstocks have the potential to dramatically improve the benefits of fuel ethanol. Their use could significantly decrease the energy (from all sources) required to produce the fuel, as well as decreasing associated greenhouse gases. However, technologies to convert cellulose to ethanol at competitive costs seem distant. For this reason, there is wide support for increased federal R&D.

Federal incentives for ethanol use — including tax incentives, the RFG oxygenate standard, and the renewable fuels standard — have promoted significant growth in the ethanol market. Annual U.S. ethanol production increased from 175 million gallons in 1980 to 6.8 billion gallons in 2007, largely as a result of these incentives. Federal incentives drive demand for the fuel, as well as making its price competitive with gasoline.

Enacted as part of the Energy Policy Act of 2005 and expanded by the Energy Independence and Security Act of 2007, the renewable fuels standard will continue to drive growth in the ethanol market, as it mandates a minimum annual amount (increasing yearly) of renewable fuel in gasoline. While other fuels will be used to some extent to meet the standard, the a large share of the mandate will be met with ethanol. The increasing demand for ethanol may lead to price pressures on motor fuel. These price pressures — and ethanol supply concerns in general — could increase interest in eliminating the tariff on imported ethanol.

Congress will likely continue to show interest in ethanol's energy and environmental costs and benefits, as well as its effects on U.S. fuel markets. Any discussion of U.S. energy policy includes promotion of alternatives to petroleum. With limited petroleum supplies, high prices, and instability in some oil-producing regions, these discussions are unlikely to end any time soon.

Order Code RL32979

CRS Report for Congress
Received through the CRS Web

Alcohol Fuels Tax Incentives

July 6, 2005

Salvatore Lazzari
Specialist in Public Finance
Resources, Science, and Industry Division

Congressional Research Service ❖ The Library of Congress

Alcohol Fuels Tax Incentives

Summary

Prior to January 1, 2005, alcohol fuel blenders qualified for a 5.2¢ tax exemption against the excise taxes otherwise due on each gallon of blended mixtures (mixtures of 10% ethanol, and 90% gasoline). This exemption, which was scheduled to decline to 5.1¢ on January 1, 2005, reduced the gasoline excise tax for "gasohol," from 18.4¢ to 13.2¢/gallon. The reduction was realized at the time when the gasoline tax was otherwise imposed: typically when the fuel was loaded from the terminal onto trucks for distribution. The 5.2¢ exemption could also be claimed later, i.e., when blenders filed their income tax return, as a 52¢ excise tax credit per gallon of alcohol used to make a fuel mixture (which was also scheduled to decline to 51¢ in tandem with the exemption on January 1, 2005). This credit, however, was not as valuable as the exemption because 1) it was taxable as income, 2) was not available instantaneously as the fuel was blended — blenders had to wait until their income tax returns were filed to reduce their tax liability by the amount of the credit, and 3) the tax credit was not refundable — it was only available to the extent of tax liability. Because the primary benefits from alcohol fuels were realized through an exemption rather than a tax credit, revenue losses from the exemption (or reduced excise taxes) accrued to the Highway Trust Fund (HTF).

The American Jobs Creation Act of 2004 (P.L. 108-357) restructured the basic tax subsidies for alcohol fuels: 1) the blender's *income* tax credits were eliminated and 2) the blender's excise tax *exemption* was replaced by an "instant" excise tax credit of the same amount — 5.1¢/gallon of a 90:10 mixture, which is also equivalent to 51¢ *per gallon of ethanol* in the mixture. These tax reforms went into effect on January 1, 2005. As before, the excise tax credit is claimed against the 18.4¢ per gallon excise tax on gasoline, so that the actual excise tax paid and remitted to the Treasury is 13.3¢ — the tax is reduced by 5.1¢/gallon just as with the exemption. When income tax effects are considered, however, the new excise tax credit has a greater economic or subsidy value than the exemption before it because income tax deductions are taken at 18.4¢ rather then 13.3¢. In other words, by labeling the tax reduction as an *excise tax credit* rather than an *excise tax exemption*, the tax law treats the blenders as paying the full excise tax of 18.4¢/ gallon rather than 13.3¢ per gallon. At a 25% marginal income tax rate, the additional 5.1¢ deduction is valued at 1.7¢/gallon of a blend or 17¢/gallon of ethanol, which means that the total after-tax subsidy for alcohol fuel mixtures is effectively 68¢/gallon of ethanol rather than the nominal rate of 51¢.

By nominally increasing the excise tax on gasohol by 5.1¢/gallon, an extra $1,500 million in FY2006 is projected to be allocated into the HTF from the general fund, which implies that HTF expenditures, and budget deficits can be expected to be higher than under the exemption. In addition to the alcohol fuel mixture excise tax credit there are three other federal tax subsidies that are available for the production and use of alcohol transportation fuels (but are little used). Comprehensive energy policy legislation H.R. 6, as passed by the Senate, includes a renewable fuels standard that would, by 2010, more than double both the use of ethanol and the revenue loss from the new alcohol fuels tax incentives.

Contents

List of Tables

Alcohol Fuels Tax Incentives After the 2004 Reforms

Introduction

President Carter's 1978 Energy Tax Act introduced the excise tax exemption for alcohol fuel blends (at 100% of the gasoline tax, which was then 4¢/gallon) to achieve energy and, more recently, certain environmental and agricultural policy objectives.[1] In 2004 this exemption, which was 5.2¢/gallon and was scheduled to decline to 5.1¢ on January 1, 2005, reduced the gasoline excise tax for "gasohol," from 18.4¢ to 13.2¢/gallon. The reduction was realized at the time when the gasoline tax was otherwise imposed: typically when the fuel was loaded from the terminal onto trucks for distribution. The 5.2¢ exemption could also be claimed later (i.e., when blenders filed their income tax return) as a 52¢ excise tax credit per gallon of alcohol used to make a fuel mixture, which was also scheduled to decline to 51¢ in tandem with the exemption on January 1, 2005. From 1978-2004, this exemption from the motor fuels excise taxes provided the major subsidy to the ethanol fuel industry — without the exemption the ethanol fuels industry would either not exist or be substantially smaller.[2]

The exemption, however, which effectively lowered the excise tax on the blended fuel, reduced revenues for the Highway Trust Fund by an estimated $14, 000 million through FY2004. As a consequence, the Congress enacted the American Jobs Creation Act of 2004 (P.L. 108-357), also known as the "Jobs Bill," which restructured the basic tax subsidy for alcohol fuels. It replaced the excise tax exemption with a new excise tax credit: the "alcohol fuel mixtures excise tax credit." Although, the two incentives are equivalent in gross, before-tax terms — they are both equal to 51¢ per gallon of ethanol — interactions between the excise tax credit and the income tax system increase the value of the incentives under this restructured or reformed system. In addition to the new "alcohol fuel mixtures excise tax credit," current federal tax law provides for a small ethanol producer credit, and several other tax incentives that, although little used, might further benefit alcohol fuels in the future.

The Senate version of H.R. 6, the Energy Policy Act of 2005, proposes to not only expand the small ethanol producer credit, but to introduce a renewable fuels

---

[1] The Energy Tax Act of 1978 (P.L. 95-618), was one part of President Carter's National Energy Plan, intended to address what was perceived as severe problems in the country's energy markets.

[2] More recently, regulatory subsidies for ethanol fuels — the federal oxygenates and reformulated fuels requirements under the Clean Air Act — as well as numerous state subsidies have been added, which have also increased the market for ethanol fuel. Still, the excise tax exemption has provided much of the economic stimulus to the industry.

standard (which is effectively an ethanol standard), a requirement that gasoline suppliers (refiners and blenders) blend at least 8 billion gallons of renewable fuels per year in producing gasoline. Such a requirement, if enacted, would more than double the amount of ethanol blended with gasoline above the current projected baseline level of ethanol use, and significantly increase federal tax revenue losses.

This report explains the provisions of the new alcohol fuels mixtures tax credit and compares the tax benefits under the new credit with the tax benefits under the old exemption. An example illustrates the mechanics of the new credit and compares it with the old exemption. The second section examines the revenue and Highway Trust Fund implications of the new tax incentive both with and without a renewable fuels standard. The final section discusses the remaining three tax subsidies for alcohol fuels, which, although little used, are nevertheless part of the current federal tax laws and might be used in the future.

The Alcohol Fuel Mixtures Excise Tax Credit

Under the new alcohol fuels mixtures tax credit, §40 of the Internal Revenue Code (IRC), gasohol blenders may claim a 51¢/gallon tax credit for alcohol used to produce a qualified mixture (a mixture of alcohol and gasoline, or a mixture of alcohol and any other special motor fuel).[3]

Unlike most tax credits, which are claimed against income tax liability (because the income tax otherwise owed is reduced or "credited" by the amount of the credit), the new alcohol mixtures credit is claimed against the motor fuels (gasoline) excise tax. However, both approaches reduce the effective excise tax burden on each gallon of ethanol used to make a 90:10 gallon of a mixture by the same amount — 5.1¢/gallon of a blend or 51¢/gallon of ethanol — regardless of whether the reduction is called a tax credit or excise tax exemption. This new excise tax credit, which became effective on January 1, 2005, will replace the old excise tax exemption as the basic tax incentive claimed on most sales of fuel ethanol — it will provide the biggest subsidy to the industry.[4]

As under the previous exemption, to qualify for the full 51¢ tax credit, the alcohol must be at least 190 proof (95% pure alcohol, determined without regard to any added denaturants or impurities). The credit is 37.78¢ per gallon if the ethanol is between 150 and 190 proof; no credit is provided for alcohol below 150 proof. This mixture credit is available only to the blender, who must not only produce the mixture but must either use the mixture as a motor fuel in a trade or business or sell it for use as a fuel. The blender may be the producer, the terminal operator, or the

---

[3] The "Jobs Bill" also introduced a parallel or equivalent system of incentives for biodiesel. Section 40A of the IRC also provides for excise tax credits against the 24.4¢/gallon tax on diesel fuel for mixtures of biodiesel. This *excise tax credit* is 50¢/gallon of recycled biodiesel and $1.00/gallon for virgin agri-biodiesel. Section 40A also provides a 50¢/gallon income tax credit ($1.00 for agri-biodiesel) for use or retail sale of pure 100% biodiesel.

[4] Moreover, the proposed renewable fuels mandate — a new regulatory subsidy — will, if enacted in H.R. 6, also provide a larger benefit than the tax subsidy.

wholesaler, distributor, or marketer. Technically, both ethanol and methanol qualify for the exemption as long as they are not derived from petroleum, natural gas, coal, or peat. In practice, however, virtually all fuel alcohol is ethanol produced from corn; very little, if any, methanol is produced from wood and other biomass (or renewable) sources because it is generally uneconomic.[5] Currently most methanol is produced from natural gas, is too expensive as a blended motor fuel and does not qualify for the tax breaks.

As did the exemption before it, the new federal tax credit for alcohol fuels also applies to certain fuel additives called oxygenates, provided they are produced from renewables such as corn and not from fossil fuels such as natural gas. ETBE is a compound derived from a chemical reaction between ethanol and isobutylene (a byproduct of both the petroleum refining process and natural gas liquids).[6] In this reaction, the ethanol is chemically transformed and is not present as a separate chemical in the final product. In 1995, the IRS ruled that blends of ETBE (ethyl tertiary butyl ether) and gasoline would also qualify for the reduced partial excise tax exemption. In effect these rulings ensured that the oxygenate required under the Clean Air Act would also qualify for the tax subsidies. Allowing ETBE to qualify for this tax exemption was intended to further stimulate the production of ethanol. Allowing ETBE to qualify for the federal tax subsidies reduces the growth of MTBE (methyl tertiary butyl ether), its main competitor. ETBE costs more to produce and therefore, without the tax subsidies, could not otherwise compete with the less costly MTBE. The recent banning and phasing-out of MTBE by many states (California, New York, and Connecticut, and others) reduces the competitive advantage of MTBE and also increases the demand for ethanol as an oxygenate.[7]

The Credit for Methanol. The mixtures credit is also available for methanol blended with gasoline at the rate of 60¢ per gallon of methanol — the equivalent of 6.0¢ on 90:10 blends — if the alcohol is at least 190 proof, and 45¢ if the methanol is between 150 and 190 proof. No credit is available for either ethanol or methanol that is less than 150 proof. As noted above, the alcohol in methanol cannot be produced from any fossil fuels — it must be produced from organic or biomass materials.

How the New Tax Credit Works

As already discussed, the new alcohol fuels mixtures excise tax credit is claimed against the excise tax, primarily the gasoline excise tax — it reduces the excise tax otherwise owed and remitted to the federal government. So, despite the reforms of

---

[5] Although blends of gasoline with biomass — derived methanol would also qualify under the tax code, such blends are disqualified under the Clean Air Act because of the associated increase of emissions of ozone — forming pollutants.

[6] Natural gas liquids are those components of wellhead gas — ethane, propane, butanes, pentanes, natural gasoline, condensate, etc. — that are liquefied at the surface in lease separators, field facilities, or gas processing plants.

[7] Currently, nineteen states have either banned, limited, or are phasing-out MTBE. See CRS Report RL32787, *MTBE in Gasoline: Clean Air and Drinking Water Issues*, by James McCarthy and Mary Tiemann.

the alcohol fuels tax incentives in 2004, the basic incentive — the 51¢/gallon tax credit — is still connected with the gasoline excise tax (and other motor fuels excise taxes); understanding how the tax credit works requires a brief explanation of the structure of the current federal excise taxes on motor fuels.

The Structure of Motor Fuels Excise Taxes. Virtually all transportation fuels are taxed under a complicated structure of tax rates and exemptions that vary by type of fuel (gasoline, diesel, propane, etc.) and transportation mode or how the fuel is used (cars, trains, buses, airplanes, etc.). Gasoline used in highway transportation — the fuel used more than any other — is taxed at a rate of 18.4¢ per gallon, composed of: an 18.3¢ Highway Trust Fund rate, which generates most of the revenue for the federal highway trust fund (HTF); and a 0.1¢ rate that is earmarked for the Leaking Underground Storage Tank Trust Fund (LUST).[8] Diesel fuel for highway use — the fuel used mostly by trucks — is taxed at 24.4¢ per gallon, also consisting of two components: a 24.3¢ rate that is allocated into the HTF, and 0.1¢ that goes into the LUST fund. In addition, special motor fuels (gasoline substitutes), jet fuel, railway diesel fuel, motorboat fuel, and virtually every other transportation motor fuel that is not specifically exempt, are also subject to tax.[9] Compressed natural gas (CNG) has, since 1993, been subject to an excise tax of 48.54¢ per MCF (thousand cubic feet) — marking the onset of the taxation of gaseous transportation fuels.[10]

Excise Tax Exemption. The tax subsidies for alcohol fuels were restructured — the excise tax exemption became an excise tax credit — under the American Jobs Creation Act of 2004 (P.L. 108-357), enacted on October 22, 2004. The reforms became effective on January 1, 2005.

Before the restructuring of the alcohol fuels tax incentives in 2004, the most important tax incentive for alcohol fuels — the one most responsible for the development of the alcohol fuels market — was the partial exemption from the otherwise standard excise tax rates on gasoline, diesel, and other transportation fuels. This exemption was 5.2¢ per gallon in 2004 and scheduled to decline to 5.1¢ beginning on January 1, 2005, before it is was replaced by the "alcohol mixtures excise tax credit." Thus, mixtures of 90% gasoline and 10% alcohol (typically called gasohol) were taxed at 13.2¢ per gallon — they were exempt from 5.2¢ of the tax [11]

---

[8] The LUST fund finances the cost of cleaning up spills from underground fuel storage tanks. All taxable transportation fuels are assessed the 0.1¢ LUST fund tax except for liquefied petroleum gas or propane.

[9] A variety of off-highway fuel uses (e.g., farming), business uses (e.g., construction equipment), and government uses (e.g., police departments and school districts) are tax exempt.

[10] Before 1993, only liquid fuels were subject to the various transportation fuels taxes — fuels that were liquid at the time they entered into the tank of the vehicle. The Omnibus Budget Reconciliation Act of 1993 (P.L. 103-66) introduced a tax on CNG.

[11] Also from Jan. 1, 1993 to Dec. 31, 2004, mixtures of 7.7% or 5.7% alcohol (either ethanol or methanol) received a prorated exemption: 7.7% ethanol blends qualified for a 4.004¢ exemption (they are taxed at 14.396¢ per gallon); and 5.7% ethanol blends qualify for a

(continued...)

CRS-5

In all these cases, the exemption equated to 52¢ *per gallon of ethanol,* for 2004, which is the same as a 52¢/gallon income tax credit.[12]

Example

Table 1 compares the net, after-tax value of the new 51¢/ gallon alcohol fuel mixture excise tax credit (column 3) as it is applied on 90:10 blends (90% gasoline, 10% ethanol), with the old 5.1¢/gallon excise tax exemption (column 2) on those same blends. For perspective, these are compared with the tax on a gallon of pure gasoline. The example assumes the ethanol is 190 proof. Subsidy value is measured in terms of its effects on the marginal costs of supplying fuel relative to the unsubsidized shift in the marginal cost curve under the 18.4¢/gallon gasoline tax. The model is described in the Appendix in greater detail.

Note first that there is no difference between the new excise tax credit and the old excise tax exemption in the actual excise tax paid or remitted to the Treasury (row 3): whether the reduction in taxes per gallon of a 90:10 blend is provided by the an excise tax exemption and or excise tax credit, the amount actually paid to the Treasury is 13.3¢/gallon. There is a difference, however, in the subsidy value between the two ways of reducing the excise taxes. In particular, **Table 1** shows that the new alcohol fuels excise tax credit has a higher subsidy value than the excise tax exemption in effect under the pre-2005 tax law. As the sixth row, third column shows, the net-after tax subsidy value of the new excise tax credit *per-gallon of a 90:10 blend* is 6.8¢/gallon as compared to 5.1¢/gallon under the exemption. Thus, the value of the excise tax credit is 1.7¢ greater, or 33.3% greater, than the value of the exemption. Per gallon of ethanol the subsidy value would be 10 times that amount or 68¢/gallon, rather than 51¢/gallon.

---

[11] (...continued)
2.964¢ per gallon exemption (they are taxed at 15.436¢ per gallon). The 5.7% and 7.7% blends correspond, respectively, to the 2.0% and 2.7% oxygen content standard for gasoline sold in ozone nonattainment areas and carbon monoxide nonattainment areas under the Clean Air Act (CAA), as amended in 1990. The CAA requires that all gasoline sold in the winter months in the 40 carbon monoxide (CO) non-attainment areas contain at least 2.7% oxygenate. Oxygenates add oxygen to gasoline and make the fuel burn more completely and more cleanly. This part of the program began on Nov. 1, 1992. The CAA also requires that all gasoline sold in 9 ozone non-attainment areas be reformulated gasoline, containing at least 2% oxygenates. Reformulated gasoline involves a more complex and extensive change to the chemical properties of fuel to 1) reduce emissions of volatile organic compounds (which form ozone), 2) reduce emissions of toxic compounds (such as formaldehyde), and 3) keep emissions of nitrogen oxide from increasing.

[12] Alcohol blended with diesel fuel or any one of the other special motor fuels is also partially exempt from tax. The exemption for "gasohol" blends also applies to blends of diesel and biomass — derived alcohol and blends of a special motor fuel and biomass — derived alcohol, whether ethanol or methanol.

CRS-6

Table 1. Comparison of the Net, After-Tax Subsidy Value of the New Mixtures Tax Credit With the Old Excise Tax Exemption

	No Credit or Exemption	Old Excise Tax Exemption	New Alcohol Fuel Mixtures Excise Tax Credit
	100% Gasoline	90:10 Blends	
Nominal (Statutory) Excise Tax Per Gallon	$0.184	$0.184	0.184
Reduction in Tax Liability Per Gallon	0	$0.051/gallon exemption	$0.051/gallon excise tax credit
Total Excise Taxes Paid to the Treasury	$0.184	$0.133	$ 0.133
Increases in Marginal Cost of Supply	$0.184	$0.133	$0.116
Difference in Marginal Cost Per Gallon Relative to the Non-Subsidized Gasoline Tax	0	- $0.051	- $0.068
Value of the Subsidy/Gallon of 90:10 Blend	0	$0.051	$0.068
Value of the Subsidy/Gallon of Ethanol	0	$0.51	$0.68

Notes: *Value of tax subsidy includes the tax benefits from the deductibility of the excise taxes against income tax liability — a higher excise tax results in greater tax deductions and a smaller tax. This figure assumes a 25% marginal income tax rate. See the Appendix for more detail.

The higher subsidy value under the excise tax credit is due to the deductibility of excise taxes as an expense against the income tax. The 2004 reforms changed the terminology of a tax subsidy: Rather than claiming an instant *exemption* of 5.1¢/gallon, blenders can now claim an instant *credit* of 5.1¢/gallon. However, the substitution of a tax credit for an equivalent exemption means that blenders are treated nominally as paying the full 18.4¢/gallon excise tax; they will get higher income tax deductions, which increases the effective amount of the new tax credit. Assuming a marginal income tax rate of 25% — the rate usually assumed by the Joint Tax Committee and other analysts — the value of the new tax credit increases by 1.7¢ per gallon (or by 33.3%). Thus, the effective value of the tax credit is not 51¢/gallon but 68¢/gallon.

CRS-7

Revenue and Highway Trust Fund Implications

From FY1932, when it was first enacted, until FY2003, the gasoline tax has generated over $400,000 million in gross tax revenues; billions more have been generated from the tax on diesel and other special highway motor fuels.[13] Net revenues have been about 25%, or about $100,000 million, less due to the deductibility of the excise taxes as a cost of doing business, leaving a net total of about $300,000 million.

In FY2003, the year for which the latest data are available, the gasoline tax alone generated $20,100 million in tax revenues for the Highway Trust Fund; $8.6 billion more were generated from the excise tax on diesel; and millions more were generated from the excise taxes on other highway motor fuels. In total, the HTF collected $33,700 billion in excise tax receipts in FY2003, about 90% from motor fuels taxes and most of that from the gasoline tax.[14] These totals include receipts generated from the taxation of gasoline blended with alcohol, almost all ethanol, which totaled in gross terms about $4,100 million in FY2003 (about $3,000 million net due to about $1,000 million in income tax offsets).

But, while the excise tax on gasoline alcohol blends generated billions in tax revenue, the lower tax rate due to the exemptions (a tax rate of 13.1¢ in 2003 due to an exemption of 5.3¢; a tax rate of 13.2¢ in 2004 due to an exemption of 5.2¢) meant that these blends did not generate as much revenue for the HTF as they would have otherwise had the ethanol blends been taxed at the same rate as pure (unblended) gasoline. From 1978, when the excise tax exemption for alcohol fuels was first enacted, to FY2004, it is estimated that the exemption cost the federal treasury approximately $14,000 million in foregone federal revenues in net terms (i.e., gross excise tax receipts less income tax receipts). Gross receipts (i.e, losses to the HTF) would have been about $18,600 million. For FY2003, for example, had ethanol blends been taxed at 18.4¢/gallon (instead of the actual 13.1¢/gallon because the exemption was 5.3¢/gallon), gross revenues for the HTF would have been about $1,400 million higher. The net revenue loss was $350 million or 25% less, again because the lower tax rate implied a smaller income tax offset.

The above effects on the HTF exclude additional losses in receipts due to a tax code provision, enacted in 1990 and repealed by the 2004 jobs bill, which allocated 2.5¢ of the taxable portion of the excise tax on ethanol blends (the 13.1¢ in 2003) to remain in the general fund. IRC §9503 (b) provided that part of the taxable portion of the tax on gasohol blends (the 13.1¢ for 90:10 blends, the 14.342¢ for 92.3:7.7 blends, and the 15.422¢ for 94.3:5.7 blends) was not allocated to the HTF in 2003, but was instead allocated into the general fund. More specifically, for the 90/10

---

[13] Brian Francis, *Gasoline Excise Taxes, 1933-2000,* Statistics of Income Bulletin; Internal Revenue Service, *Federal Excise Taxes Reported to or Collected by the IRS, Alcohol and Tobacco Tax and Trade Bureau, and Customs Service: 1997-2004,* Table 21.

[14] Three of the six separate excise taxes that finance the HTF are imposed on highway motor fuels. The other three excise taxes, which generated $3.1 billion in tax receipts in FY2003, are: the retail excise tax on heavy trucks, tractors, and trailers, the tax on truck tires, and the heavy vehicle use tax.

blends, the law in 2003 provided that 3.1¢ of the 13.1¢ tax remain in the general fund; for blends containing less than 10% ethanol, 2.5¢ remains in the general fund and is not allocated into the HTF. For 2003 about $700 million was allocated to the general fund instead of the HTF under this provision. Thus the total HTF revenue loss resulting from the alcohol fuels exemption was about $2,100 million for FY2003.

Revenue Effects of the New Alcohol Fuels Mixtures Tax Credit

The substitution of a tax credit for an exemption against the excise taxes will increase revenues to the HTF, and reduce revenues by the same amount for the general fund. Based on current projections by both the Office of Management and Budget (OMB) and the Joint Committee on Taxation, foregone revenues from the excise tax credit is projected to be about $1,5000 million for FY2006. **Table 2** shows the two revenue loss projections.

Table 2. Projections of Revenue Losses to the General Fund (and Increases to the HTF) from the New Alcohol Fuels Mixtures Excise Tax Credit ($millions)

Year	OMB Projections	JCT Projections
2004	$1,450	$1,100
2005	$1,490	$1,400
2006	$1,550	$1,400
2007	$1,590	$1,400
2008	$1,620	$1,500
2009	$1,650	$1,500
2010	$1,680	$1,500

Sources: Executive Office of the President. Office of Management and Budget. *Budget Document of the United States: Analytical Perspectives, FY2006,* p. 323; Joint Committee on Taxation. *Estimates of Federal Tax Expenditures for Fiscal years 2005-2009.* (JCS-1-05), Jan. 12, 2005.

However, because the income tax offset or deductibility takes place at the 18.4¢ rate, the net income offsets are higher relative to the non-subsidized full gasoline tax rate — there are additional net revenue loss due to this higher income tax deduction. At the assumed 25% marginal income tax rate, these offsets are estimated at 1/4 of the gross revenue losses in **Table 2**. Thus, for FY2006, using the OMB estimated gross revenue losses of $1,550 million, income tax offsets (i.e., foregone income tax receipts into the general fund) are estimated at $387.5 million.

CRS-9

Revenue Losses Under the Proposed Renewable Fuels Standard

Combined with the renewable fuels standard (RFS) proposed in the Senate's comprehensive energy policy bill, the new tax credit for alcohol fuels mixtures is projected to raise revenue losses significantly over those baselines already projected by the OMB. This is illustrated in **Table 3**. As shown in column (3), revenue losses from the new excise tax credit are projected to more than double by 2010 if the RFS under the Senate's version of H.R. 6 is enacted.

Table 3. Projected Ethanol Use and Corresponding Revenue Losses to the General Fund (and Increases to the HTF), Baseline vs. the Renewable Fuels Standard (RFS), FY2006-2012

	Baseline OMB Data		Renewable Fuels Standard		
Year	**Baseline Ethanol Use (mil. gals.)**	**Baseline Revenue Losses ($mil.)**	**Ethanol Use Under Senate's RFS (mil. gals.)**	**Revenue Losses Under Senate's RFS ($mil.)**	**RFS Revenue Losses Over Baseline ($mil.)**
2006	3039	1550	4000	2040	490
2007	3118	1590	4700	2397	807
2008	3176	1620	5400	2754	1134
2009	3235	1650	6100	3111	1461
2010	3294	1680	6800	3468	1788
2011	N.A.		7400	3774	
2012	N.A.		8000	4080	

Sources: Author's calculations based on OMB data and data in H.R. 6. OMB data is from Executive Office of the President, Office of Management and Budget, *Budget Document of the United States: Analytical Perspectives, FY2006*, p. 323;

Other Possible Tax Subsidies for Alcohol Fuels

Although the new alcohol fuels mixtures excise tax credit will, as the exemption before it, become the major tax incentive for ethanol fuels, there are also others.

Tax Credits for Pure Alcohol Fuels

An income tax credit is also available for straight (or "neat") alcohol (known as E85 or, in the case of methanol, M85) used as fuel — fuels that contain a minimum of 85% alcohol, and are thus not mixtures. The amounts of tax credits is 60¢/gallon of either ethanol and methanol. This credit is available only to the user directly (who must use it in a trade or business), or to the seller who must sell it at retail to the ultimate user (as long as it is placed in the fuel tank of the buyer's vehicle). In all these cases, the alcohol may be either ethanol or methanol but must not be produced from fossil fuels, effectively limiting the tax credit to ethanol from corn. The market for these straight, or neat fuels, is very small. The credit for straight alcohol fuels was not amended by the jobs bill.

Small Ethanol Producer Tax Credit

Current law provides for an income tax credit of 10¢ per gallon ($4.20 per barrel) for up to 15 million gallons of annual ethanol production by a small ethanol producer, defined as one with ethanol production capacity of 30 million gallons per year or less (about 2,000 barrels per day). This credit, which was enacted as part of the Omnibus Budget Reconciliation Act of 1990 (P.L. 101-508), is strictly a production tax credit available only to the manufacturer who sells the alcohol to another person for blending into a qualified mixture in the buyer's trade or business, for use as a fuel in the buyer's trade or business, or for sale at retail where such fuel is placed in the fuel tank of the retail customer. Casual off-farm production of ethanol does not qualify for this credit. The small ethanol producer credit is limited in the same way as the blender's tax credit. The amount of the credit is reduced to take into account any excise tax exemption claimed on ethanol output and sales. The 2004 "Jobs Bill" allowed the flow- or pass-through of the small ethanol producer credit to the patrons of a cooperative, thus allowing farmer cooperatives that produce ethanol to also benefit from the provision.[15]

Income Tax Deduction for Alcohol-Fueled Vehicles

Individuals or businesses that purchase alternative fuel vehicles (AFVs) can claim a tax deduction from adjusted gross income for the incremental costs of new vehicles and upgrades to existing conventionally fueled vehicles. The maximum tax deduction for cars is $2,000, but for trucks it can go as high as $50,000. A tax deduction is also available, up to $100,000, for investments in any equipment needed in dispensing the alternative fuels — for storing and dispensing the clean fuel and otherwise refueling clean fuel burning vehicles. For both of these tax incentives, alternative fuels are defined as compressed natural gas, liquefied petroleum gas, liquefied natural gas, hydrogen, and electricity, and they include 85% (neat) alcohol fuels, ether, or any combination of these produced from biomass. The deduction is reduced by 75% in 2006 — the $2,000 deduction would become $500 — for clean

---

[15] Under IRC§521 farmers cooperatives are exempt from income taxes as long as any cooperative income flows through to the patrons as "patronage dividends" — basically net income (profits) are allocated back to the cooperative patrons (members), and in effect, federal tax law treats cooperatives as partnerships (rather than corporations). Prior to the 2004 "Jobs Bill," the only credits that were allowed to be passed through to the patrons were the rehabilitation credit, the energy credit, and the reforestation tax credit. The 2004 amendments added the small ethanol producer credit to the list of credits allocable to the patrons, which effectively reduces each patron's dividends by a proportionate share of the credits based on the patron's share of ethanol production or total business done with the cooperative (or some other allocation criteria).

fuel vehicles (or refueling property) purchased during 2006, and by 100% thereafter i.e., it is eliminated after December 31, 2006. This deduction is currently little used.

Section 29 Production Tax Credit

An income tax credit is also available for the production of a broad variety of fuels derived from various alternative energy resources (such as oil from tar sands or shale, gas from coalbeds, brine or tight formations, synthetic fuels, etc.). This is the alternative fuels production tax credit, also known as the §29 tax credit (because it is part of Internal Revenue Code section 29), which in 2004 was $6.47 per barrel of oil equivalent. Certain types of alcohol fuels — either ethanol or methanol produced synthetically from coal or lignite — could qualify for this non-refundable tax credit. Alcohol fuels produced from biomass do not qualify for this credit, although gas produced from biomass does qualify. There is little if any production of liquid synthetic fuels from coal in the United States so that, based on current information, this credit is not claimed on alcohol fuels used in transportation.[16]

[16] For a more detailed description and an analysis see CRS Report 97-679, *Economic Analysis of the §29 Tax Credit for Unconventional Fuels,* by Salvatore Lazzari.

Technical Appendix

This technical appendix calculates the economic or subsidy value of the new alcohol fuels excise tax credit as compared to the old excise tax exemption and the non-subsidized 100% gasoline. That subsidy value is measured as the difference in marginal costs for a profit-maximizing gasoline supplier.

Pre-income tax profits, with a unit excise tax, such as a gasoline tax, is:

$$(1)\ \pi^b = P \cdot Q - C(Q) - T^1 \cdot Q$$

where P is price, Q is output, C (Q) is a total cost function, and T^1 is a unit excise tax.

Maximizing before-tax profits π^b with respect to output Q leads to the following condition:

$$(2)\ \delta\pi^b / \delta Q = P - C'(Q) - T^1 = 0$$

which implies that,

$$(3)\ P = C'(Q) + T^1$$

i.e., profit maximization leads the firm to produce at that point where price P is equal to the marginal cost C'(Q) and the excise tax T^1. In other words, in a competitive market, a profit maximizing firm facing a unit tax on gasoline faces a higher marginal cost curve by the amount of the tax — the marginal cost curve shifts up (or is higher than the non-taxed marginal costs) by 18.4¢/gallon.

A pure income or profits tax t is completely neutral (it does not affect the profit maximizing level of output) under these conditions, as is shown next.

$$(4)\ \pi^a = [P \cdot Q - C(Q) - T^1 \cdot Q](1 - t)$$

$$(5)\ \delta\pi^a / \delta Q = [P - C'(Q) - T^1](1 - t) = 0$$

which implies that,

$$(6)\ P = C'(Q) + T^1$$

Equation (6) is the same as equation (3) because the (1 - t) term cancels — the equilibrium level of output and price is unaffected by the pure business income tax even with the imposition of a unit tax, such as the 18.4¢ gasoline tax.

CRS-13

Now, however, assume that there are two unit excise taxes, T^1 is the excise tax per gallon actually paid, and T^2 is the excise tax rate deductible for purposes of the income tax. Then, the after-tax profit function becomes:

$$(7)\ \pi^a = [P \cdot Q - C(Q) - T^1 \cdot Q] - t[P \cdot Q - C(Q) - T^2 \cdot Q]$$

$$(8) = [P \cdot Q - C(Q)](1 - t) - Q[T^1 - t \cdot T^2]$$

Finding the level of Q that maximizes profits lead to the following condition:

$$(9)\ \delta\pi^a / \delta Q = [P - C'(Q)](1 - t) - [T^1 - t \cdot T^2] = 0$$

which reduces to:

$$(10)\ P = C'(Q) + [T^1 - t \cdot T^2] / (1 - t)$$

Where P and C'(Q) are, as before, price and marginal costs, and $[T^1 - t \cdot T^2]/(1 - t)$ is the shift in the marginal cost curve after allowance for the deductibility of excise taxes as an expense against the income tax.

To illustrate, when a gasoline distributor is paying 18.4¢/ gallon (T^1 = 18.4¢), and deducting the same amount against the income tax (T^2 = 18.4¢) then the tax terms in equation (10) become:

$$(11)\ T^1(1 - t) / (1 - t) = T^1$$

Note that because $T^1 = T^2$ the (1 - t) terms cancel and the shift in marginal cost is simply T^1 = 18.4¢. This is also true when a gasoline blender is paying 13.3¢ because of the exemption — the blender is also deducting 13.3¢ against the income tax and the shift or increase in marginal costs equals the amount of the tax, 13.3¢.

However, when T^1 and T^2 differ, as they do under the new restructured alcohol fuels tax incentives, then the shift in marginal costs is defined by equation (10). Under present law, i.e., since January 1, 2005, T^1 = 13.3¢, but T^2 = 18.4¢. Thus, assuming the marginal corporate income tax rate is 25% (t = .25), the tax term in equation (10) — the shift or increase in marginal costs — becomes:

(12)13.3¢ - .25 (18.4¢)/ (1 - .25)

(13) = 13.3¢-4.6¢/.75

(14) = 11.6¢

That represents the increase in marginal costs under the new alcohol fuels excise tax credit (row 4, column 3 in **Table 1**). This is 6.8¢ less than the increase under the gasoline tax, and 1.7¢ less then the increase in costs under the old exemption (the subsidy value is 6.8¢/gallon rather than 5.1¢/gallon)

Intermediate-Level Blends of Ethanol in Gasoline, and the Ethanol "Blend Wall"

Brent D. Yacobucci
Specialist in Energy and Environmental Policy

January 28, 2010

Congressional Research Service
7-5700
www.crs.gov
R40445

CRS Report for Congress
Prepared for Members and Committees of Congress

Summary

On March 6, 2009, Growth Energy (on behalf of 52 U.S. ethanol producers) applied to the Environmental Protection Agency (EPA) for a waiver from the current Clean Air Act limitation on ethanol content in gasoline. Currently, ethanol content in gasoline is capped at 10% (E10); the application requests an increase in the maximum concentration to 15% (E15). If granted, the waiver would allow the use of significantly more ethanol in gasoline than is currently permitted. The existing limitation leads to an upper bound of roughly 15 billion gallons of ethanol in all U.S. gasoline. This "blend wall" could limit the fuel industry's ability to meet an Energy Independence and Security Act (EISA, P.L. 110-140) requirement to blend increasing amounts of renewable fuels (including ethanol) into motor fuels—thus the interest among ethanol producers in the waiver.

On November, 30, 2009, EPA sent a letter to Growth Energy neither granting nor denying the waiver, stating that studies necessary for the agency to make a decision have not been completed, and that some of that data may be available in May or June of 2010. To meet the high volumes of renewable fuels mandated by EISA, EPA recognized that "it is clear that ethanol will need to be blended into gasoline at levels greater than the current limit of 10 percent."

Under EISA, the EPA Administrator must grant or deny the waiver request within 270 days of receipt (December 1, 2009). The Clean Air Act is silent on the consequences if EPA does not grant or deny the waiver within the 270-day window, as is the case in the Growth Energy petition.

To grant the waiver, the petitioner must establish to EPA that the increased ethanol content will not "cause or contribute to a failure of any emission control device or system" to meet emissions standards. EPA is to consider short- and long-term (full useful life) effects on evaporative and exhaust emissions from various vehicles and engines, including cars, light trucks, and non-road engines (e.g., lawnmowers). In its November 30 letter, EPA noted that long-term testing on newer vehicles has not been completed, but that the agency expects that model year 2001 and newer vehicles "will likely be able to accommodate higher ethanol blends, such as E15." In the letter the agency made no statements about older vehicles or non-road engines, but stated that EPA could "be in a position to approve E15 for 2001 and newer vehicles in the mid-year timeframe."

In addition to the emissions control concerns, other factors affecting consideration of the blend wall include vehicle and engine warranties and the effects on infrastructure. Currently, no automaker warranties its vehicles to use gasoline with higher than 10% ethanol. Small engine manufacturers similarly limit the allowable level of ethanol. In addition, most gasoline distribution systems (e.g., gas pumps) are designed to dispense up to E10. While some of these vehicle and fuel distribution systems may be able to operate effectively on E15 or higher, their warranties/certifications would likely need to be updated.

If EPA were to grant a waiver only for newer vehicles, a key question is how fuel pumps might be labeled to keep owners from using E15 in older vehicles and other equipment. A related question is whether fuel suppliers would even be willing to sell E15 if some of their customers may not use it. Further, it is unclear whether existing fuel distribution systems which were designed to dispense E10 can handle the higher-level ethanol blends.

Contents

Figures

Contacts

Background

There is growing interest in the potential for ethanol to displace petroleum as a transportation fuel.[1] In 2008, the United States consumed roughly 9 billion gallons of fuel ethanol, representing about 6% of all U.S. gasoline consumption (by volume).[2] Fuel ethanol consumption has grown from roughly 1 billion gallons per year in the early 1990s, largely as a result of federal policies promoting its use, including tax incentives and mandates for the use of renewable fuels.[3]

Arguably the most significant incentive for ethanol's use is the renewable fuel standard (RFS) established in the Energy Policy Act of 2005[4] and expanded in the Energy Independence and Security Act of 2007.[5] The RFS mandates the use of 9.0 billion gallons of renewable fuel in 2008, increasing steadily through 2022 (**Figure 1**). While the RFS is not an explicit ethanol mandate, the vast majority of the requirement has been met using corn-based ethanol. Going forward, there are limitations on the amount of corn-based ethanol that may be used to meet the mandate, although it is likely that much of the additional mandate for "advanced biofuels"[6] will be met using ethanol derived from sugarcane and from cellulosic feedstocks such as perennial grasses, fast-growing trees, and agricultural wastes.[7] By 2022, EISA requires the use of 36 billion gallons of renewable fuels, and much of this will likely be ethanol from a variety of feedstocks.

However, there is a key obstacle to the use of so much ethanol in gasoline. Currently, although some ethanol is sold as an alternative fuel (E85),[8] most is sold as an additive in conventional and reformulated gasoline. At present, the amount of ethanol that may be blended in gasoline is limited to 10% by volume (E10) by guidance developed by the Environmental Protection Agency (EPA) under the Clean Air Act, as well as by vehicle and engine warranties, and certification procedures for fuel dispensing equipment.

Under the RFS, assuming that most of the mandate is met using ethanol, gasoline blenders are likely to hit a limit in the next few years. In 2012, the RFS will require over 15 billion gallons of renewable fuel, while projected gasoline consumption in 2012 is slightly less than 150 billion gallons.[9] After 2012, the renewable fuel mandate will continue to increase. However, a limit of 10% ethanol means that ethanol for gasoline blending (not including E85) likely cannot exceed

[1] For more information on fuel ethanol, see CRS Report RL33290, *Fuel Ethanol: Background and Public Policy Issues*, by Brent D. Yacobucci.

[2] Renewable Fuels Association (RFA), Industry Statistics, http://www.ethanolrfa.org/industry/statistics/, accessed September 10, 2008.

[3] CRS Report R40110, *Biofuels Incentives: A Summary of Federal Programs*, by Brent D. Yacobucci

[4] EPAct 2005, P.L. 109-58.

[5] EISA, P.L. 110-140.

[6] Biofuels produced from feedstocks other than corn starch and with 50% lower lifecycle greenhouse gas emissions compared to gasoline.

[7] For more information on the RFS, see CRS Report R40155, *Selected Issues Related to an Expansion of the Renewable Fuel Standard (RFS)*, by Brent D. Yacobucci and Randy Schnepf.

[8] A blend of 85% ethanol and 15% gasoline. Ethanol-gasoline blends are designated with an "E" followed by a number—the percentage ethanol concentration by volume. For example, a blend of 10% ethanol and 90% gasoline is referred to as "E10."

[9] U.S. Energy Information Administration (EIA), *Annual Energy Outlook 2010 Early Release*, December 2009, Reference Case Table 11.

15 billion gallons per year.[10] This "blend wall" is the maximum possible volume of ethanol that can be blended into U.S. motor gasoline. It is likely that the actual limit is lower since older fuel tanks and pumps at some retail stations may not be equipped to handle ethanol-blended fuel.

Figure 1. Renewable Fuel Standard Under the Energy Independence and Security Act

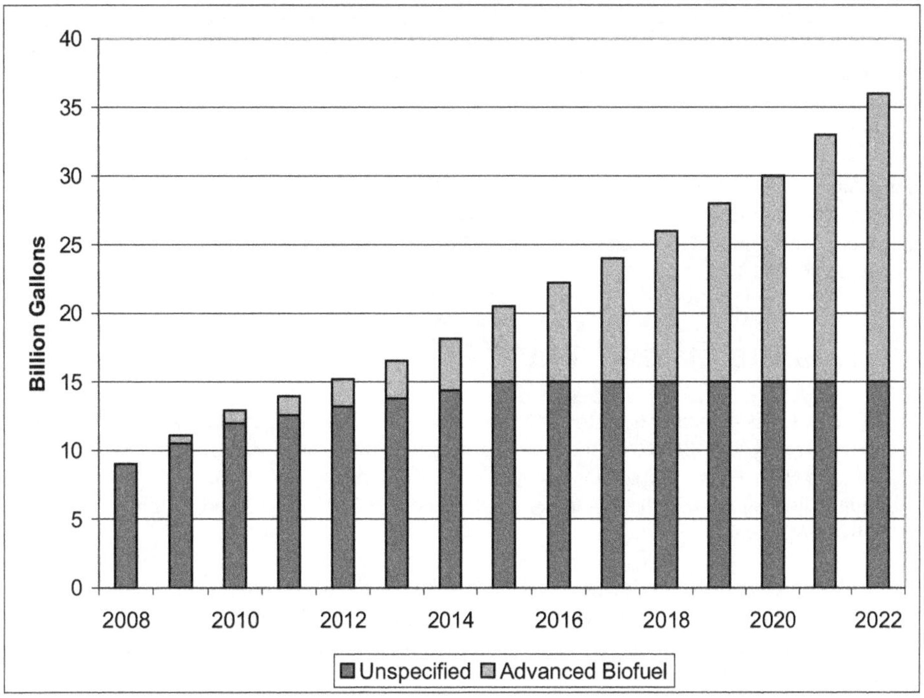

Source: CRS Analysis of P.L. 110-140.

Because of this "blend wall," there is interest, especially among ethanol producers, in increasing the allowable concentration of ethanol in gasoline. Research is ongoing on intermediate-level blends including 15%, 20%, 30%, and 40% ethanol (E15, E20, E30, and E40, respectively).

On March 6, 2009, Growth Energy (on behalf of 52 U.S. ethanol producers) applied to the Environmental Protection Agency (EPA) for a waiver from the current Clean Air Act E10 limit.[11] The application requests an increase in the maximum concentration to 15% (E15). If granted, the waiver would allow the use of significantly more ethanol in gasoline than is currently permitted.[12]

[10] If gasoline demand were to increase, the maximum amount of ethanol that could be blended into that gasoline would increase proportionally. Likewise, if gasoline demand were to decrease, the maximum amount of ethanol that could be blended would decrease proportionally.

[11] Growth Energy on Behalf of 52 United States Ethanol Manufacturers, *Application for a Waiver Pursuant to Section 211(f)(4) of the Clean Air Act for E-15*, March 6, 2009.

[12] Instead of the roughly 15 billion gallon limit in 2012 discussed above, ethanol could represent up to roughly 22 billion gallons of gasoline content.

Under EISA, EPA had 270 days (December 1, 2009) to grant or deny the waiver. On November, 30, 2009, EPA sent a letter to Growth Energy neither granting nor denying the waiver, stating that studies necessary for the agency to make a decision have not been completed, and that some of that data may be available in May or June of 2010. To meet the high volumes of renewable fuels mandated by EISA, EPA recognized that "it is clear that ethanol will need to be blended into gasoline at levels greater than the current limit of 10 percent."[13] In the letter, EPA noted that long-term testing on newer vehicles has not been completed, but that the agency expects that model year 2001 and newer vehicles "will likely be able to accommodate higher ethanol blends, such as E15."[14] In the letter the agency made no statements about older vehicles or non-road engines, but stated that EPA could "be in a position to approve E15 for 2001 and newer vehicles in the mid-year timeframe."

If EPA were to grant Growth Energy's petition to increase gasoline ethanol content to 15%, that would address only one component of the blend wall. The other impediments—vehicle and engine warranties, distribution infrastructure, and the ability of older vehciles and non-road equipment to use the fuel—would likely still need to be addressed before ethanol use in gasoline were taken beyond 10%.

What Is the "Blend Wall"?

The "blend wall" is the upper limit to the total amount of ethanol that can be blended into U.S. gasoline. Currently, gasoline ethanol content is limited to 10% by volume, and in 2008 gasoline consumption was roughly 140 billion gallons.[15] Therefore, the current blend wall is roughly 14 to 15 billion gallons of ethanol that could be blended into gasoline.[16] The blend wall is largely driven by three factors.

First, under the Clean Air Act, it is currently unlawful to sell gasoline that contains additives at levels higher than those approved by EPA. For ethanol, that limit is 10% by volume. To allow a higher percentage, a fuel manufacturer would need to petition EPA for a waiver. (See "Approval of New Fuels and Fuel Additives.")

Second, automakers currently warranty their vehicles to operate on ethanol/gasoline blends up to 10%. While there is data to suggest that newer vehicles could be operated reliably on higher levels of ethanol without modification, no automaker has yet approved those higher blends for

[13] Gina McCarthy, Notice of a Receipt of a Clean Air Act Waiver Application to Increase the Allowable Ethanol Content of Gasoline to 15 Percent; Letter to Petitioners, U.S. Environmental Protection Agency, Office of Air and Radiation, Washington, DC, November 30, 2009, p. 1, http://www.epa.gov/otaq/regs/fuels/additive/lettertogrowthenergy11-30-09.pdf.

[14] Ibid. p. 2.

[15] Beyond the blend wall, more ethanol could be used in transportation as E85 in flexible fuel vehicles (FFVs) specially designed for its use. However, there are far fewer FFVs than conventional vehicles, and most of these are currently operated on gasoline. If all of the roughly 6 to 7 million FFVs were operated on E85 all of the time, that could represent an additional 5 to 6 billion gallons of ethanol use beyond the limits of the blend wall.

[16] Technically, the blend wall is a number slightly higher than 10% of gasoline consumption. Ethanol has a lower energy content than gasoline, thus one gallon of ethanol does not displace an entire gallon of gasoline. As the share of ethanol in gasoline increases, the total volume must increase to provide an equivalent amount of energy. For example, 140 billion gallons of gasoline would have the equivalent energy to roughly 145 billion gallons of a 10% ethanol blend.

use.[17] Further, small engine manufacturers generally advise against using gasoline with more than 10% ethanol in machines such as lawnmowers, trimmers, and snowmobiles. Even if EPA were to approve higher ethanol blends for sale, it is unclear whether vehicle and machine owners would be willing to use the new fuel without explicit approval from the engine/vehicle manufacturer.[18]

Third, most existing infrastructure (e.g., underground gasoline storage tanks, fuel pumps) are designed and certified to deliver blends up to E10. It is unclear whether they can tolerate higher ethanol concentrations. Underwriters Laboratories (UL), an independent testing and certification company, recently announced guidance supporting the use of ethanol blends up to a maximum of 15% in existing fuel pumps currently certified to dispense E10.[19] However, according to the same announcement, UL stated that "under normal business conditions, E10 at the dispenser can vary from about seven to 13 percent ethanol."[20] Assuming a similar variance would exist for E15, it is likely that under normal conditions, ethanol concentrations would exceed the 15% limit. Therefore, a higher maximum level, perhaps 18%, would be necessary to allow those pumps to be certified to deliver E15.

While all three of these components of the blend wall are relevant, this report focuses on the process for addressing the first component, the current Clean Air Act restriction on ethanol concentration in gasoline.

Approval of New Fuels and Fuel Additives

For a blend of gasoline and gasoline additives to be approved under Section 211(f)(1)(A) of the Clean Air Act, it must be "substantially similar" to unleaded gasoline.[21] EPA has defined "gasoline" to have an upper limit of 2.7% oxygen content (by weight), effectively limiting the ethanol concentration to roughly 7.5% (by volume).[22] However, Section 211(f)(4) of the Clean Air Act (as amended by EISA) allows manufacturers of fuels and fuel additives to apply for a waiver from the "substantially similar" requirement if they can prove that the use of the fuel or additive will not "cause or contribute to" a vehicle not meeting applicable emissions standards over its useful life.

> The EPA Administrator, upon application of any manufacturer of any fuel or fuel additive, may waive the prohibitions established under paragraph (1) or (3) of this subsection or the limitation specified in paragraph (2) of this subsection, if he determines that the applicant has established that such fuel or fuel additive or a specified concentration thereof, and the

[17] For example, in Brazil all gasoline contains between 20% and 25% ethanol. While specific vehicle requirements differ between the United States and Brazil, especially emissions control standards, there is reason to believe that somewhat higher blends could be used in the United States.

[18] This was a key concern in the state of Minnesota. That state mandates that all gasoline in the state contain 10% ethanol. That mandate is set to increase to 20% in 2013, but only if EPA approves the use of the fuel, and automakers warranty their vehicles to operate on the fuel.

[19] Underwriter's Laboratories, *Underwriter's Laboratories Announces Support for Authorities Having Jurisdiction Who Decide to Permit the Use of Existing UL Listed Gasoline Dispensers with Automotive Fuel Containing up to a Maximum of 15% Ethanol*, Northbrook, IL, February 19, 2009.

[20] Ibid.

[21] 42 U.S.C. 7545(f)(1)(A).

[22] EPA, "Regulation of Fuels and Fuel Additives; Definition of Substantially Similar," Revised Interpretive Rule, Final Action, February 11, 1991. 56 Federal Register 5332-5336.

> emission products of such fuel or fuel additive or specified concentration thereof, will not cause or contribute to a failure of any emission control device or system (over the useful life of the motor vehicle, motor vehicle engine, nonroad engine or nonroad vehicle in which such device or system is used) to achieve compliance by the vehicle or engine with the emission standards with respect to which it has been certified pursuant to Sections 206 and 213(a) of this title. The Administrator shall take final action to grant or deny an application submitted under this paragraph, after public notice and comment, within 270 days of the receipt of such an application.[23]

EPA has twice granted waivers for 10% ethanol under Section 211(f). The first was granted in 1978 to Gas Plus, Inc. for blends of ethanol up to 10%.[24] The second was in 1982 to Synco 76 Fuel Corp. for a blend of 10% ethanol plus a proprietary additive.[25] To allow the use of E15 or E20, EPA would need to revise its definition of "substantially similar" to allow a higher oxygen content, or a manufacturer would need to petition EPA for a waiver under Section 211(f), as Growth Energy has done.

What Studies or Data Must Accompany a Section 211(f) Waiver Request?

According to EPA, there are no specific guidelines for what data must accompany a waiver application. However, based on communication between EPA's Office of Transportation and Air Quality (OTAQ) and the Minnesota Department of Agriculture,[26] as well as a presentation made by a member of OTAQ staff to the American Petroleum Institute Technology Committee,[27] a submission must include data on both evaporative and exhaust emissions. The data must be comprehensive, assessing the emissions effects both short-term and over the full useful life of the vehicle.[28] EPA expects that the aging of the vehicles will occur under both controlled (e.g., dynamometer) and on-road environmental (e.g., hot and cold weather) conditions. Further, the tests must be done for a variety of vehicles (e.g., new and used, car, truck, and motorcycle), and the selection of vehicles should reflect their frequency on the road.

According to EPA, the application must include an assessment of the health effects of the fuel (e.g., inhalation exposure studies), and should also include data assessing the durability of vehicles and vehicle parts using the fuel. These include assessments of the compatibility of the new fuel (or blend level) with engine materials, and the effects on operability and performance.

Because gasoline is also used in other engines (e.g., lawnmowers, snowmobiles, boats, etc.), the long-term effects on emissions and engine durability for these engines must also be studied, according to EPA. In the case of higher-level ethanol blends, this may be a key concern. While newer automobiles have complex fuel systems, including computers that can measure and adjust fuel/air ratios in real time, most small non-road engines have much simpler carburetor systems

[23] 42 U.S.C. 7545(f).

[24] 44 *Federal Register* 20777.

[25] 47 *Federal Register* 22404.

[26] Margo Oge, Environmental Protection Agency (EPA), Office of Transportation and Air Quality (OTAQ), Letter to Gene Hugoson, Director, Minnesota Department of Agriculture, March 6, 2008.

[27] Karl Simon, EPA, OTAQ, "Mid Level Ethanol Blend Experimental Framework – EPA Staff Recommendations," Presentation to the American Petroleum Institute Technology Committee Meeting, Chicago, June 4, 2008.

[28] Full useful life for modern cars and light trucks is 150,000 miles. For motorcycles, full useful life is 20,000 miles.

with set fuel/air ratios. One potential problem is that ethanol contains oxygen:[29] by increasing the oxygen content in the fuel—increasing the ethanol content from 10% to 20% effectively doubles the oxygen content—while keeping the amount of air coming into the engine constant, the engine will run much leaner. This could cause the engine to misfire, and/or to run much hotter than originally designed, especially in the case of air-cooled engines.

What Actions Are Federal Agencies Such as the Department of Energy and EPA Taking to Study the Compatibility of Higher Blends of Fuel in Non-flex Fuel Vehicles? What Are the Timelines of These Studies, and Will They Be Comprehensive Enough to Support a Section 211(f) Waiver Request?

Preliminary research has been completed or is ongoing on many of the above data requirements. Much of the preliminary research has been conducted by or for the state of Minnesota. Minnesota has a state law requiring the use of E10 across the state. Assuming E20 is approved as a motor fuel, the state will mandate its use starting in 2013. Therefore, Minnesota has headed much of the research that led to the Growth Energy waiver application. According to the Minnesota Department of Agriculture, some of the preliminary research has been completed or is ongoing on materials compatibility and driveability.[30] Preliminary research has also been conducted on exhaust and evaporative emissions, but this research will likely take the most time, since data must eventually be collected for the test vehicles' full useful life.

In a presentation to EPA's Clean Air Act Advisory Committee's Mobile Sources Technical Review Committee, representatives of Chrysler and Honda highlighted key research areas in assessing mid-level ethanol blends.[31] For cars and trucks they categorized the research into seven main topics: durability, tailpipe emissions, evaporative emissions, driveability, materials compatibility, emissions inventory, and on-board diagnostic (OBD) integrity. For most of these topics, they show that fuel producers, automakers, EPA and/or DOE have completed "preliminary, partial or screening" assessments, but that comprehensive testing has just started in some areas, while other areas may still need to be addressed. According to their timeline, much of the comprehensive research would not be completed before the end of 2009. Similar research must be completed for non-road engines. However, their timeline showed that the planning for that research was incomplete as of mid-2008.

In its November 30, 2009, letter to Growth Energy, EPA noted that durability testing is ongoing at the Department of Energy (DOE). According to the letter, DOE is testing a total of 19 newer

[29] In modern vehicles, additional oxygen is generally a benefit in that it promotes more complete combustion, potentially reducing carbon monoxide and other pollutant emissions.

[30] Ralph Groschen, Minnesota Department of Agriculture, "What's Happening With E20?," Presentation to the Iowa Renewable Fuels Summit, Des Moines, IA, January 31, 2008. In her letter to Gene Hugoson of the Minnesota Department of Agriculture, OTAQ Director Oge stated that "the draft reports presented to us ... are a good first step in beginning the evaluation of the effects of E20. We understand that you consider these test results to be preliminary and not sufficient for a complete waiver application, and we agree with that determination." Margo Oge, op. cit.

[31] Reg Modlin, Chrysler Corporation, and Dave Rainey, Honda Motor Company, "Assessing Effects of Mid-Level Ethanol Blends," Presentation to the Mobile Sources Technical Review Subcommittee Meeting, Arlington, VA, May 8, 2008.

vehicles, has completed testing of two of those vehicles, and expects "testing will be completed on an additional 12 vehicles by the end of May. As a result EPA expects to have a significant amount of the total data being generated through this testing program available to us by mid-June."[32] The letter made no comment on the status of testing for older vehicles or for non-road engines.

What Are the Potential Outcomes of a Waiver Request?

Under Section 211(f)(4), as amended by Section 251 of EISA, the Administrator must grant or deny the waiver request within 270 days of receipt. Before being amended by EISA, the language in Section 211(f)(4) stated that "if the Administrator has not acted to grant or deny an application under this paragraph within one hundred and eighty days of receipt of such application, the waiver authorized by this paragraph shall be treated as granted." The amended section does not now specify the status of a waiver request if EPA neither grants nor denies the request within 270 days, as is the case with Growth Energy's current request.

A question that has been raised is whether EPA can grant a partial waiver. For example, some contend that it is possible for EPA to quickly grant a waiver to allow E12 or E13, and take more time to review Growth Energy's application for E15. In recent press reports, Agriculture Secretary Tom Vilsack supported this strategy.[33] It is unclear whether EPA has authority to grant a partial waiver under the Clean Air Act, although EPA does have the authority to determine that E12 or E13 is "substantially similar" to gasoline.

What Entity Can Make a Request?

According to Section 211(f)(4), the EPA Administrator may waive the limitations "upon application of any manufacturer of any fuel or fuel additive." Therefore, presumably any gasoline or ethanol producer may petition EPA for the waiver, provided they can demonstrate to EPA that the new additive or (in this case) specified concentration of an existing additive will meet the criteria set out in Section 211(f)(4).[34] In the case of the current waiver application, Growth Energy filed the application on behalf of 52 U.S. ethanol manufacturers,[35] in partnership with the

[32] Gina McCarthy, *Notice of a Receipt of a Clean Air Act Waiver Application to Increase the Allowable Ethanol Content of Gasoline to 15 Percent; Letter to Petitioners*, U.S. Environmental Protection Agency, Office of Air and Radiation, Washington, DC, November 30, 2009, pp. 1-2.

[33] Christopher Doering, "USDA Chief Backs Boost of Ethanol Blend in Gasoline," *Reuters*, March 9, 2009.

[34] There has been some confusion over whether the petition would have to come from a gasoline producer (since the waiver is for increased ethanol content in gasoline), but the language in Section 211(f)(4) states that the application may come from "any manufacturer of any fuel or fuel additive."

[35] While the cover page of the application mentions 52 for the number of ethanol manufacturers represented, the text of the application lists 54 companies: Absolute Energy, LLC; Agri-Energy LLC/Dakota Renewable; Amazing Energy, LLC; Arizona Grain Inc.; Arkalon Energy, LLC; Big River Resources, LLC; Cardinal Ethanol, LLC; Castle Rock Renewable Fuels, LLC; Conestoga Energy; DENCO; Didion Ethanol, East Kansas Agri Energy, LLC; Front Range Energy LLC; Golden Grain Energy, LLC; Granite Falls Energy, LLC; Green Plains Renewable Energy, Inc.; Hawkeye Renewables LLC, IBEC Ethanol; ICM; Kansas Ethanol, LLC; LifeLine Foods, Inc.; Little Sioux Corn Processors, LLC; Marquis Energy, LLC; Nesika Energy, LLC; Patriot Renewable Fuels, LLC; Pinal Energy, Poet Biorefining – Alexandria; Poet Biorefining – Ashton; Poet Biorefining – Big Stone; Poet Biorefining – Caro; Poet Biorefining – Chancellor; Poet Biorefining – Coon Rapids; Poet Biorefining – Corning; Poet Biorefining – Emmetsburg; Poet Biorefining – Glenville; Poet Biorefining – Gowrie; Poet Biorefining – Groton; Poet Biorefining – Hanlontown; Poet Biorefining – Hudson; Poet Biorefining – Jewell; Poet Biorefining – Laddonia; Poet Biorefining – Lake Crystal; Poet (continued...)

American Coalition for Ethanol, the Renewable Fuels Association, and the National Ethanol Vehicle Coalition.

Other Than a Successful Section 211 Waiver Request, Are There Other Means to Approve Higher Blends of Ethanol, Such as an Executive Order or Other Administrative Action?

The provisions of Section 211 are explicit, and there seem to be few options outside of the Section 211(f)(4) waiver process for E15 or other intermediate blends to be approved. While there may be no administrative action that could permit the use of E15 other than an EPA waiver or a determination that E15 is "substantially similar" to gasoline, there are potential legislative options. These include

- amending the Clean Air Act to explicitly allow the use of E15 (or some other level of ethanol);
- amending the Clean Air Act to provide expedited approval of higher levels of previously-approved fuel additives; and
- mandating the production and sale of flexible fuel vehicles (since intermediate blends between E85 and E0—straight gasoline with no ethanol—are already approved for use in these vehicles), and promoting (or mandating) the use of E85 fuel.

Growth Energy's Waiver Application

As stated above, on March 6, 2009, Growth Energy petitioned EPA for a waiver to allow the use of up to 15% ethanol in gasoline. Under the Clean Air Act, EPA had up to 270 days (December 1, 2009) to approve or deny the waiver request, but on November 30, 2009, EPA sent a letter to Growth Energy stating that it will continue to evaluate the petition and that it may "be in a position to approve E15 for 2001 and newer vehicles in the mid-year [2010] timeframe."[36] In their application, Growth Energy states that "recent and extensive research demonstrates that use of higher ethanol blends will significantly benefit the environment by reducing greenhouse gas emissions, reducing harmful tailpipe emissions, reducing smog, using less energy for an equivalent amount of fuel, and protecting natural resources."[37] Growth Energy contends that

> available data and multiple recent studies[38] regarding the impact of various intermediate blends [of ethanol] on emissions, materials compatibility, durability, and driveability, were

(...continued)

Biorefining – Leipsic; Poet Biorefining – Macon; Poet Biorefining – Mitchell; Poet Biorefining – Portland; Poet Biorefining – Preston; POET Ethanol Products; Prairie Horizon Agri-Energy LLC; Quad County Corn Processors; Renew Energy; Siouxland Ethanol LLC; Sire; and Western Plains Energy, LLC.

[36] Gina McCarthy, op. cit., p. 2.

[37] Growth Energy, op. cit., p. 2.

[38] Growth Energy cites seven studies completed by or for the U.S. Department of Energy, the American Coalition for Ethanol, the State of Minnesota, the Coordinating Research Council, the Rochester Institute of Technology, the Minnesota Center for Automotive Research, and Stockholm University.

> completed on extensive and representative test fleets, provide a reliable comparison to certification conditions, and demonstrate that use of E-15 will not cause or contribute to failure of any emission control device or system to meet its certification emissions standards.[39]

Growth Energy cites a Department of Energy (DOE) study[40] that found a statistically significant decrease in carbon monoxide emissions using E15, and a marginally significant decrease in non-methane hydrocarbon emissions.[41] The same study also found a statistically significant increase in acetaldehyde emissions, and a marginally significant increase in formaldehyde emissions.[42] Both formaldehyde and acetaldehyde are regulated as toxic air pollutants under Sections 202 and 211 of the Clean Air Act. However, the fact that emissions increased using the fuel is not enough for EPA to deny the waiver: EPA would need to prove that that increase in emissions is enough to cause the vehicle or engine to fall out of compliance with emissions standards. Growth Energy asserts that the DOE study and other studies have found that the use of E15 results in emissions within applicable limits.[43] In its November 30 letter to Growth Energy, EPA stated that "we want to make sure we have all necessary science to make the right decision," including more long-term testing data.[44]

Other Issues

As stated above, EPA approval is not the only hurdle in enabling the use of intermediate-level ethanol blends. A key non-vehicle issue is whether existing infrastructure can support ethanol blends above E10. Like automobiles, most existing gasoline tanks and pumps are designed and certified to handle up to E10.[45] Even if the fuel is approved by EPA for use in motor vehicles, presumably fuel suppliers would be unwilling to sell the fuel unless they are confident that it will not damage their existing systems or lead to liability issues in the future. Otherwise, it seems doubtful that fuel suppliers would voluntarily upgrade their systems to handle the new fuel.

In addition to fuel supply concerns, for vehicle and machine owners to accept the new fuel, engine and auto manufacturers would likely need to convince their customers that both new and existing equipment would not be damaged by using the new fuel, and that its use would not void vehicle and equipment warranties. This may be especially difficult for small-engine manufacturers and users who are currently concerned about the effects on their engines from E10, let alone higher blends of ethanol.

---

[39] Ibid. p 6.

[40] U.S. Department of Energy, *Effects of Intermediate Ethanol Blends on Legacy Vehicles and Small Non-Road Engines*, Report 1, Oak Ridge, TN, October 2008.

[41] Growth Energy, op. cit., p. 16.

[42] Ibid.

[43] Ibid., p. 18.

[44] Gina McCarthy, op. cit., p. 1.

[45] Potential concerns that have been raised include whether the higher ethanol concentration will corrode seals and other components, and whether the higher concentration would lead to stress cracking of tanks and other metal components.

Author Contact Information

Brent D. Yacobucci
Specialist in Energy and Environmental Policy
byacobucci@crs.loc.gov, 7-9662

Selected Issues Related to an Expansion of the Renewable Fuel Standard (RFS)

Randy Schnepf
Specialist in Agricultural Policy

Brent D. Yacobucci
Specialist in Energy and Environmental Policy

January 28, 2010

Congressional Research Service
7-5700
www.crs.gov
R40155

CRS Report for Congress
Prepared for Members and Committees of Congress

Summary

Biofuels have grown significantly in the past few years as a component of U.S. motor fuel supply. Current U.S. biofuels supply relies primarily on ethanol produced from Midwest corn. Today, ethanol is blended in more than half of all U.S. gasoline (at the 10% level or lower in most cases). Federal policy has played a key role in the emergence of the U.S. biofuels industry in general, and the corn ethanol industry in particular. U.S. biofuels production is supported by federal and state policies that include minimum usage requirements, blending and production tax credits, an import tariff to limit importation of foreign-produced ethanol, loans and loan guarantees to facilitate the development of biofuels production and distribution infrastructure, and research grants.

Congress first established a Renewable Fuel Standard (RFS)—a mandatory minimum volume of biofuels to be used in the national transportation fuel supply—with the enactment of the Energy Policy Act of 2005 (EPAct, P.L. 109-58). This initial RFS mandated that a minimum of 4 billion gallons be used in 2006, and that this minimum usage volume rise to 7.5 billion gallons by 2012. Two years later, the Energy Independence and Security Act of 2007 (EISA, P.L. 110-140) superseded and greatly expanded the biofuels blending mandate. EISA required the annual use of 9 billion gallons of biofuels in 2008 and expanded the mandate to 36 billion gallons annually in 2022, of which only 15 billion gallons can be ethanol from corn starch. The remaining 21 billion gallons are to be so-called "advanced biofuels"—that is, biofuels produced from feedstocks other than corn starch, including sugarcane, oil crops, and cellulose.

In the long term, the expanded RFS is likely to play a dominant role in the development of the U.S. biofuels sector, but with considerable uncertainty regarding potential spillover effects in other markets and on other important policy goals. Emerging resource constraints related to the rapid expansion of U.S. corn ethanol production have provoked questions about the sustainability of continued growth in biofuels production and about possible unintended consequences, including potential market and environmental effects of a major shift in U.S. agricultural land use patterns, increased energy demand to grow feedstocks and process them into liquid fuel, and near-term barriers to developing the infrastructure needed to deliver increasing volumes of biofuels to the market.

Questions also exist about the ability of the U.S. biofuels industry to meet the expanding mandate for biofuels from non-corn sources such as cellulosic biomass materials (e.g., grasses, trees, and agricultural and municipal wastes). Although cellulosic feedstock sources appear promising, technological barriers make them very expensive to produce relative to conventional ethanol and keep their future uncertain. Another important biofuel, biodiesel, is produced from soybeans and other oil crops, but also remains expensive to produce owing to the relatively high prices of its feedstocks.

Key policy questions are whether a renewable fuel mandate is the most effective policy to promote the above goals, if government intervention in the industry is appropriate, and, if so, what level is appropriate.

This report describes the nature of the RFS mandate, its implementation, and some emerging issues regarding the long-term growth of the U.S. biofuels sector.

Contents

Figures

Tables

Contacts

Introduction

High petroleum and gasoline prices, concerns over global climate change, and the desire to promote domestic rural economies have raised interest in biofuels as an alternative to petroleum in the U.S. transportation sector. In response to this interest, U.S. policymakers have enacted an increasing variety of policies, at both the state and federal levels, to directly support U.S. biofuels production and use. Policy measures include blending and production tax credits, an import tariff to limit importation of foreign-produced ethanol, loans and loan guarantees to facilitate the development of biofuels production and distribution infrastructure, research grants, and minimum usage requirements.[1] As a result of expanding policy support, biofuels (primarily corn-based ethanol and biodiesel) have grown significantly in the past few years as a component of U.S. motor fuels, comprising about 6% of total transportation fuel consumption in 2009.[2]

Arguably, the most significant federal programs for biofuels have been tax credits for the production or sale of ethanol and biodiesel. However, with the establishment of the Renewable Fuel Standard (RFS) under the Energy Policy Act of 2005 (EPAct, P.L. 109-58), Congress mandated biofuels use. Then, two years later Congress passed the Energy Independence and Security Act of 2007 (EISA, P.L. 110-140) which replaced the original biofuels blending mandate with an expanded version that rises from 9 billion gallons in 2008 to 36 billion gallons in 2022. In the long term, the expanded mandate may prove even more significant than tax incentives in promoting the use of these fuels.

This report focuses specifically on the RFS and some of the potential long-term implications. It describes the general nature of the biofuels RFS and its implementation, and outlines some of the emerging issues related to the sustainability of continued growth in biofuels production as well as the emergence of potential unintended consequences. This report does not address the broader public policy issue surrounding how best to support U.S. energy policy.

Biofuels Defined

Any fuel produced from biological materials (e.g., food crops, agricultural residues, or municipal waste) qualifies as a "biofuel"; however, the term is often used to refer to liquid transportation fuels. Ethanol is generally blended into gasoline at the 10% level (E10) or lower. It can be used in purer forms such as E85 (85% ethanol and 15% gasoline) in flex-fuel vehicles specially designed for higher ethanol content, but E85 presently represents less than 1% of U.S. ethanol consumption.

Current U.S. biofuels supply relies almost exclusively on ethanol produced from Midwest corn (**Table 1**). Approximately 10.7 billion gallons of ethanol were produced in the United States in 2009, about 99% from corn. Other domestic feedstocks for ethanol include grain sorghum and sweet sorghum. After ethanol, biodiesel is the next most significant biofuel in the United States.

[1] For more information on incentives (both tax and non-tax) for ethanol, see CRS Report R40110, *Biofuels Incentives: A Summary of Federal Programs*, by Brent D. Yacobucci.

[2] In gasoline-equivalent shares with 5.27% for ethanol and 0.77% for biodiesel. CRS estimates based on extrapolating from EIA/DOE, "Table C1. Estimated Consumption of Vehicle Fuels in the United States, by Fuel Type, 2003-2007," with recent data for 2008 and 2009.

Biodiesel is produced from any type of organic-based oil including vegetable oils, animal fats, and waste restaurant grease and oils. In the United States and Brazil, biodiesel is primarily made from soybean oil. In the European Union, rapeseed oil is the primary feedstock. The U.S. biodiesel industry has suffered from unfavorable market conditions in recent years. U.S. biodiesel production in 2009 was only 470 million gallons,[3] down from 700 million gallons in 2008. In 2009 U.S. biodiesel production represented about 1% of roughly 44 billion gallons of national on-road diesel fuel use.[4] Other biofuels with the potential to play a role in the U.S. market include diesel fuel substitutes and other alcohols (e.g., methanol and butanol) produced from biomass.

Table 1. U.S. Production of Biofuels from Various Feedstocks

Fuel	Feedstock	U.S. Production in 2009
Ethanol	Corn	10.6 billion gallons
	Sorghum, wheat, barley, and brewery waste	less than 100 million gallons
	Cane sugar	1.5 million gallons (plus a projected 200 million gallons imported from Brazil and Caribbean countries)
	Cellulose	No commercial production
Biodiesel	Soybean oil	approximately 470 million gallons
	Other vegetable oils	less than 10 million gallons
	Recycled grease	less than 10 million gallons
	Cellulose	No commercial production
Methanol	Cellulose	No commercial production
Butanol	Cellulose, other biomass	No commercial production

Source: Renewable Fuels Association; National Biodiesel Board; CRS analysis.

Because of concerns over the significant expansion in corn-based ethanol supply, interest has grown in spurring the development of motor fuels produced from cellulosic biomass materials, including grasses, trees, and agricultural and municipal wastes. However, cellulosic biofuels are currently expensive to produce relative to conventional ethanol, and many uncertainties remain concerning both the viability and the speed of commercial development.[5]

In addition to expanding domestic production of biofuels, there is some interest in expanding imports of sugar-based ethanol—usually produced from sugar cane in Brazil. In 2009 the United States imported a projected 200 million gallons of sugar-ethanol.[6] However, ethanol from Brazil is currently subject to a 2.5% ad valorem tariff and a most-favored-nation duty of $0.54 per gallon of ethanol (for fuel use)[7] that in most years is a significant barrier to direct Brazilian

[3] National Biodiesel Board, "Biodiesel Production Estimates, 2005-current." Dec. 14, 2009; however, other sources suggest the figure is much lower in the 300 to 350 million gallons range.

[4] U.S. Energy Information Administration, *U.S. Product Supplied for Crude Oil and Petroleum Products*, Washington, DC, July 28, 2008, http://tonto.eia.doe.gov/dnav/pet/pet_cons_psup_dc_nus_mbbl_a.htm.

[5] For more information, see CRS Report RL34738, *Cellulosic Biofuels: Analysis of Policy Issues for Congress*, by Kelsi Bracmort et al.

[6] Mostly from Brazil and Caribbean Basin Initiative (CBI) countries.

[7] CRS Report R40110, *Biofuels Incentives: A Summary of Federal Programs*, by Brent D. Yacobucci.

imports. Some Brazilian ethanol can be brought into the United States duty free if it is dehydrated (reprocessed) in Caribbean Basin Initiative (CBI) countries.[8] Up to 7% of the U.S. ethanol market could be supplied duty-free in this fashion; historically, however, ethanol dehydrated in CBI countries has only represented about 2% of the total U.S. market.

The Renewable Fuel Standard (RFS)

Congress first established a Renewable Fuel Standard (RFS)—a mandatory minimum volume of biofuels to be used in the national transportation fuel supply—in 2005 with the enactment of the Energy Policy Act of 2005 (EPAct, P.L. 109-58). This initial RFS (sometimes referred to as RFS1) mandated that a minimum of 4 billion gallons be used in 2006, and that this minimum usage volume rise to 7.5 billion gallons by 2012 (**Table 2**). Two years later, the Energy Independence and Security Act of 2007 (EISA, P.L. 110-140) superseded and greatly expanded the biofuels blending mandate. This expanded RFS is sometimes referred to as RFS2.

Energy Independence and Security Act of 2007 (EISA; P.L. 110-140)

Section 202 of EISA requires the use of 9 billion gallons of renewable fuels in 2008, increasing annually to reach 36 billion gallons in 2022. Although the RFS has been called an ethanol mandate, there is no explicit requirement to use ethanol. It is more correctly called a biofuels mandate with limits on corn-ethanol inclusion. There are also specific requirements for the use of biodiesel and cellulosic biofuels; however, it is expected that, in the early years, the vast majority of the RFS will be met using ethanol produced from corn starch.

In order to encourage the use of non-corn-based biofuels, the RFS includes a cap on the volume of ethanol derived from corn starch that can be counted under the RFS. The corn-ethanol cap rises until 2015 when it is fixed at 15 billion gallons. Any corn-starch ethanol blended in excess of the annual mandate (while still eligible for the tax credit of $0.45/gallon of ethanol produced) is not credited toward the annual RFS mandate. In 2022, the RFS mandates the use of 21 billion gallons of so-called "advanced biofuels"[9] with special carve-outs for cellulosic biofuels and biodiesel. Currently, production of advanced biofuels is limited to ethanol derived from sugar and biodiesel.

---

[8] For more information on CBI imports, see CRS Report RS21930, *Ethanol Imports and the Caribbean Basin Initiative (CBI)*, by Brent D. Yacobucci.

[9] The term "advanced biofuels" comes from legislation in the 110th Congress, and is defined in Section 201 of the Energy Independence and Security Act of 2007 (EISA). In many cases, the definition of "advanced biofuels" includes mature technologies and fuels that are currently produced in large amounts. For example, the EISA definition of "advanced biofuels" potentially includes ethanol from sugar cane, despite the fact that Brazilian sugar growers have been producing fuel ethanol for decades. EISA defines "advanced biofuels" as biofuels other than ethanol derived from corn starch (kernels) having 50% lower lifecycle greenhouse gas emissions relative to gasoline. Possible fuels include biodiesel from oil seeds, ethanol from sugar cane, and ethanol from cellulosic materials (including non-starch parts of the corn plant, such as the stalk).

The Expanded RFS Defined

Usage Volume Requirements

The expanded RFS (or RFS2) includes all motor fuel, as well as heating oil (**Table 2**). It rises from 9 billion gallons (bgal.) in 2008 to 36 bgal. in 2022. However, the corn-based ethanol share of the expanded RFS is capped at 15 bgal. in 2015, and all subsequent annual increases are to be derived entirely from advanced biofuels—defined as biofuels derived from feedstocks other than corn starch. The advanced biofuels volume under the RFS2, which includes specific carve-outs for cellulosic biofuels and biodiesel, reaches 21 bgal. by 2022.

Table 2. EISA 2007 Expansion of the Renewable Fuel Standard

(in billions of gallons)

		RFS2: Biofuel mandate for motor fuel, home heating oil, and boiler fuel					
				Portion to be from advanced biofuels			
Year	RFS1 in EPAct of 2005	Total RFS2	Cap on corn starch-derived ethanol	Total non-corn starch	Cellulosic	Biodiesel	Other
2006	4.0	—	—	—	—	—	—
2007	4.7	—	—	—	—	—	—
2008	5.4	9.00	9.0	0.00	0.00	0.00	0.00
2009	6.1	11.10	10.5	0.60	0.00	0.50	0.10
2010	6.8	12.95	12.0	0.95	0.10	0.65	0.20
2011	7.4	13.95	12.6	1.35	0.25	0.80	0.30
2012	7.5	15.20	13.2	2.00	0.50	1.00	0.50
2013	7.6 (est.)	16.55	13.8	2.75	1.00	1.00	0.75
2014	7.7 (est.)	18.15	14.4	3.75	1.75	1.00	1.00
2015	7.8 (est.)	20.50	15.0	5.50	3.00	1.00	1.50
2016	7.9 (est.)	22.25	15.0	7.25	4.25	1.00	2.00
2017	8.1 (est.)	24.00	15.0	9.00	5.50	1.00	2.50
2018	8.2 (est.)	26.00	15.0	11.00	7.00	1.00	3.00
2019	8.3 (est.)	28.00	15.0	13.00	8.50	1.00	3.50
2020	8.4 (est.)	30.00	15.0	15.00	10.50	1.00	3.50
2021	8.5 (est.)	33.00	15.0	18.00	13.50	1.00	3.50
2022	8.6 (est.)	36.00	15.0	21.00	16.00	1.00	4.00

Source: RFS1 is from EPAct (P.L. 109-58), Section 1501; RFS2 is from EISA (P.L. 110-140), Section 202.

Required Reduction in Greenhouse Gas Emissions

The RFS2 requires that renewable fuels produced in facilities that commence operation after enactment of EISA (December 19, 2007) must achieve at least a 20% reduction in life-cycle greenhouse gas emissions relative to gasoline to count under the total RFS2. This requirement rises to 50% for the advanced biofuels RFS, and 60% for the cellulosic biofuels RFS.

Figure 1. Renewable Fuels Standard (RFS2) vs U.S. Ethanol Production Since 1995
(in billions of gallons)

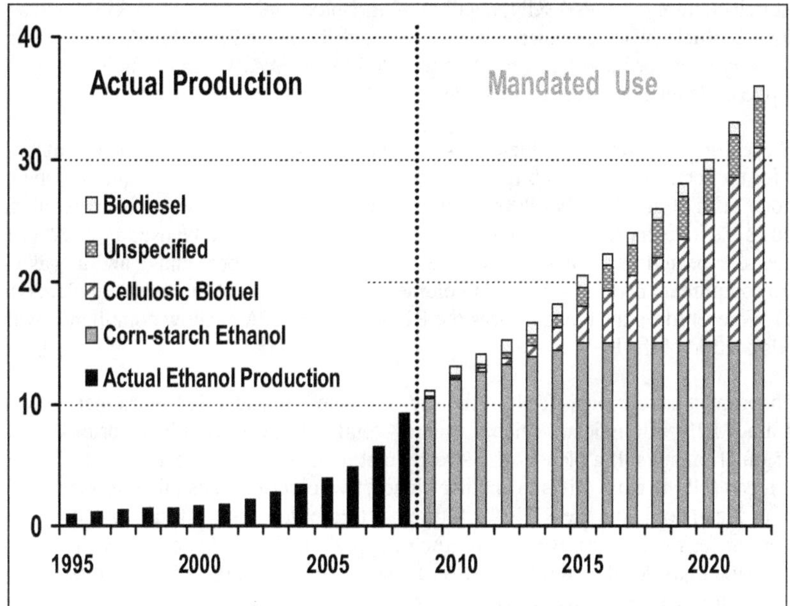

Source: Actual ethanol production data for 1995-2008 is from Renewable Fuels Association; the RFS2 by category is from EISA (P.L. 110-140).

Implementation of the RFS

Under the Energy Independence and Security Act of 2007, the U.S. Environmental Protection Agency (EPA) is responsible for revising and implementing regulations to ensure that gasoline sold in the United States contains a minimum volume of renewable fuel.[10] On May 1, 2007, EPA issued a final rule on the original RFS program (RFS1) detailing compliance standards for fuel suppliers, as well as a system to trade renewable fuel credits between suppliers. On May 26, 2009, EPA proposed rules for the expanded program (RFS2), including life-cycle analysis methods necessary to categorize fuels as advanced biofuels, and new rules for credit verification and trading.

EPA has not finalized its rules for implementing the RFS2 of EISA. However, it is expected that the RFS2 mandates will be implemented in a manner similar to that of RFS1, using a tracking system based on renewable identification numbers (RINs).[11]

[10] For more information, see the EPA website for "Renewable Fuel Standard Program," at http://www.epa.gov/otaq/renewablefuels/index.htm#regulations.

[11] The following discussion is drawn from two resources: "Renewable Identification Numbers (RINs) and Government Biofuels Blending Mandates," Robert Wisner, *AgMRC Renewable Energy Newsletter*, Agricultural Marketing Research Center, Iowa State University, April 2009; available at http://www.agmrc.org/renewable_energy/agmrc_renewable_energy_newsletter.cfm; and "Renewable Identification Numbers are the Tracking Instrument and Bellwether of US Biofuel Mandates," Wyatt Thompson, Seth Meyer, and Pat Westhoff, *EuroChoices 8(3)*, published (continued...)

A RIN is a unique 38-character number that is issued (in accordance with EPA guidelines) by the biofuel producer or importer at the point of biofuel production or the port of importation. A gallon of corn-starch ethanol receives 1 RIN; a gallon of agri-biodiesel receives 1.5 RINs, and a gallon of cellulosic biofuels (when commercially available) will receive 2.5 RINs. After a RIN is created by a biofuel producer or importer, it must be reported to the EPA. When biofuels change ownership (e.g., from a producer to a blender), the RINs are also transferred.

The RFS is enforced on retail fuel blenders (not on biofuels producers). Companies that blend gasoline for the retail market are obligated to include a quantity of biofuels equal to a percentage of their total sales of gasoline. For 2009, EPA has set the RFS at 10.21%.[12] This percentage is computed as the industry total amount of biofuels that is mandated to be used in 2009 as a percentage of expected total gasoline use. The company uses this percentage and its gasoline volume to compute its mandated biofuels volume. When the biofuels are actually blended into gasoline (or diesel fuel), the blender turns the RINs in to the EPA to show compliance with the company's portion of the RFS.

Each blender must have enough RINs at the end of the year to show that it has met its share of the mandate. Each RIN will indicate which mandate—total, advanced, cellulosic, or biodiesel—if any, the biofuel satisfies. If a blender has already met its mandated share and has blended surplus biofuels, it can sell the extra RINs to another blender (who perhaps has failed to meet its blending mandate) or it can hold onto the RINs for future use (either to satisfy a future blending requirement or for future sale). Because biofuels supply and demand varies over time and across regions, a market has developed for RINs. As a result, the tradability of RINs offers flexibility for how the mandates can be met by blenders.

RFS as Public Policy

Proponents' Viewpoint

Supporters of an RFS claim it serves several public policy interests in that it:

- reduces the risk of investing in renewable biofuels by guaranteeing biofuels demand for a projected period (such risk would otherwise keep significant investment capital on the sidelines);
- enhances U.S. energy security via the production of liquid fuel from a renewable domestic source resulting in decreased reliance on imported fossil fuels (the U.S. currently imports over half of its petroleum, two-thirds of which is consumed by the transportation sector);
- provides an additional source of demand—renewable biofuels—for U.S. agricultural output that would have significant agricultural and rural economic

(...continued)

online June 15, 2009, pp. 43-50.

[12] "Renewable Fuel Standard for 2009, Issued Pursuant to Section 211(o) of the Clean Air Act," EPA, *Federal Register*, vol. 73, no. 226, Nov. 21, 2008, at http://www.epa.gov/fedrgstr/EPA-AIR/2008/November/Day-21/a27613.pdf

benefits via increased farm and rural incomes and substantial rural employment opportunities;[13]

- underwrites the environmental benefits of renewable biofuels over fossil fuels (most biofuels are non-toxic, biodegradable, and produced from renewable feedstocks), and
- responds to climate change concerns because agricultural-based biofuels emit substantially lower volumes of direct greenhouse gases (GHGs) than fossil fuels when produced, harvested, and processed under the right circumstances.

Critics' Viewpoints

Critics of an RFS, particularly of the EISA expansion of the original RFS, have taken issue with many specific aspects of biofuels production and use, including the following:

- By picking the "winner," policymakers may exclude or retard the development of other, potentially preferable alternative energy sources.[14] Critics contend that biofuels are given a huge advantage via billions of dollars of annual subsidies that distort investment markets by redirecting venture capital and other investment dollars away from competing alternative energy sources. Instead, these critics have argued for a more "technology-neutral" policy such as a carbon tax, a cap-and-trade system of carbon credits, or a floor price on imported petroleum.
- Continued large federal incentives in support of ethanol production are no longer necessary, particularly since the sector is no longer in its "economic infancy" and would have been profitable during much of 2006 and 2007 without federal subsidies.[15]
- The expanded mandate could have substantial unintended consequences in other areas of policy importance, including energy/petroleum security, pollutant and greenhouse gas emissions, agricultural commodity and food markets, land use patterns, soil and water quality, conservation, the ability of the gasoline-marketing infrastructure and auto fleet to accommodate higher ethanol concentrations in gasoline, the likelihood of modifications in engine design, and other considerations.
- Taxpayers are being asked to finance ever-increasing biofuels subsidies that have the potential to affect future federal budgetary choices.

[13] For example, see John M. Urbanchuk (Director, LECG LLC), *Contribution of the Ethanol Industry to the Economy of the United States*, white paper prepared for National Corn Growers Assoc., February 21, 2006.

[14] For example, see Bruce A. Babcock, "High Crop Prices, Ethanol Mandates, and the Public Good: Do They Coexist?" *Iowa Ag Review*, Vol. 13, No. 2, Spring 2007; and Robert Hahn and Caroline Cecot, "The Benefits and Costs of Ethanol," Working Paper 07-17, AEI-Brookings Joint Center for Regulatory Studies, November 2007.

[15] Chris Hurt, Wally Tyner, and Otto Doering, Department of Agricultural Economics, Purdue University, *Economics of Ethanol*, December 2006, West Lafayette, IN.

The Increasing Cost of Biofuels Policy

A 2007 survey of federal and state government subsidies in support of ethanol production reported that total annual federal support fell somewhere in the range of $5.4 to $6.6 billion per year—nearly $1 per gallon.[16] In 2009, federal and state subsidies were roughly in the range of $6 to $8 billion.[17] Federal and state subsidies include tax credits, infrastructure development grants, special infrastructure depreciation allowances, loans and loan guarantees, feedstock development funds, research funds, and others. The major direct federal costs associated with the implementation of the RFS are the federal tax credits available to the various biofuels that are blended to meet the RFS mandate (**Table 3**).

Table 3. Federal Tax Credits Available for Qualifying Biofuels

Biofuel	Tax Credit: $/gallon	Details	Expiration Date
Volumetric Ethanol Excise Tax (VEET) Credit	$0.45	Available in unlimited amount to all qualifying biofuels.	Dec. 31, 2010
Small Ethanol Producer Credit	$0.10	Available on the first 15 million gallons (mgal) of any producer with production capacity below 60 mgal.	Dec. 31, 2010
Biodiesel Tax Credit: virgin oil	$1.00	Available in unlimited amount to all qualifying biodiesel.	Dec. 31, 2009
Biodiesel Tax Credit: recycled oil	$0.50	Available in unlimited amount to all qualifying biodiesel.	Dec. 31, 2009
Small Agri-Biodiesel Producer Credit	$0.10	Available on the first 15 mgal of any producer with production capacity below 60 mgal.	Dec. 31, 2009
Cellulosic Biofuels Production Tax Credit	$1.01	Available in unlimited amount to all qualifying biofuels.	Dec. 31, 2012

Source: CRS Report R40110, *Biofuels Incentives: A Summary of Federal Programs*, by Brent D. Yacobucci

Under the RFS2, federal tax credits alone will expand dramatically during the life of the program. Based on CRS calculations, federal biofuels tax credit subsidies will grow from about $6.7 billion in 2010 to over $27 billion in 2022 if the RFS requirements are fully met. The total liability from 2008 through 2022 (under the assumption that the RFS is fully met during that period) is estimated at nearly $200 billion.

[16] Ronald Steenblik, *Biofuels—At What Cost? Government Support for Ethanol and Biodiesel in the United States*, Global Subsidies Initiative of the International Institute for Sustainable Development, Geneva, Switzerland, September 2007, p. 37; available at http://www.globalsubsidies.org.

[17] CRS projection based on available data.

Figure 2. Annual Minimum Liability for Biofuel Tax Credits Under the RFS2

($ billion)

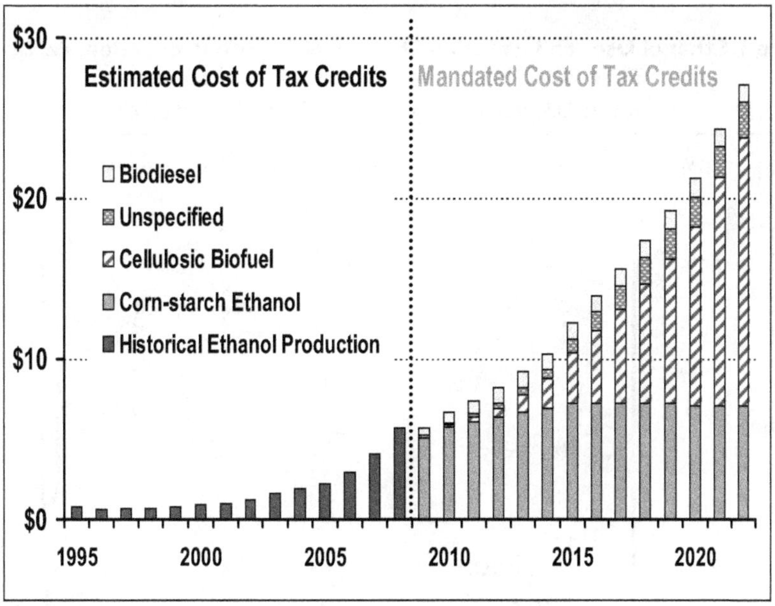

Source: CRS projections are based on the extension of current law: EISA (P.L. 110-140) and 2008 farm bill (P.L. 110-246). Actual data on tax credits is held by the IRS and is not publically available.

Notes: Assumes that all expiring tax credits are extended through 2022 and that the RFS2 mandate is fully filled. CRS has made simplifying assumptions concerning the share of small producer tax credits in order to derive these projections.

Potential Issues with the Expanded RFS

As shown earlier (**Table 1**), most U.S. biofuels production is ethanol produced from corn starch. As a result, as the U.S. ethanol industry has grown over the years, so too has its usage share of the annual corn crop. In 2001, national ethanol production was using about 7% of the U.S. corn crop; by 2005 it was using over 14%. Then, in late 2005, a series of events combined to produce extremely favorable economic returns in the U.S. ethanol sector—MTBE was phased out as a gasoline oxygenate; Hurricane Katrina severely damaged Gulf Coast petroleum importing and refining infrastructure, putting them offline for several months and driving petroleum product prices sharply higher; and meanwhile, corn prices remained relatively low at about $2 per bushel. These factors were followed closely by the passage of EPAct in December 2005, and the establishment of the first Renewable Fuel Standard (RFS), which—by guaranteeing a market for new ethanol production—removed much of the investment risk from the sector.

As a result of these mutually reinforcing factors, investment money flowed into the construction of new ethanol plants, and U.S. ethanol production escalated sharply, reaching a forecast 10.7 billion gallons in 2009. U.S. ethanol production is projected to consume roughly 32% of the 2009 U.S. corn crop. Under the expanded RFS, the 2015 corn ethanol cap of 15 billion gallons would place a call on as much as 40% of the volume of U.S. corn production based on yield and area trends.[18] Such a shift towards greater corn use for biofuels implies higher prices for other corn users, including both the livestock and export sectors (**Figure 3**).

Figure 3. Ethanol Uses an Increasing Share of U.S. Corn Production, Particularly Since 2005, While Feed Use Has Fallen Sharply

(annual U.S. corn disappearance as a percent of total use)

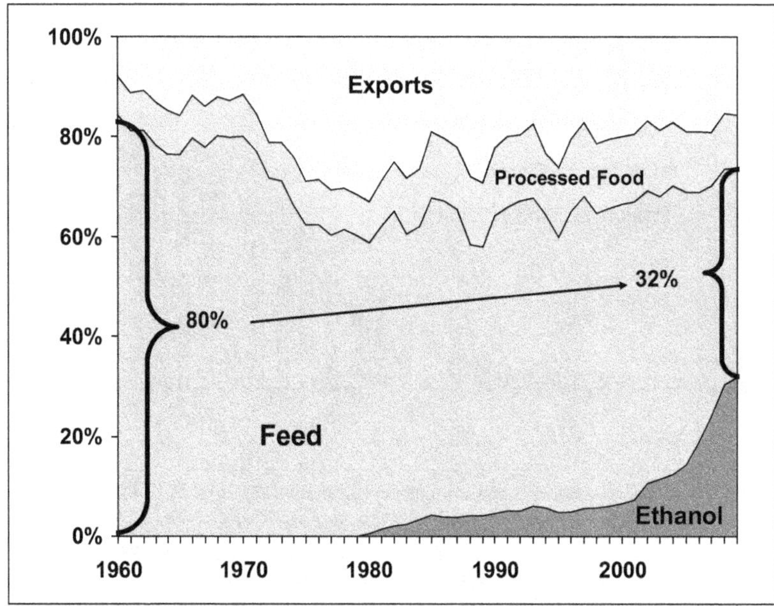

Source: USDA, PSD database, January 12, 2010.

An RFS-driven expansion in biofuels feedstocks (especially corn for grain and stover) is likely to heighten competition for available cropland between biofuels feedstocks and other field crops, as well as to engender an intensification of agricultural activity on U.S. cropland to meet growing demand for food, feed, and fuel resources. This could have consequences for several important agricultural markets including:

- grains—because corn would compete with other grains for land;
- livestock—because the cost of animal feed would likely increase with the price of corn;
- agricultural inputs—because corn is more input-intensive (in terms of fertilizers and pesticides) than other major field crops; and
- land—because the value of cropland, as well as total harvested acreage, would both likely increase.

[18] CRS projection based on the *FAPRI August 2009 Baseline Update for US Agricultural Markets*, FAPRI-MU Report #06-09, August 2009.

In addition to agricultural effects, an increase in corn-based ethanol production would likely have other market effects, including effects on:

- energy markets—because natural gas is a key input in both corn and ethanol production (and should production of biofuels exceed the mandate, then they will compete with traditional petroleum fuels for transportation fuel demand);
- water quality—because the expansion of corn-based ethanol production is likely to involve heavier use of farm chemicals with increased potential for run-off or leaching;
- soil fertility—because several potential biofuels activities (including intensive year-over-year corn production, diversion of corn stover to cellulosic biofuels production and away from field retention as a soil amendment under low-till cultivation, and the expansion of biofuels feedstock cultivation on marginal land) could result in diminished soil fertility and/or increased erosion;
- wildlife habitat—because expanding biofuels feedstock production on marginal lands traditionally left fallow under a conserving practice could compete with wildlife and fowl habitat; and
- federal budget exposure—because applying the federal biofuels production tax credits to the RFS requirements produces a budget liability of nearly $200 billion for the 2009-2022 period.

Overview of Long-Run Corn Ethanol Supply Issues

The ability of the U.S. corn industry to continue to expand production and satisfy the steady growth in demand depends, first and foremost, on continued productivity gains. U.S. corn yields have shown strong, steady growth since the late 1940s, with some acceleration occurring since the mid-1990s as bio-engineered advances in seed technology have heightened drought and pest resistance in corn plants (**Figure 4**). In addition, U.S. cropland planted to corn has increased in recent years from the 1983 low of 60.3 million acres to as high as 93.5 million acres in 2007.

Corn Prices

However, expanding U.S. corn production has only partially offset the rapid growth in demand following the rapid expansion of the U.S. ethanol industry that has occurred since 2005. As a result, corn prices have trended steadily upward in direct relation to the added growth in demand from the ethanol sector (**Figure 5**). Both USDA and the Food and Agricultural Policy Research Institute (FAPRI), in their annual agricultural baseline reports, project corn prices to remain in the $3.75 to $4.00 per bushel range through 2018, compared with an average farm price of $2.15 per bushel during the previous 10-year period (1997-2006).[19]

[19] *USDA Agricultural Projections to 2018*, Long-Term Projections Report, OCE-2009-1, Office of the Chief Economist, February 2009; and *FAPRI August 2009 Baseline Update for US Agricultural Markets*, FAPRI-MU Report #06-09, August 2009.

Figure 4. U.S. Annual Corn Planted Acres and Yield

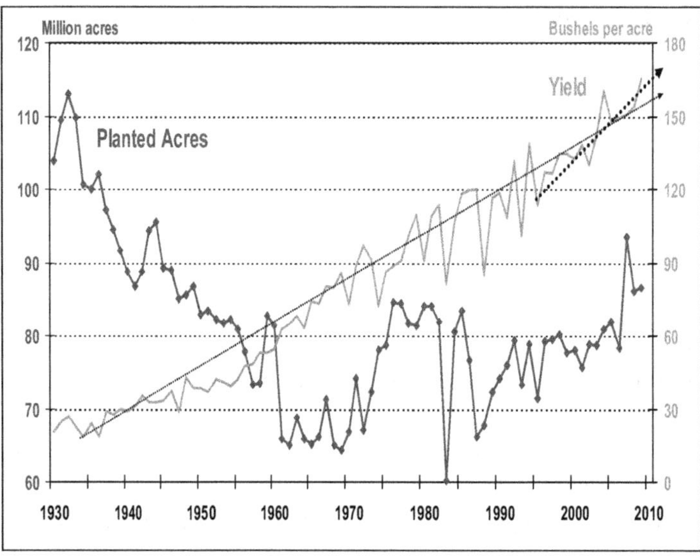

Source: USDA, PSD database, as of January 12, 2010.

Figure 5. Monthly U.S. Corn Prices Have Trended Upward Since Late 2005

(central Illinois cash price for no. 2, yellow corn)

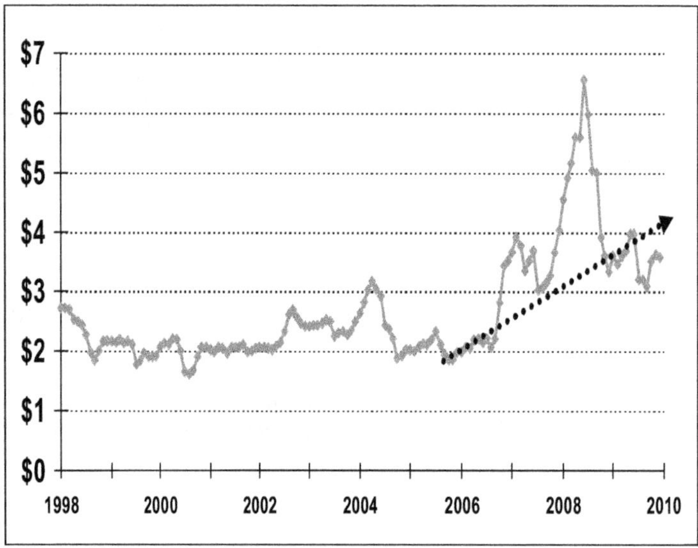

Source: USDA, ERS, Feed Grains Database, at http://www.ers.usda.gov/Data/feedgrains/.

Corn Yields

It is likely that the upward-trending farm prices will encourage continued research investments to move corn yields steadily higher in the future. However, even slight differences in the long-run growth rate portend large impacts in the price outlook. Some economists think that yield increases will slow in coming decades because of land degradation and the impact of climate change. Others suggest that dramatic developments in bio-engineering and seed technology will push corn yields sharply higher. A prime example of the differences in U.S. corn yield outlooks is the contrast between USDA, whose economists project U.S. corn yields to reach about 240 bushels per acre by 2050, compared with the scientists of the biotech seed company Monsanto, who predict that corn yields will be much higher—as much as 300 bushels per acre—by 2030 (20 years earlier than USDA's 240 bushels per acre is achieved).[20] According to USDA, achieving "300-bushel corn" by 2030 would require an extraordinary deviation (a tripling) from both projected and accelerated corn yield trends, and would be historically unprecedented.[21]

Corn Area

Prospects for further expansion in crop area are far less certain, as corn is an energy-intensive crop that prefers deep, fertile soils and timely precipitation. Within the prime corn-growing regions of the Corn Belt, per-acre returns for corn easily dwarf other field crops that vie for the same acreage. Recent seed developments have allowed corn production to expand dramatically into the central and northern Plains states. However, the risk of investing up front in high operating costs to be offset at harvest by strong returns is higher as production moves into less traditional regions, such as the northern Plains, the Delta, and the Southeast.

Corn-Soybean Rotation

The most likely source of new corn acreage will come from shifts in crop rotation from soybeans to corn.[22] However, crop intensification also has its limits. Corn (of the grass family) is traditionally planted in an annual rotation with soybeans (a broad-leaf legume) that offers important agronomic benefits including pest and disease control, as well as enhanced soil fertility.[23] When farmers shift away from this rotation, corn yields tend to suffer. Planting corn-on-corn in two consecutive years usually results in a 10% to 20% yield decline in the second year. As a result, the corn-to-soybean price ratio would have to tilt fairly strongly in favor of corn for this practice to be profitable. Given the limitations on corn area expansion and rotational intensification, it is likely that the sustainable long-run corn planted area is probably in the range of 90 to 95 million acres. If this is the case, then it would mean that future growth in U.S. corn production will be increasingly dependent on yield growth.

---

[20] Philip Brasher, "2050 Corn Harvest Will Affect Food, Fuel Policies," *Des Moines Register*, Nov. 15, 2009.

[21] Paul W. Heisey, "Science, Technology, and Prospects for Growth in U.S. Corn Yields," *Amber Waves*, vol. 7, no. 4, Economic Research Service, USDA, December 2009.

[22] Chad E. Hart, "Feeding the Ethanol Boom: Where Will the Corn Come From?" *Iowa Ag Review*, vol. 12, no. 4 (Fall 2006), pp. 4-5.

[23] Bruce A. Babcock and David A. Hennessy, "Getting More Corn Acres from the Corn Belt," *Iowa Ag Review*, vol. 12, no. 4 (Fall 2006), pp. 6-7.

Overview of Non-Corn-Starch-Ethanol RFS Issues

EISA defines "advanced biofuels" very broadly as biofuels other than corn-starch ethanol; however, the principal focus of advanced biofuels is on biofuels based on cellulosic biomass. Under the RFS2, advanced biofuels use is mandated to reach a minimum of 21 billion gallons by 2022, of which at least 16 billion gallons must be some type of cellulosic biofuels. The cellulosic biofuels RFS mandate begins in 2010 with an initial 100 million gallon requirement.

Potential Advantages of Cellulosic Biofuels

Biofuels produced from cellulosic feedstocks, such as prairie grasses and fast-growing trees or agricultural waste, have the potential to improve the energy and environmental effects of U.S. biofuels while offering significant cost savings on the feedstock production side (e.g., because they are high-yielding, grown on marginal land, perennial rather than annual). Further, moving away from feed and food crops to dedicated energy crops could avoid some of the agricultural supply and price concerns associated with corn ethanol. However, many obstacles must first be overcome before commercially competitive cellulosic biofuels production occurs.[24]

In the near term, it is likely that corn stover[25] will be the primary biomass of choice for cellulosic biofuels production. This is because many ethanol plants already exist in corn production zones and an extension of those plants to include cellulosic biofuels production from stover would offer some scale economies. However, stover-to-biofuel conversion has its own set of potential environmental trade-offs, paramount of which is the dilemma of sacrificing soil fertility gains by harvesting the stover rather than returning it to the soil under no- or minimum-tillage practices.

Cellulosic Biofuels Production Uncertainties

There are substantial uncertainties regarding both the costs of producing cellulosic feedstocks and the costs of producing biofuels from those feedstocks. Dedicated perennial crops are often slow to establish, and it can take several years before a marketable crop is produced. Crops heavy in cellulose tend to be bulky and represent significant problems in terms of harvesting, transporting, and storing. New harvesting machinery would need to be developed to guarantee an economic supply of cellulosic feedstocks.[26] Seasonality issues involving the operation of a biofuels plant year-round based on a four- or five-month harvest period of biomass suggest that bulkiness is likely to matter a great deal. In addition, most marginal lands (i.e., the low-cost biomass production zones) are located far from major urban markets, making it difficult to reconcile plant location with the cost of fuel distribution.

Under current technologies, the cost of the physical conversion process for cellulosic biofuel (including physical, chemical, enzymatic, and microbial treatment and conversion of the biomass feedstocks into motor fuel) remains significantly higher than for corn ethanol or other alternative

[24] For more information, see CRS Report RL34738, *Cellulosic Biofuels: Analysis of Policy Issues for Congress*, by Kelsi Bracmort et al.

[25] Stover is the above-soil part of the corn plant excluding the kernels.

[26] To economically supply field residues to biofuels producers, farm equipment manufacturers likely would need to develop one-pass harvesters that could collect and separate crops and crop residues at the same time.

fuels. Many scientists still suggest that commercialization of cellulosic ethanol is several years down the road.[27]

These uncertainties, plus the financial crisis of 2008 and the ensuing recession and credit crunch, have severely curtailed new investment in the biofuels sector.[28] Some initial investments have been made in small-scale (generally less than 5 million gallons per year) cellulosic ethanol plants, but as of early 2010 no commercial-scale cellulosic biofuel plant is yet online in the United States. An unofficial CRS estimate of operational U.S. cellulosic plant capacity by mid-2010 falls far short of the RFS mandate.[29] As a result, the Environmental Protection Agency may be compelled to issue a waiver of the 2010 cellulosic mandate.[30]

Unintended Policy Outcomes of the "Advanced Biofuels" Mandate

Because the advanced biofuels mandate in the RFS is a fixed mandate, irrespective of prices, the above uncertainties about the production of cellulosic ethanol could have significant implications for fuel supply and fuel prices. If cellulosic ethanol production is unable to advance rapidly enough to meet the RFS mandate for non-corn-starch ethanol, then other unexpected biofuels sources may be forced to step in and fill the void:

- Production of domestic sorghum-starch ethanol may expand across the prairie states and in other regions less suitable for corn production.
- Costly domestic sugar-beet ethanol or biodiesel production may be undertaken to fill the mandate.
- Imports of Brazilian sugar-cane ethanol could expand.

Energy Supply Issues

Biofuels are not primary energy sources. Energy stored in biological material (through photosynthesis) must be converted into a more useful, portable fuel. This conversion requires energy. The amount and types of energy used to produce biofuels (e.g., coal versus natural gas), and the feedstocks for biofuels production (e.g., corn versus cellulosic biomass), are critical in determining a biofuels net energy balance and the environmental benefits of a biofuel.

Energy Balance

To analyze the net energy consumption of ethanol, the entire fuel cycle must be considered. The fuel cycle consists of all inputs and processes involved in the development, delivery and final use of the fuel. For corn-based ethanol, these inputs include the energy needed to produce fertilizers, operate farm equipment, transport corn, convert corn to ethanol, and distribute the final product.

[27] For example, the Department of Energy's goal is to make cellulosic biofuels cost-competitive with corn ethanol by 2012. Other groups are less optimistic.

[28] Robert Wisner, "Cellulosic Ethanol: Will the Mandates be Met?" *AgMRC Renewable Energy Newsletter*, Agricultural Marketing Research Center, Iowa State University, September 2009.

[29] Based on various news media reports.

[30] For more information, see CRS Report RS22870, *Waiver Authority Under the Renewable Fuel Standard (RFS)*, by Brent D. Yacobucci.

USDA estimated an energy output/input ratio of 1.67 based on 2001 survey data (and assuming the then-most-advanced technology for corn and ethanol production)—in other words, the energy contained in a gallon of corn ethanol was 67% higher than the amount of energy needed to produce and distribute it.[31] A 2006 review of research studies (from the 1990s and early 2000s) on ethanol's energy balance and greenhouse gas emissions found that most studies give corn-based ethanol a slight positive energy balance of about 1.2.[32] Ethanol industry sources argue that recent technology and innovation have continued to improve corn ethanol's energy balance.

If feedstocks other than corn are used to produce biofuels, it is expected that lower nitrogen fertilizer use would greatly improve the energy balance. Further, if biomass were used to provide process energy at the biofuels refinery (rather than coal or natural gas), the energy savings would be even greater.[33] Some estimates are that cellulosic ethanol could have an energy balance of 8.0 or more.[34] Similarly high energy balances have been calculated for sugar-cane ethanol and certain types of biodiesel.

Natural Gas Demand

As biofuels production increases, the energy needed to process biomass into liquid fuel can be expected to increase. The resultant increase in energy demand will likely support higher energy prices. The two principal processing fuels used in the United States are natural gas and coal. Other fuels include electricity and biomass.

The United States has been a net importer of natural gas since the early 1980s. A significant increase in its use as a processing fuel in the production of ethanol—and a feedstock for fertilizer production—would likely increase U.S. demand for natural gas, implying higher prices that would reach all natural gas consumers. In the longer run, the U.S. natural gas supply situation is in flux, as recent technological breakthroughs in accessing gas shale have the potential to alter long-run U.S. natural gas supplies.[35]

The EISA RFS proposal boosts corn ethanol production to 15 billion gallons by 2015, requiring an increase in natural gas and/or fertilizer consumption. If the entire 15 billion gallons of corn ethanol were processed using natural gas, the energy requirements would be equivalent to approximately 746 billion cu. ft. of natural gas[36] or slightly more than 3% of total U.S. natural gas consumption, which was an estimated 23.2 trillion cu. ft. in 2008.[37] After 2015, annual eligible

---

[31] Hosein Shapouri et al., "The 2001 Net Energy Balance of Corn-Ethanol," Office of the Chief Economist, USDA, 2004.

[32] Alexander E. Farrell, Richard J. Plevin, Brian T. Turner, Andrew D. Jones, Michael O'Hare, and Daniel M. Kammen, "Ethanol Can Contribute to Energy and Environmental Goals," *Science*, Jan. 27, 2006, pp. 506-508.

[33] "Ethanol Energy Balance," Alternative Fuels & Advanced Vehicles Data Center, Dept. of Energy, available at http://www.afdc.energy.gov/afdc/ethanol/balance.html.

[34] David Andress, *Ethanol Energy Balances*, November 2002.

[35] CRS Report R40894, *Unconventional Gas Shales: Development, Technology, and Policy Issues* , coordinated by Anthony Andrews.

[36] CRS calculations based on energy usage rates of 49,733 Btu/gal of ethanol from Shapouri (2004), roughly 60,000 Btu/gal from Farrell (2006). Hosein Shapouri and Andrew McAloon, *The 2001 Net Energy Balance of Corn-Ethanol*, USDA, Office of the Chief Economist,Washington, 2004; Farrell et al., "Ethanol Can Contribute to Energy and Environmental Goals."

[37] U.S. Department of Energy (DOE), Energy Information Administration (EIA), *Natural Gas Consumption by End* (continued...)

corn-starch ethanol under the RFS is capped at 15 billion gallons and advanced biofuels account for increases in renewable fuel use. At that point, demand for natural gas in the biofuels sector will likely stabilize along with ethanol production.

Energy Security[38]

Despite the fact that ethanol displaces gasoline, the benefits to energy security from ethanol remain relatively small. While roughly 32% of the U.S. corn crop was used for ethanol in 2009, the resultant ethanol only accounts for about 6% of gasoline consumption on an energy-equivalent basis.[39] Expanding corn-based ethanol production to levels needed to significantly promote U.S. energy security is likely to be infeasible. If the entire 2009 record U.S. corn crop of 13.151 billion bushels were used as ethanol feedstock, the resultant 37 billion gallons of ethanol (24.2 billion gasoline-equivalent gallons, or GEG) would represent about 18% of estimated national gasoline use of approximately 138 billion gallons.[40] In contrast, the import share of U.S. liquid fuel consumption (crude oil and other petroleum products) is estimated at 71% in 2007.[41]

An expanded RFS would certainly displace petroleum consumption, but the overall effect on life-cycle fossil fuel consumption is questionable, especially if there is a large reliance on corn-based ethanol. Under the EISA RFS mandate, by 2022 biofuels will likely represent less than 25% of gasoline energy demand.

The specific definition of "advanced biofuels" also affects the overall energy security picture for biofuels. For example an expanded RFS provides an incentive to increase imports of sugar-cane ethanol, especially from Brazil. The expanded RFS also provides an incentive for imports of biodiesel and other renewable diesel substitutes from tropical countries. This would represent a "diversification" of fuel sources, not the "domestication" that some claim is true energy security.

Energy Prices

The effects of the expanded RFS on energy prices are uncertain. If wholesale biofuels prices remain higher than gasoline prices (after all economic incentives are taken into account), then mandating higher and higher levels of biofuels would likely lead to higher gasoline pump prices. However, if petroleum prices—and thus gasoline prices—are high, the use of some biofuels might help to mitigate high gasoline prices.

(...continued)

Use; 2008 data from http://tonto.eia.doe.gov/dnav/ng/ng_cons_sum_dcu_nus_a.htm.

38 A key question in evaluating the energy security benefits or costs of an expanded RFS is "what is the definition of energy security." For many policymakers, "energy security" and "energy independence" (i.e., producing all energy within our borders) are synonymous. For others, "energy security" means guaranteeing that we have reliable supplies of energy regardless of their origin. For this section, the former definition is used.

39 By volume, ethanol accounted for approximately 8% of gasoline consumption in the United States in 2009, but a gallon of ethanol yields only about 68% of the energy of a gallon of gasoline.

40 This estimate is based on USDA's January 12, 2010, *World Agricultural Supply and Demand Estimates (WASDE) Report*, using comparable conversion rates.

41 DOE, EIA, Annual Energy Review 2010, Table A1, "Total Energy Supply and Disposition Summary," Washington, December 14, 2009, at http://www.eia.doe.gov/oiaf/aeo/pdf/appa.pdf.

Current production costs are so high for some biofuels, especially cellulosic biofuels and biodiesel from algae, that significant technological advances—or significant increases in petroleum prices—are necessary to lower their production costs to make them competitive with gasoline. Without cost reductions, mandating large amounts of these fuels would likely raise fuel prices. If a price were placed on greenhouse gas emissions—perhaps through the enactment of a cap and trade bill—then the economics could shift in favor of these fuels despite their high production costs, as they have lower fuel-cycle and life-cycle greenhouse gas emissions (see below).

Greenhouse Gas Emissions

Biofuels proponents argue that a key benefit of biofuels use is a decrease in greenhouse gas (GHG) emissions. However, some question the overall GHG benefit of biofuels, especially corn-based ethanol. There is a wide range of fuel-cycle estimates for greenhouse gas reductions from corn-based ethanol. However, most studies have found that corn-based ethanol reduces fuel-cycle GHG emissions by 10%-20% per mile relative to gasoline.[42] These estimates vary depending on several factors including the cultivation practice (e.g., minimum-tillage versus normal tillage) used to grow the corn and the fuel used to process the corn into ethanol (e.g., natural gas versus coal). These same studies find that biofuels produced from sugar cane or cellulosic biomass could reduce fuel-cycle GHG emissions by as much as 90% per mile relative to gasoline.

However, fuel-cycle analyses generally do not take changes in land use into account. For example, if a previously uncultivated piece of land is tilled to plant biofuels crops, some of the carbon stored in the field could be released. In that case, the overall GHG benefit of biofuels could be compromised.[43] One study estimates that taking land use into account (a life-cycle analysis, as opposed to a fuel-cycle analysis), the GHG reduction from corn ethanol is less than 3% per mile relative to gasoline,[44] while cellulosic biofuels have a life-cycle reduction of 50%.[45] Other recent studies indicate even smaller GHG reductions.

Biofuels produced at facilities commencing operations after the date of enactment must have a 20% life-cycle emissions reduction to qualify under the EISA expanded RFS. However, it is expected that this provision may not be relevant to a large share of conventional ethanol since much of the capacity to meet the 15 billion gallon cap currently exists or will come from expansions of existing plants.

Ethanol Infrastructure and Distribution Issues

In addition to the above concerns about raw material supply for ethanol production (both feedstock and energy), there are issues involving ethanol distribution and infrastructure. Expanding ethanol production likely will strain the existing supply infrastructure. Further,

[42] EPA, Greenhouse Gas Impacts of Expanded Renewable and Alternative Fuels Use, April 2007; Farrell et al.

[43] See Timothy Searchinger, Ralph Heimlich, and R. A. Houghton, et al., "Use of U.S. Croplands for Biofuels Increases Greenhouse Gases Through Emissions from Land-Use Change," *Science*, vol. 319, no. 5867 (February 2008).

[44] Mark A. Delucchi, Draft Report: Life Cycle Analyses of Biofuels, 2006.

[45] While a 50% life-cycle reduction is still significant, it is far less than the 90% reduction suggested some by fuel-cycle analyses.

expansion of ethanol use beyond the current 10% blend will require investment in entirely new infrastructure that would be necessary to handle an increasing percentage of ethanol in gasoline. If biomass-based diesel substitutes are produced in much larger quantities, some of these infrastructure issues may be mitigated.

Distribution Issues

Ethanol-blended gasoline tends to separate in pipelines due to the presence of water in the lines. Further, ethanol is corrosive and may damage existing pipelines and storage tanks. Therefore, unlike petroleum products, ethanol and ethanol blended gasoline cannot be shipped by pipeline in the United States. Another issue with pipeline transportation is that corn ethanol must be moved from rural areas in the Midwest to more populated areas, which are often located along the coasts. This shipment is in the opposite direction of existing pipeline transportation, which moves gasoline from refiners along the coast to other coastal cities and into the interior of the country. While some studies have concluded that shipping ethanol or ethanol-blended gasoline via pipeline could be feasible, no major U.S. pipeline has made the investments to allow such shipments.[46]

Thus, the current distribution system for ethanol is dependent on rail cars, tanker trucks, and barges. These deliver ethanol to fuel terminals where it is blended with gasoline before shipment via tanker truck to gasoline retailers. However, these transport modes lead to prices higher than for pipeline transport, and the supply of current shipping options (especially rail cars) is limited. For example, according to industry estimates, the number of ethanol carloads has tripled between 2001 and 2006, and the number is expected to increase by another 30% in 2007, although final data is not yet available.[47] A significant increase in corn-based ethanol production would further strain this tight transport situation.

Because of these distribution issues, some pipeline operators are seeking ways to make their systems compatible with ethanol or ethanol-blended gasoline. These modifications could include coating the interior of pipelines with epoxy or some other, corrosion-resistant material. Another potential strategy could be to replace all susceptible pipeline components with newer, hardier components. However, even if such modifications are technically possible, they likely will be expensive, and could further increase ethanol transportation costs.

As non-corn biofuels play a larger role, as required in EISA, some of the supply infrastructure concerns may be alleviated. Cellulosic biofuels potentially can be produced from a variety of feedstocks, and may not be as dependent on a single crop from one region of the country. For example, municipal solid waste is ubiquitous across the United States, and could serve as a ready feedstock for biofuels production if the technology were developed to convert it economically to fuel. Further, increased imports of biofuels from other countries could allow for greater use of biofuels, especially along the coasts. Moreover, some biofuels, especially some diesel substitutes, may be able to be mixed with petroleum fuels at the refinery and placed directly into the pipeline.

---

[46] Some small, proprietary ethanol pipelines do exist. American Petroleum Institute, *Shipping Ethanol Through Pipelines*. Available at http://www.api.org/aboutoilgas/sectors/pipeline/upload/pipelineethanolshipment-2.doc.

[47] Ilan Brat and Daniel Machalaba, "Can Ethanol Get a Ticket to Ride?," *The Wall Street Journal*, Feb. 1, 2007, p. B1.

Higher-Level Ethanol Blends

More than half of all U.S. gasoline contains some ethanol (mostly blended at the 10% level or lower). U.S. ethanol consumption in 2009 is estimated at 10.7 billion gallons, which was blended into roughly 138 billion gallons of gasoline. This represents only about 8% of annual gasoline demand on a volume basis, and only about 5% on an energy basis (since ethanol contains roughly 68% of the energy content of petroleum-based gasoline).

One key benefit of gasoline-ethanol blends up to 10% ethanol is that they are compatible with existing vehicles and infrastructure (fuel tanks, retail pumps, etc.). All automakers that produce cars and light trucks for the U.S. market warranty their vehicles to run on gasoline with up to 10% ethanol (E10). This 10% currently is an upper bound (sometimes referred to as the "blend wall") to the amount of ethanol that can be introduced into the gasoline pool. If most or all gasoline in the country contained 10% ethanol, this would allow only for roughly 15 billion gallons, far less than the amount of biofuels mandated in EISA. In response to the impending "blend wall," on March 6, 2009, Growth Energy (a biofuels advocacy consultancy) and 54 ethanol manufacturers submitted a waiver application to the Environmental Protection Agency (EPA) to increase the allowable ethanol content of gasoline to 15%.[48] The waiver request remains under review by EPA, pending vehicle test results. A decision may be forthcoming by mid-June.

As a major producer of ethanol for its domestic market, Brazil has a mandate that all of its gasoline contain 20%-25% ethanol. For the United States to move to E20 (20% ethanol, 80% gasoline), it may be that few (if any) modifications would need to be made to existing vehicles and infrastructure. Vehicle testing, however, would be necessary to determine whether new vehicle parts would be required, or if existing vehicles are compatible with E20. Similar testing would be necessary for terminal tanks, tanker trucks, retail storage tanks, pumps, and the like. In addition, EPA would need to certify that the fuel will not lead to increased air quality problems.

There is also interest in expanding the use of E85 (85% ethanol, 15% gasoline). Current E85 consumption represents only about 1% of ethanol consumption in the United States. A key reason for the relatively low consumption of E85 is that relatively few vehicles operate on E85. The National Ethanol Vehicle Coalition estimates that there are approximately 6 million E85-capable vehicles on U.S. roads,[49] as compared to approximately 230 million gasoline- and diesel-fueled vehicles.[50] Most E85-capable vehicles are "flexible fuel vehicles" or FFVs. An FFV can operate on any mixture of gasoline and between 0% and 85% ethanol. However, ethanol has a lower per-gallon energy content than gasoline. Therefore, FFVs tend to have lower fuel economy when operating on E85. For the use of E85 to be economical, the pump price for E85 must be low enough to make up for the decreased fuel economy relative to gasoline. Generally, to have equivalent per-mile costs, E85 must cost 20% to 30% less per gallon at the pump than gasoline. Owners of a large majority of the FFVs on U.S. roads choose to fuel them exclusively with gasoline, largely due to higher per-mile fuel cost and lower availability of E85.

E85 capacity is expanding rapidly, with the number of E85 stations nearly tripling between January 2006 and January 2008. As of early 2009, there were an estimated 1,900 retail E85

[48] For more information on the waiver request, see EPA at http://www.epa.gov/otaq/additive.htm.

[49] National Ethanol Vehicle Coalition, *Frequently Asked Questions*, accessed February 3, 2006, at http://www.e85fuel.com/e85101/faq.php.

[50] Federal Highway Administration, *Highway Statistics 2003*, November 2004, Washington.

stations in the United States (out of 170,000 stations nationwide).[51] E85 stations represent slightly more than 1% of U.S. gasoline retailers. Further expansion will require significant investments, especially at the retail level. Installation of a new E85 pump and underground tank can cost as much as $100,000 to $200,000.[52] However, if existing equipment can be used with little modification, the cost could be less than $10,000.

Vehicle Infrastructure Issues

As was stated above, if a large portion of any increased RFS is met using ethanol, then the United States likely does not have the vehicles to consume the fuel. The 10% blend wall on ethanol in gasoline for conventional vehicles poses a significant barrier to expanding ethanol consumption beyond 15 billion gallons per year.[53] To allow more ethanol use, vehicles will need to be certified and warranted for higher-level ethanol blends, or the number of ethanol FFVs will need to increase. Turnover of the U.S. automobile fleet is likely to slow during the recession, making it more difficult to integrate FFVs into the fleet.

Conclusion

There is continuing interest in expanding the U.S. biofuels industry as a strategy for promoting energy security and achieving environmental goals. However, it is possible that increased biofuel production may place desired policy objectives in conflict with one another. There are limits to the amount of biofuels that can be produced from current feedstocks and questions about the net energy and environmental benefits they might provide. Further, rapid expansion of biofuels production may have many unintended and undesirable consequences for agricultural commodity costs, fossil energy use, and environmental degradation. Owing to these concerns, alternative strategies for energy conservation and alternative energy production are widely seen as warranting consideration.

Author Contact Information

Randy Schnepf
Specialist in Agricultural Policy
rschnepf@crs.loc.gov, 7-4277

Brent D. Yacobucci
Specialist in Energy and Environmental Policy
byacobucci@crs.loc.gov, 7-9662

[51] Renewable Fuels Association, at http://www.ethanolrfa.org/resource/e85/.

[52] David Sedgwick, *Automotive News*, January 29, 2007. p. 112.

[53] Note that 15 billion gallons is the corn starch ethanol limit for the expanded RFS in the EISA.

Regulatory Announcement

EPA Finalizes Regulations for the National Renewable Fuel Standard Program for 2010 and Beyond

The U.S. Environmental Protection Agency is finalizing revisions to the National Renewable Fuel Standard program (commonly known as the RFS program). This rule makes changes to the Renewable Fuel Standard program as required by the Energy Independence and Security Act of 2007 (EISA). The revised statutory requirements establish new specific annual volume standards for cellulosic biofuel, biomass-based diesel, advanced biofuel, and total renewable fuel that must be used in transportation fuel. The revised statutory requirements also include new definitions and criteria for both renewable fuels and the feedstocks used to produce them, including new greenhouse gas (GHG) emission thresholds as determined by lifecycle analysis. The regulatory requirements for RFS will apply to domestic and foreign producers and importers of renewable fuel used in the U.S.

Key Actions

This final action lays the foundation for achieving significant reductions of greenhouse gas emissions from the use of renewable fuels, reductions of imported petroleum and further development and expansion of our nation's renewable fuels sector.

This action is also setting the 2010 RFS volume standard at 12.95 billion gallons (bg). Further, for the first time, EPA is setting volume standards for specific categories of renewable fuels including cellulosic, biomass-based diesel, and total advanced renewable fuels. For 2010, the cellulosic standard is being set at 6.5 million gallons (mg); the biomass-based diesel standard is being set at 1.15 bg, (combining the 2009 and 2010 standards as proposed).

United States
Environmental Protection
Agency

Office of Transportation and Air Quality
EPA-420-F-10-007
February 2010

Regulatory Announcement

In order to qualify for these new volume categories, fuels must demonstrate that they meet certain minimum greenhouse gas reduction standards, based on a lifecycle assessment, in comparison to the petroleum fuels they displace.

For its final determinations, EPA is using the best available models and has incorporated updated information based on:

- significant new scientific data available to the agency
- rigorous independent peer review
- extensive public comments

For the fuel pathways modeled, the following meet or exceed the respective required minimum GHG reduction standards:

- corn based ethanol plants using new efficient technologies
- soy based biodiesel
- biodiesel made from waste grease, oils, and fats
- sugarcane based ethanol

Fuels derived from cellulosic materials meet, and generally significantly exceed, the minimum GHG reduction standard.

Additional information on these and other key changes can be found below.

New Renewable Volume Standards

This final rule revises the annual renewable fuel standards (RFS2) and makes the necessary program modifications as set forth in EISA. Of these modifications, several are notable. First, the required renewable fuel volume continues to increase under RFS2, reaching 36 bg by 2022. The following chart shows the volume requirements from EISA:

2

Regulatory Announcement

EISA Renewable Fuel Volume Requirements (billion gallons)

Year	Cellulosic biofuel requirement	Biomass-based diesel requirement	Total Advanced biofuel requirement	Total renewable fuel requirement
2008	n/a	n/a	n/a	**9.0**
2009	n/a	0.5	0.6	**11.1**
2010	0.1	0.65	0.95	**12.95**
2011	0.25	0.80	1.35	**13.95**
2012	0.5	1.0	2.0	**15.2**
2013	1.0	a	2.75	**16.55**
2014	1.75	a	3.75	**18.15**
2015	3.0	a	5.5	**20.5**
2016	4.25	a	7.25	**22.25**
2017	5.5	a	9.0	**24.0**
2018	7.0	a	11.0	**26.0**
2019	8.5	a	13.0	**28.0**
2020	10.5	a	15.0	**30.0**
2021	13.5	a	18.0	**33.0**
2022	16.0	a	21.0	**36.0**
2023+	b	b	b	**b**

[a] *To be determined by EPA through a future rulemaking, but no less than 1.0 billion gallons.*

[b] *To be determined by EPA through a future rulemaking.*

EISA Expands Coverage to Include Diesel and Nonroad Fuels

EISA expanded the RFS program beyond gasoline to generally cover all transportation fuel. This now includes gasoline and diesel fuel intended for use in highway vehicles and engines, and nonroad, locomotive and marine engines. These provisions continue to apply to refiners, blenders, and importers of transportation fuel (with limited flexibilities for small refiners), and their percentage standards apply to the total amount of gasoline and diesel they produce for such use.

2010 Standards

For 2010, EISA set a total renewable fuel standard of 12.95 billion gallons. This total volume, presented as a fraction of a refiner's or importer's gasoline and diesel volume, must be renewable fuel. The final 2010 standards are shown in below.

3

Regulatory Announcement

Standards for 2010

Fuel Category	Percentage of Fuel Required to be Renewable	Volume of Renewable Fuel (in billion gal)
Cellulosic biofuel	0.004%	0.0065
Biomass-based diesel	*1.10%	*1.15
Total Advanced biofuel	0.61%	0.95
Renewable fuel	8.25%	12.95

**Combined 2009/2010 Biomass-Based Diesel Volumes Applied in 2010*

Setting the 2010 Cellulosic Standard

EISA requires the Administrator to evaluate and make an appropriate market determination for setting the cellulosic standard each year. Based on an updated market analysis considering detailed information from pilot and demonstration scale plants, an Energy Information Administration analysis, and other publically and privately available market information, we are setting the 2010 cellulosic biofuel standard at 6.5 million ethanol-equivalent gallons. While this volume is significantly less than that set forth in EISA for 2010, a number of companies and projects appear to be poised to expand production over the next several years. Since the cellulosic standard is lower than the level otherwise required by EISA, we will also make cellulosic credits available to obligated parties for end-of-year compliance, should they need them, at a price of $1.56 per gallon (gallon-RIN). In addition, while we have lowered the cellulosic standard below the level otherwise required in the Act, we have maintained the advanced biofuel and total renewable standards as that set in EISA for 2010. We are continuing to assess the growth of the cellulosic biofuel industry and intend to issue a notice of proposed rulemaking (NPRM) each spring and a final rule by November 30 of each year to set the renewable fuel standards for each ensuing year.

Treatment of Biomass-based Diesel in 2010

This rule also includes special provisions to account for the 2009 biomass-based diesel volume requirements in EISA. As described in the final rule, in November 2008 we used the new total renewable fuel volume of 11.1 billion gallons from EISA as the basis for the 2009 total renewable fuel standard that we issued under the RFS1 regulations. While this approach ensured that the total mandated renewable fuel volume required by EISA for 2009 was used, the RFS1 regulatory structure did not provide a mechanism for implementing the 0.5 billion gallon 2009 requirement for biomass-based diesel. We are addressing this issue in this rule combining the 2010 biomass-based diesel requirement of 0.65 billion gallons with the 2009 biomass-based diesel requirement of 0.5 billion gallons to require that obligated parties meet a combined 2009/2010 requirement of 1.15 billion gallons by the end of the 2010 compliance year.

Greenhouse Gas Reduction Thresholds

EISA established new renewable fuel categories and eligibility requirements, including setting the first mandatory GHG reduction thresholds for the various categories of fuels. A significant aspect of the RFS2 program is the requirement that the lifecycle GHG emissions of a qualifying

Regulatory Announcement

renewable fuel must be less than the lifecycle GHG emissions of the 2005 baseline average gasoline or diesel fuel that it replaces. Four different levels of reductions are required for the four different renewable fuel standards. These lifecycle performance improvement thresholds are listed in the table below:

Lifecycle GHG Thresholds Specified in EISA

(Percent reduction from 2005 baseline)

Renewable fuel[a]	**20%**
Advanced biofuel	**50%**
Biomass-based diesel	**50%**
Cellulosic biofuel	**60%**

[a] The 20% criterion generally applies to renewable fuel from new facilities that commenced construction after December 19, 2007.

Compliance with each threshold requires a comprehensive evaluation of renewable fuels, as well as the baseline for gasoline and diesel, on the basis of their lifecycle emissions. As mandated by EISA, the greenhouse gas emissions assessments must evaluate the aggregate quantity of greenhouse gas emissions (including direct emissions and significant indirect emissions such as significant emissions from land use changes) related to the full lifecycle, including all stages of fuel and feedstock production, distribution and use by the ultimate consumer.

EPA's lifecycle methodology required breaking new scientific ground and using analytical tools in new ways. Throughout the development of EPA's lifecycle analysis, the Agency employed a collaborative, transparent, and science-based approach. EPA recognizes that as the state of scientific knowledge continues to evolve in this area, the lifecycle GHG assessments for a variety of fuel pathways are likely to be updated. Therefore, while EPA is using its current lifecycle assessments to inform the regulatory determinations for fuel pathways in this final rule, as required by the statute, the Agency is also committing to further reassess these determinations and lifecycle estimates.

Based on the Agency's current modeling of specific fuel pathways, which incorporated comments received through the third-party peer review process, and data and information from new studies and public comments, EPA has determined that:

- Ethanol produced from corn starch at a new (or expanded capacity from an existing) natural gas-fired facility using advanced efficient technologies that we expect will be most typical of new production facilities complies with the 20% GHG emission reduction threshold
- Biobutanol from corn starch complies with the 20% GHG threshold
  Ethanol produced from sugarcane complies with the applicable 50% GHG reduction threshold for the advanced fuel category
- Biodiesel from soy oil and renewable diesel from waste oils, fats, and greases complies with the 50% GHG threshold for the biomass-based diesel category
- Diesel produced from algal oils complies with the 50% GHG threshold for the biomass-based diesel category
- Cellulosic ethanol and cellulosic diesel (based on currently modeled pathways) comply with the 60% GHG reduction threshold applicable to cellulosic biofuels

5

Regulatory Announcement

In addition to finalizing a threshold compliance determination for those pathways shown above that we specifically modeled, our technical judgment indicates certain other pathways are likely to be similar enough to modeled pathways that we are also assured these similar pathways qualify. Further, for other fuels we are establishing a process whereby a biofuel producer can petition the Agency to consider whether their product would be eligible for use in complying with an EISA standard. For additional information on the lifecycle GHG emissions methodology and results for renewable fuel pathways, and details on the petition process, please refer to the Lifecycle GHG Analysis Fact Sheet, EPA420-F-10-006 or the RFS2 preamble.

Requirements for Feedstock Producers

EISA changed the definition of renewable fuel to require that it be made from feedstocks that qualify as "renewable biomass." EISA's definition of the term "renewable biomass" limits the types of biomass as well as the types of land from which the biomass may be harvested. The definition generally applies restrictions to two feedstock sectors: the agricultural sector (planted crops and crop residues) and the non-agricultural sector (planted trees and tree residues, animal waste material and byproducts, slash and pre-commercial thinnings). These definitions affect feedstock use for production of compliant renewable fuels.

In the RFS2 rule, EPA is finalizing details applicable to renewable fuel producers which are necessary to implement this requirement. For both domestic and foreign non-agricultural sector feedstocks, renewable fuel producers can comply with specific recordkeeping and reporting requirements for their individual facilities by collecting and maintaining appropriate records from their feedstock suppliers that their feedstocks comply with the renewable biomass requirement. Producers may also, as an alternative to these individual recordkeeping and reporting requirements, opt to form a consortium to fund an independent third party to conduct annual renewable biomass quality-assurance surveys, based on a plan approved by EPA.

For agriculturally-based feedstocks produced in the U.S., renewable fuel producers will be in compliance based on EPA's aggregate compliance determination. EPA will monitor agricultural land data yearly and should the baseline level of approved agricultural land be exceeded, the individual recordkeeping and reporting requirements imposed on the non-agricultural sector would then be required. The program also provides an option for a similar, future aggregate determination for renewable fuel produced from foreign-based agricultural feedstocks, if the source region can provide sufficient data to support an effective aggregate analysis and monitoring program. Otherwise, foreign producers must verify using one of the approaches applied in the non-agricultural sector.

Overview of Impacts of Increasing Volume Requirements in the RFS2 Program

The increased use of renewable fuels required by the RFS2 standards is expected to reduce dependence on foreign sources of crude oil, increase domestic sources of energy, while at the same time providing important reductions in greenhouse gas emissions that contribute to climate change.

6

Regulatory Announcement

Petroleum Consumption, Energy Security and Fuel Costs

We estimate that the increased use of renewable fuels needed to reach the 36 billion gallons mandated by 2022 relative to market projections in the absence of the mandate will displace about 13.6 billion gallons of petroleum-based gasoline and diesel fuel. This represents about 7 percent of expected annual gasoline and diesel consumption in 2022. Furthermore, we expect the rule to decrease oil imports by $41.5 billion, and to result in additional energy security benefits of $2.6 billion. By 2022, the increased use of renewable fuels is expected to decrease gasoline costs by 2.4 cents per gallon and to decrease diesel costs by 12.1 cents per gallon.

Greenhouse Gas Emissions

The expanded use of renewable fuels is expected to reduce greenhouse gas emissions by 138 million metric tons when the program is fully implemented in 2022. The reductions would be equivalent to taking about 27 million vehicles off the road.

Emissions and Air Quality

The increased use of renewable fuels will also impact emissions with some emissions such as hydrocarbons, nitrogen oxides (NOx), acetaldehyde and ethanol expected to increase and others such as carbon monoxide (CO) and benzene expected to decrease. However, the impacts of these emissions on criteria air pollutants are highly variable from region to region. Overall the emission changes are projected to lead to increases in population-weighted annual average ambient PM and ozone concentrations, which in turn are anticipated to lead to up to 245 cases of adult premature mortality.

Agriculture Sector and Related Impacts

In 2022, the increased use of renewable fuels is expected to expand the market for agricultural products such as corn and soybeans and open new markets for advanced biofuels. We estimate that the RFS2 program would increase net farm income by $13 billion dollars – or more than 36 percent -- in 2022. We also expect corn exports to decrease by 8 percent, and soybean exports to decrease by 14 percent.

The rule is expected to increase the cost of food $10 per person in 2022.

For More Information

For more information on the final RFS2 rule please visit the RFS website at: http://www.epa.gov/otaq/renewablefuels/index.htm

Contact EPA's Office of Transportation and Air Quality, Assessment and Standards Division information line at: asdinfo@epa.gov, or (734) 214-4636

7

Technical Highlights

E85 and Flex Fuel Vehicles

Ethanol is a renewable fuel made from plants. Essentially non-drinkable grain alcohol, ethanol is produced by fermenting plant sugars. It can be made from corn, sugar cane, and other starchy agricultural product. The cellulose in agricultural wastes such as waste woods and corn stalks (also know as "cellulosic ethanol") can also be used as a base. In the United States, most ethanol is currently made from corn, although because of rapidly developing research, cellulosic ethanol may soon become a larger part of the market.

E85 Fuel

While pure ethanol is rarely used for transportation fuel, there are several ethanol-gasoline blends in use today. E85 is a blend of 85 percent denatured ethanol and 15 percent gasoline. In certain areas, higher percentages of gasoline will be added to E-85 during the winter to ensure that vehicles are able to start at very cold temperatures.

E85 cannot be used in a conventional, gasoline-only engine. Vehicles must be specially designed to run on it. The only vehicles currently available to U.S. drivers are known as flex fuel vehicles (FFVs), because they can run on E85, gasoline, or any blend of the two. Much like diesel fuel, E85 is available at specially-marked fueling pumps. Today, nearly 700 fueling stations offer it.

Another common mix is E10, a blend of 10 percent ethanol and 90 percent gasoline. E10 is available in many areas across the United States and can be used in any gasoline vehicle manufactured after 1980.

Office of Transportation and Air Quality
EPA-420-F-09-065
October 2009

Technical Highlights

Flex Fuel Vehicles

Ethanol-fueled vehicles date back to the 1880s when Henry Ford designed a car that ran solely on ethanol. He later built the first flex fuel vehicle: a 1908 Model T designed to operate on either ethanol or gasoline.

Today's FFVs feature specially-designed fuel systems and other components that allow a vehicle to operate on a mixture of gasoline and ethanol that can vary from 0 percent to 85 percent ethanol. These cars and trucks have the same power, acceleration, payload, and cruise speed as conventionally fueled vehicles. Maintenance for ethanol-fueled vehicles is very similar to that of regular cars and trucks. However, owners should identify the car as an FFV when ordering replacement parts.

Today, the United States has more than 6 million FFVs on the road. These vehicles are available in a range of models, including sedans, pick-up trucks, and minivans. Additionally, several auto manufacturers have announced plans to greatly expand the number of FFV models they will offer. In fact, you may even be driving one now. To find out, check the inside of your gas tank door for an identification sticker.

Affordability

FFVs are priced the same as gasoline-only vehicles, offering drivers the opportunity to buy an E85 capable vehicle at no additional cost.

In general, E85 reduces fuel economy and range by about 20-30 percent, meaning an FFV will travel fewer miles on a tank of E85 than on a tank of gasoline. This is because ethanol contains less energy than gasoline. Vehicles can be designed to be optimized for E85--which would reduce or eliminate this tendency. However, no such vehicles are currently on the market. The pump price for E85 is often lower than regular gasoline; however, prices vary depending on supply and market conditions.

E85 & Conventional Vehicles

Consumers should never use E85 in a conventional, gasoline-only vehicle. This can lead to a range of problems, including not being able to start the engine, damage to engine components, illumination of the check engine light, and emissions increases.

It is technically possible to convert a conventional gasoline vehicle to run on E85; however, such conversions would likely be illegal unless they are certified by the U.S. Environmental Protection Agency (EPA). In addition, converting a conventional vehicle to E85 may violate the terms of the vehicle warranty. For more information on the vehicle conversion processs, please visit EPA's Web site at: www.epa.gov/otaq/cert/dearmfr/cisd0602.pdf.

Technical Highlights

Benefits

Much of the increased interest in ethanol as a vehicle fuel is due to its ability to replace gasoline from imported oil. The United States is currently the world's largest ethanol producer, and most of the ethanol we use is produced domestically from corn grown by American farmers.

E85 also provides important reductions in greenhouse gas (GHG) emissions. When made from corn, E85 reduces lifecycle GHG emissions (which include the energy required to grow and process corn into ethanol) by 15-20% as compared to gasoline. E85 made from cellulose can reduce emissions by around 70 percent as compared to gasoline.

EPA's stringent Tier II vehicle emission standards require that FFVs achieve the same low emissions level regardless of whether E85 or gasoline is used. However, E85 can further reduce emissions of certain pollutants as compared to conventional gasoline or lower volume ethanol blends. For example, E85 is less volatile than gasoline or low volume ethanol blends, which results in fewer evaporative emissions. Using E85 also reduces carbon monoxide emissions and provides significant reductions in emissions of many harmful toxics, including benzene, a known human carcinogen. However, E85 also increases emissions of acetaldehyde--a toxic pollutant. EPA is conducting additional analysis to expand our understanding of the emissions impacts of E85.

3

Renewable Energy Programs in the 2008 Farm Bill

Megan Stubbs
Analyst in Agricultural Conservation and Natural Resources Policy

September 30, 2009

Congressional Research Service
7-5700
www.crs.gov
RL34130

CRS Report for Congress
Prepared for Members and Committees of Congress

Summary

The Food, Conservation, and Energy Act of 2008 (P.L. 110-246, the 2008 farm bill) extends and expands many of the renewable energy programs originally authorized in the 2002 farm bill. The bill also continues the emphasis on the research and development of advanced and cellulosic bioenergy authorized in the 2007 Energy Independence and Security Act (P.L. 110-140).

Farm bill debate over U.S. biomass-based renewable energy production policy focused mainly on the continuation of subsidies for ethanol blenders, continuation of the import tariff for ethanol, and the impact of corn-based ethanol on agriculture. The enacted bill requires reports on the economic impacts of ethanol production, reflecting concerns that the increasing share of corn production being used for ethanol had contributed to high commodity prices and food price inflation.

Title VII, the research title of the 2008 farm bill, contains numerous renewable energy related provisions that promote research, development, and demonstration of biomass-based renewable energy and biofuels. The Sun Grant Initiative coordinates and funds research at land grant institutions on biobased energy technologies. The Agricultural Bioenergy Feedstock and Energy Efficiency Research and Extension Initiative provides support for on-farm biomass energy crop production research and demonstration.

Title IX, the energy title of the farm bill, authorizes mandatory funds (not subject to appropriations) of $1.1 billion, and discretionary funds (subject to appropriations) totaling $1.0 billion, for the FY2008-FY2012 period. Energy grants and loans provided through initiatives such as the Bioenergy Program for Advanced Biofuels promote the development of cellulosic biorefinery capacity. The Repowering Assistance Program supports increasing efficiencies in existing refineries. Programs such as the Rural Energy for America Program (REAP) assist rural communities and businesses in becoming more energy-efficient and self-sufficient, with an emphasis on small operations. The Biomass Crop Assistance Program, the Biorefinery Assistance Program, and the Forest Biomass for Energy Program provide support to develop alternative feedstock resources and the infrastructure to support the production, harvest, storage, and processing of cellulosic biomass feedstocks. Cellulosic feedstocks—for example, switchgrass and woody biomass—are given high priority both in research and funding.

Title XV of the 2008 farm bill contains tax and trade provisions. It continues current biofuels tax incentives, reducing those for corn-based ethanol but expanding tax credits for cellulosic ethanol. The tariff on ethanol imports is extended.

Implementation of the farm bill's energy provisions is underway. The stimulus package passed by Congress in February 2009 provided additional funds for some energy programs. President Obama, in May 2009, directed the U.S. Department of Agriculture (USDA) and the Department of Energy (DOE) to accelerate implementation of renewable energy programs. Notices have appeared in the *Federal Register* soliciting applications for the Biorefinery Program, the Rural Energy for America Program, and the Biomass Crop Assistance Program.

Contents

Tables

Appendixes

Contacts

Background

Renewable energy policy in the 2008 farm bill builds on earlier programs, many of which were established in the 2002 farm bill. That bill, the Farm Security and Rural Investment Act of 2002 (P.L. 107-171, FSRIA), was the first omnibus farm bill to explicitly include an energy title (Title IX). The energy title authorized grants, loans, and loan guarantees to foster research on agriculture-based renewable energy, to share development risk, and to promote the adoption of renewable energy systems.[1] For example, the U.S. Department of Agriculture's (USDA's) Bioenergy Program (Sec. 9006 of P.L. 107-171)—whose funding expired in FY2006—provided direct incentives to expand actual production of bioenergy.

Since enactment of the 2002 farm bill, interest in renewable energy has grown rapidly, due in large part to a strong rise in domestic and international petroleum prices and a dramatic acceleration in domestic biofuels production (primarily corn-based ethanol).[2] Many policymakers view agriculture-based biofuels as both a catalyst for rural economic development and a response to growing energy import dependence. Ethanol and biodiesel, the two most widely used biofuels, receive significant federal support in the form of tax incentives, loans and grants, and regulatory programs.[3]

The 2008 farm bill became law six months after the enactment of the Energy Independence and Security Act of 2007 (EISA, P.L. 110-140), and many of its provisions also build on the goals of EISA.[4] The emphasis on facilitating production of biofuels derived from cellulosic feedstocks reflects the goals of the renewable fuels standard (RFS) in EISA. EISA includes a significant expansion of the RFS to 36 billion gallons by 2022, with carve-outs for biodiesel (1 billion gallons by 2012) and cellulosic ethanol (16 billion gallons by 2022) and an implicit cap on corn starch ethanol (15 billion gallons by 2015). Provisions in the farm bill reflect the increased role for biofuels mandated by the expansion of the RFS and its likely impact on the U.S. agriculture sector.[5]

The emphasis on cellulosic ethanol also reflects increasing concerns about the economic and environmental issues associated with corn starch-based ethanol.[6] Record high commodity prices in 2007 and mid-2008, combined with high energy costs, resulted in sharp increases in livestock feed costs, export prices, and domestic food price inflation. For the first time, an agricultural commodity is directly competing with petroleum in the marketplace. Ethanol production, the profitability of which depends directly on both petroleum and corn prices, accounts for about a third of U.S. corn production. The increase in corn used for U.S. ethanol production exceeds the

---

[1] For more information, see CRS Report RL32712, *Agriculture-Based Renewable Energy Production*, by Randy Schnepf. For details of previous law, see USDA, *2002 Farm Bill*, "Title IX—Energy," available at http://www.ers.usda.gov/Features/Farmbill/titles/titleIXenergy.htm.

[2] For more information on agriculture and bioenergy, see CRS Report RL32712, *Agriculture-Based Renewable Energy Production*; and CRS Report RL33290, *Fuel Ethanol: Background and Public Policy Issues*.

[3] For a listing of federal incentives in support of biofuels production, see CRS Report R40110, *Biofuels Incentives: A Summary of Federal Programs*.

[4] For more information, see CRS Report RL34239, *Biofuels Provisions in the 2007 Energy Bill and the 2008 Farm Bill: A Side-by-Side Comparison*.

[5] For more information, see CRS Report RL34265, *Selected Issues Related to an Expansion of the Renewable Fuel Standard (RFS)*.

[6] For more information, see CRS Report RL34738, *Cellulosic Biofuels: Analysis of Policy Issues for Congress*.

increase in corn produced during the past three years. When petroleum prices rise, so does demand for ethanol as a substitute, which in turn increases both the demand for and price of corn. The "food versus fuel" debate intensified during the farm bill debate as food price inflation accelerated both in the U.S. and globally—highlighting some of the potential problems associated with replacing even a small share of the nation's gasoline consumption with corn-based ethanol. Competition for limited corn supplies between livestock producers, ethanol refiners, exporters, and other domestic users have resulted in calls for at least a partial waiver of the RFS in 2009.[7]

Several of the federal programs that currently support renewable energy production in general, and agriculture-based energy production in particular, are outside the purview of USDA and have legislative origins outside of the farm bill.[8] For example, the RFS mandates the inclusion of an increasing volume of biofuels in the national fuel supply. This originated with the Energy Policy Act of 2005 (P.L. 109-58) and was more recently expanded in EISA. Similarly, the federal tax credits available to biofuel blenders and wind energy producers were initially contained in the American Jobs Creation Act of 2004 (P.L. 108-357), although they have been incorporated in the farm bill.

Major Energy Provisions in the 2008 Farm Bill

The Food, Conservation, and Energy Act of 2008 (P.L. 110-246) significantly expands existing programs to promote biofuels. Like the previous farm bill, it contains a distinct energy title (Title IX) that covers a wide range of energy and agricultural topics with extensive attention to biofuels, including corn starch-based ethanol, cellulosic ethanol, and biodiesel. Research provisions relating to renewable energy are found in Title VII and tax and trade provisions are found in Title XV.

The enacted 2008 farm bill keeps the structure of Title IX as it was in the Senate-passed version of the farm bill. Title IX serves as a substitute amendment to the 2002 farm bill Title IX and consists of 3 sections. The first section, 9001, contains 13 new provisions which effectively replace the provisions of the 2002 bill. (See the **Appendix** for a side-by-side comparison of previous law with the energy provisions of the 2008 farm bill.)

Key biofuels-related provisions in the enacted 2008 farm bill include:

- emphasis on cellulosic ethanol production through new blender tax credits, promotion of cellulosic feedstocks production, feedstocks infrastructure and refinery development;
- grants and loan guarantees for biofuels (especially cellulosic) research, development, deployment, and production;
- studies of the market and environmental impacts of increased biofuel use;

[7] In April 2008, Texas Gov. Rick Perry wrote the Environmental Protection Agency seeking a waiver from the federal ethanol mandate, noting its contribution to higher food prices and dire impact on the cattle industry. The waiver request was denied. Also, see CRS Report RS22870, *Waiver Authority Under the Renewable Fuel Standard (RFS).*

[8] For more information see CRS Report RL32712, *Agriculture-Based Renewable Energy Production*. For a complete listing of renewable energy legislative proposals introduced during the 110th Congress, see CRS Report RL33831, *Energy Efficiency and Renewable Energy Legislation in the 110th Congress.*

- expansion of biofuel feedstock availability;
- expansion of the existing biobased marketing program to encourage federal procurement of biobased products;
- support for rural energy efficiency and self-sufficiency;
- reauthorization of biofuels research programs within the U.S. Department of Agriculture and Environmental Protection Agency;
- an education program to promote the use and understanding of biodiesel;
- reduction of the blender tax credit for corn-based ethanol;
- continuation and expansion of the federal bio-products certification program;
- environmental safeguards through greenhouse gas emission requirements on new biofuel production; and
- continuation of the import duty on ethanol.

Energy Policy Issues in the 2008 Farm Bill

Cellulosic Biofuels

The 2008 farm bill energy title provides $1 billion in financial incentives and support to encourage the production of advanced (mainly cellulosic) biofuels.[9] Grants and loan guarantees leverage industry investments in new technologies and the production of cellulosic feedstocks. For instance, the Biomass Crop Assistance Program supports the production of dedicated crop and forest cellulosic feedstocks and provides incentives for harvest and post-production storage and transport. Advanced biofuels refinery capacity construction is assisted under the Biorefinery Program for Advanced Biofuels through grants and loans for the development, construction, and retrofitting of commercial-scale refineries to produce advanced biofuels. These programs are supported by increased funding for advanced biofuels research under the Agricultural Bioenergy Feedstock and Energy Efficiency Research and Extension Initiative, and the Sun Grant Program which support and coordinate advanced biofuels research, extension, and development between government agencies, universities, and research institutions.

Cellulosic ethanol is produced from cellulose, hemicellulose, or lignin derived from the structural material that provides much of the mass of plants. Besides corn, several other agricultural products are viable feedstock and appear to offer attractive long-term supply potential—particularly cellulose-based feedstock such as prairie grasses and fast-growing woody crops such as hybrid poplar and willow trees, as well as waste biomass materials (logging residues, wood processing mill residues, urban wood wastes, and selected agricultural residues such as sugar cane bagasse and rice straw). Some cellulosic feedstock, such as native prairie grasses (e.g.,

[9] Advanced biofuels include biofuels derived from cellulosic feedstocks; sugar and starch other than corn kernel-starch; waste material including crop residue, animal, plant, or food waste; diesel fuel produced from renewable biomass including vegetable oil and animal fat; butanol or other alcohols produced through the conversion of organic matter; and other fuels derived from cellulosic biomass. For more information, see CRS Report RL34738, *Cellulosic Biofuels: Analysis of Policy Issues for Congress.*

switchgrass), appear to offer environmental benefits over corn-based ethanol because they thrive on marginal lands (as well as on prime cropland) and need little water and no fertilizer.

Currently, cellulosic ethanol is not produced on a commercial scale. As of July 2008, there was one commercial scale refinery under construction, two demonstration-scale plants and three pilot-scale plants completed. Industry sources expect commercial production of cellulosic ethanol to begin in 2010 or 2011. In January 2009, USDA announced funding for a cellulosic biofuels plant under the Biorefinery Program with projected output of 20 million gallons annually, beginning in 2010. The RFS mandates cellulosic ethanol production of 100 million gallons in 2010. For more information on cellulosic ethanol see CRS Report RL34738, *Cellulosic Biofuels: Analysis of Policy Issues for Congress.*

Tax Credits and Tariffs

Title XV of the farm bill contains provisions which extend and modify tax credits and tariffs on ethanol. In keeping with the promotion of cellulosic ethanol, a blender credit of $1.01 per gallon applies to ethanol produced from qualifying cellulosic feedstocks. This tax credit is intended to spur investment in cellulosic ethanol production. The ethanol blender tax credit of $0.51 per gallon (which applies to all ethanol blended, including imports) was reduced to $0.45 per gallon in January 2009. Section 15331 of the farm bill requires the reduction starting the first year following that year in which U.S. ethanol production and imports exceed 7.5 billion gallons. Production and imports in 2008 were estimated to have exceeded 9 billion gallons.

The $0.54 per gallon import tariff for ethanol benefits the U.S. ethanol industry by protecting U.S. ethanol from lower-cost imports and also keeps imported ethanol from benefitting from the blender tax credit when it is blended into gasoline in the United States. The tariff was set to expire at the end of 2008 but has been extended to the end of 2010 by the farm bill (P.L. 110-246, Section 15333). With the decrease in the blender tax credit to $0.45, the tariff now exceeds the blender tax credit by nine cents, and so more than offsets the benefit of the blender tax credit.

Economic Impacts of Ethanol Production

The impact of increased ethanol production on agricultural and rural economies was a subject of debate during the farm bill process. As a result, the farm bill includes provisions requiring a series of reports assessing how ethanol production may be impacting the farm economy, the environment, and consumer food prices. Among these are the Comprehensive Study of Biofuels (to be conducted by the U.S. Department of Agriculture (USDA), the Environmental Protection Agency (EPA), the Department of Energy (DOE), and the National Academy of Sciences) and the Biofuels Infrastructure Study by USDA, DOE, Department of Transportation (DOT), and EPA. The Biomass Crop Assistance Program requires an assessment of the economic impacts of expanded cellulosic biomass production on local economies and infrastructures. Likewise, the Biomass Research and Development Program requires a report on the economic impacts of rural economies of biorefinery expansion and conversion by USDA.

Funding for Energy Programs

Table 1 illustrates mandatory and discretionary spending levels for renewable energy programs authorized in the 2008 farm bill. Mandatory funding is through USDA's Commodity Credit

Corporation (CCC).[10] Programs identified as receiving mandatory funds are funded at these levels unless Congress limits funding to a lower amount through the appropriations or legislative process. Discretionary programs are funded each year through the annual appropriations process.

Title IX authorizes $1.1 billion in mandatory funding for FY2008 through FY2012, compared with $800 million in the 2002 farm bill (FY2002-FY2007). Mandatory authorization in the 2008 farm bill included $320 million to the Biorefinery Assistance Program, $300 million to the Bioenergy Program for Advanced Biofuels, and $255 million to the Rural Energy for America Program (REAP). Authorizations for appropriations in the 2008 farm bill total $1 billion, four times the $245 million in the 2002 farm bill. Most of the increase is for the Biorefinery Assistance Program, which has an authorization $600 million higher than in the 2002 farm bill. **Table 2** provides a list of provisions in the 2008 farm bill's energy title, and selected energy programs in the research title, for FY2008 through FY2010, along with their funding levels (as requested by the President, and as authorized and provided by Congress) where available.

[10] The CCC is the funding mechanism for the mandatory payments that are administered by various agencies of USDA, including all of the farm commodity price and income support programs and selected conservation programs.

Table 1. 2008 Farm Bill (P.L. 110-246): Authorized Funding for Energy Provisions, FY2008-FY2012

($ millions)

2008 Farm Bill Section	Provision Name	Funding Type	FY2008	FY2009	FY2010	FY2011	FY2012	Total: FY2008-FY2012
Title VII Research								
Sec. 7205	Nutrient Management Research and Extension Initiative	Discretionary	SSAN	SSAN	SSAN	SSAN	SSAN	SSAN
Sec. 7207	Bioenergy Feedstock and Energy Efficiency Research and Extension Initiative	Discretionary	50	50	50	50	50	250
Sec. 7526	Sun Grant Program	Discretionary	75	75	75	75	75	375
Title IX Energy								
Sec. 9002[a]	Biobased Markets Program	Mandatory	1	2	2	2	2	9
		Discretionary	0	2	2	2	2	8
Sec. 9003[a]	Biorefinery Assistance	Mandatory	0	75	245	0	0	320
		Discretionary	0	150	150	150	150	600
Sec. 9004[a]	Repowering Assistance	Mandatory	0	35	0	0	0	35
		Discretionary	0	15	15	15	15	60
Sec. 9005[a]	Bioenergy Program for Advanced Biofuels	Mandatory	0	55	55	85	105	300
		Discretionary	0	25	25	25	25	100
Sec. 9006[a]	Biodiesel Fuel Education Program	Mandatory	1	1	1	1	1	5
Sec. 9007[a]	Rural Energy for America Program	Mandatory	0	55	60	70	70	255
		Discretionary	0	25	25	25	25	100
Sec. 9008[a]	Biomass Research and Development Act	Mandatory	0	20	28	30	40	118
		Discretionary	0	35	35	35	35	140
Sec. 9009[a]	Rural Energy Self-Sufficiency Initiative	Discretionary	0	5	5	5	5	20
Sec. 9010[a]	Feedstock Flexibility Program for Bioenergy Producers	Mandatory	SSAN	SSAN	SSAN	SSAN	SSAN	SSAN

CRS-6

2008 Farm Bill Section	Provision Name	Funding Type	FY2008	FY2009	FY2010	FY2011	FY2012	Total: FY2008-FY2012
Sec. 9011[a]	Biomass Crop Assistance Program	Mandatory	SSAN	SSAN	SSAN	SSAN	SSAN	SSAN
Sec. 9012[a]	Forest Biomass for Energy	Discretionary	0	15	15	15	15	60
Sec. 9013[a]	Community Wood Energy Program	Discretionary	0	5	5	5	5	20
Sec. 9003	Renewable Fertilizer Study	Discretionary	0	1	0	0	0	1
	Total Discretionary Funding	**Discretionary**	**$125**	**$403**	**$402**	**$402**	**$402**	**$1,734**
	Total Mandatory Funding	**Mandatory**	**$2**	**$243**	**$391**	**$188**	**$218**	**$1,042**

Source: P.L. 110-246 (Food and Energy Security Act of 2008).

Notes: "SSAN" = Such sums as necessary.

a. Section 9001 of the 2008 farm bill (P.L. 110-246) amends title IX of the 2002 farm bill (P.L. 107-171). Sections 9001 through 9013 of the table are the amended section numbers.

CRS-7

Table 2. 2008 Farm Bill Energy Funding (Presidential Request, Authorization, and Enactment), by Provision, FY2008 to FY2010

($ millions)

2008 Farm Bill Provision		FY2008			FY2009			FY2010		
		Pres. Req.	Auth.	Enact.[a]	Pres. Req.	Auth.	Enact.[a]	Pres. Req.	Auth.	Enact.[a]
Sec. 7205	Nutrient Management Research and Extension Initiative	0	SSAN	0	0	SSAN	0	0	SSNA	na
Sec. 7207	Bioenergy Feedstock and Energy Efficiency Research and Extension Initiative	0	50	0	0	50	0	0	50	na
Sec. 7526	Sun Grant Program	0	75	0	0	75	0	0	75	na
Sec. 9002[b]	Federal Biobased Markets Program	0	1	1	0	2	2	2	2	na
Sec. 9003[b]	Biorefinery Assistance	0	0	0	0	225	75	17	395	na
Sec. 9004[b]	Repowering Assistance	0	0	0	0	50	35	0	50	na
Sec. 9005[b]	Bioenergy Program for Advanced Biofuels[c]	0	0	0	55	80	55	55	80	na
Sec. 9006[b]	Biodiesel Education Program	0	1	1	0	1	1	0	1	na
Sec. 9007[b]	Rural Energy for America Program (REAP)[c]	0	0	38	5	80	65	128	85	na
Sec. 9008[b]	Biomass Research and Development[c, d]	0	0	0	0	55	20	0	63	na
Sec. 9009[b]	Rural Energy Self-Sufficiency Initiative	0	0	0	0	5	0	0	5	na
Sec. 9010[b]	Feedstock Flexibility Program for Bioenergy Producers	0	SSAN	0	0	SSAN	116	0	SSAN	na
Sec. 9011[b]	Biomass Crop Assistance Program		SSAN	0		SSAN	SSAN	0	SSAN	na
Sec. 9012[b]	Forest Biomass for Energy	0	0	0	0	15	0	0	15	na
Sec. 9013[b]	Community Wood Energy Program	0	0	0	0	5	0	0	5	na
Sec. 9003	Renewable Fertilizer Study	0	0	0	0	1	0	0	0	na
Total Energy Programs		**0**	**127**	**40**	**60**	**644**	**369**	**202**	**826**	**na**

Note: "Pres. Req." = Presidential request; "Auth." = Farm bill authorized level; "Enact." = Enacted; "SSAN" = Such sums as necessary; na = not available.

a. Represents a combination of the amount authorized by the 2008 Farm Bill, less any reductions in annual appropriations acts, plus discretionary funding provided where applicable.

b. Section 9001 of the 2008 farm bill (P.L. 110-246) amends title IX of the 2002 farm bill (P.L. 107-171). Sections 9001 through 9013 of the table are the amended section numbers.

c. Joint program with the Department of Energy.

d. Section 310 (7 U.S.C. 8609) of the Biomass Research and Development Act of 2000 (P.L. 106-224) provides discretionary authorization of $200 million for each of FY2006 through FY2015.

Appendix. Comparison of the Enacted 2008 Farm Bill (P.L. 110-246) with Previous Law

2002 Farm Bill (FSRIA, P.L. 107-171) or Other Law (as indicated)	Enacted Farm Bill (P.L. 110-246)
Title VII: Agriculture Research and Extension	
Nutrient Management Research and Extension Initiative	
Section 1673(h) of the Food, Agriculture, Conservation, and Trade Act of 1990 (P.L. 101-624) authorizes matching grants under the farm bill nutrient management research and extension initiative for finding innovative methods and technologies for economic use or disposal of animal waste. Extendes through 2007 in section 7120 of the 2002 farm bill. Such sums as necessary are appropriated annually for FY1999-FY2007. [7 U.S.C. 5925a]	Extends the nutrient Management research and extension initiative through FY2012 and adds dairy cattle waste as a type of waste to be studied. Also adds an amendment to include the production of renewable energy from animal waste as an eligible activity to receive grants under this section. Authorizes such sums as necessary annually for FY2007-FY2012. [Sec. 7205]
Agricultural Bioenergy Feedstock and Energy Efficiency Research and Extension Initiative	
No provision.	Establishes the Agricultural Bioenergy Feedstock and Energy Efficiency Research and Extension Initiative, a program to award competitive matching (up to 50%) grants for projects with a focus on supporting on-farm biomass crop research and the dissemination of results to enhance the production of biomass energy crops and the integration of such production with the production of bioenergy. Discretionary appropriations of $50 million annually are authorized for FY2008-FY2012. [Sec. 7207]
Sun Grant Program	
Section 9011. This provision was added subsequent to the 2002 farm bill under the Sun Grant Research Initiative Act of 2003. Establishes 5 national Sun Grant research centers based at land-grant universities, each covering a different national region, to enhance coordination and collaboration between USDA, DOE, and land-grant universities in the development, distribution, and implementation of biobased energy technologies. Competitive grants are available to land-grant schools within each region. Authorized appropriations of $25 million in FY2005, $50 million in FY2006, and $75 million for each of FY2007 through FY2010 for total discretionary funding of $375 million during FY2005-FY2010. [7 U.S.C. 8109]	Reauthorizes the Sun Grant Program through FY2012 and establishes a 6th regional center—Western Insular Pacific Sub-Center—at the University of Hawaii. Authorizes discretionary funding of $75 million annually for FY2008-FY2012. [Sec. 7526]
Title IX: Energy	
Definitions	
Sec. 9001. Defines Administrator, Biomass, Biobased Product, Procuring Agency, Renewable Energy, Rural Small Business, and Secretary. [7 U.S.C. 8101]	New section 9001 of FSRIA. Adds several definitions including "Advanced Biofuels," which excludes any fuel derived from corn starch, but includes ethanol derived from other plant starches (e.g., sorghum), sugar, as well as cellulosic biomass or organic waste; it also includes organically-derived biogas, butanol or other alcohols; and, notably, biodiesel. Other definitions are biobased product; biomass conversion facility; biorefinery; intermediate ingredient or feedstock; renewable biomass; and renewable energy. Adopts the Senate definitions with amendments. Advanced biofuels include aviation, jet, and heating fuels made from cellulosic biomass. [Sec. 9001]

2002 Farm Bill (FSRIA, P.L. 107-171) or Other Law (as indicated)	Enacted Farm Bill (P.L. 110-246)
Biobased Markets Program	
Sec. 9002. Requires federal agencies to purchase biobased products under certain conditions and authorize a voluntary biobased labeling program. USDA regulations define biobased products, identify biobased product categories, and specify the criteria for qualifying those products for preferred procurement. Mandatory Commodity Credit Corporation (CCC) funding of $1 million is authorized for each of FY2002 through FY2007 for testing biobased products. [7 U.S.C. 8101]	New section 9002 of FSRIA. Renames program as the Biobased Markets Program. Requires procuring agencies to establish a program and specifications for procuring biobased products (excluding motor vehicle fuels, heating oil, or electricity). Establishes the voluntary labeling program: "USDA Certified Biobased Product." Requires USDA to establish a national registry of biobased testing centers and a report on implementation. Mandatory CCC funding of $1 million in 2008, and $2 million annually for FY2009-FY2012. Discretionary funding of $2 million annually is authorized for FY2009-FY2012. [Sec. 9001]
Biorefinery Assistance	
No provision.	New section 9003 of FSRIA. Biorefinery Assistance Program. Assists in the development of new and emerging technologies for the development of advanced biofuels. Provides competitive grants and loan guarantees for construction and retrofitting of biorefineries for the production of advanced biofuels. Biorefinery grants provided for up to 30% of total cost. Each loan guarantee is limited to $250 million or 80% of project cost. Mandatory funding of $75 million in FY2009 and $245 million in FY2010, available until expended for loan guarantees. Discretionary funding of $150 million annually is authorized for FY2009-FY2012. [Sec. 9001]
Repowering Assistance	
Section 9003. Establishes a grant program to help finance the cost of developing and constructing biorefineries and biofuel production plants to carry out projects to demonstrate the commercial viability of converting biomass to fuels or chemicals. Mandatory funding is not authorized and discretionary funding has not been appropriated for the program. Therefore, no implementation regulations have been developed. [7 U.S.C. 8103]	New section 9004 of FSRIA. Provides payments to encourage biorefineries in existence on the date of enactment to convert from fossil fuel to renewable energy power sources. Encourages new production of energy for refineries from renewable biomass. Mandatory funding of $35 million for FY2009, to remain available until expended. Discretionary funding of $15 million annually for FY2009-FY2012 is authorized. [Sec. 9001]
Energy Program for Advanced Biofuels	
Section 9010. Originally created by a 1999 Executive Order during the Clinton Administration, the bioenergy program provides mandatory CCC incentive payments to biofuels producers based on year-to-year increases in the quantity of biofuel produced. Mandatory CCC funding of $150 million is available for each of FY2002 through FY2006. No funding is authorized for FY2007. [7 U.S.C. 8108]	New section 9005 of FSRIA. Establishes the Bioenergy Program for Advanced Biofuels to encourage production of advanced biofuels. Not more than 5% of the funds can go to facilities with total refining capacity exceeding 150 million gallons per year. Producers of advanced biofuels contracts with USDA to receive payments based on the quantity and duration of production of advanced biofuels, the net renewable energy content of the biofuel, and other factors. Payments limited to ensure equitable distribution. Mandatory funding of $55 million for 2009, $55 million for FY2010, $85 million for FY2011, and $105 million for FY2012. Discretionary funding of $25 million annually is authorized for FY2009-FY2012. [Sec. 9001]
Biodiesel Fuel Education Program	
Section 9004. Administered by USDA's Cooperative State Research, Education, and Extension Service, the program awards competitive grants to nonprofit organizations that educate governmental and private entities operating vehicle fleets, and educates the public about the benefits of biodiesel	New section 9006 of FSRIA. Extends the Biodiesel Fuel Education Program through 2012. Mandatory CCC funds of $1 million are provided annually for FY2008-FY2012. [Sec. 9001]

2002 Farm Bill (FSRIA, P.L. 107-171) or Other Law (as indicated)	Enacted Farm Bill (P.L. 110-246)
fuel use. Mandatory CCC funding of $1 million is authorized for each of FY2003 through FY2007. [7 U.S.C. 8104]	
Energy Audit and Renewable Energy Development Program	
Section 9005. A competitive grant program for eligible entities to provide energy audits and technical assistance to agricultural producers and rural small businesses to assist them in becoming more energy efficient and in using renewable energy technology and resources. Authorized appropriations of such sums as are necessary to carry out the program for each of FY2002 through FY2007. [7 U.S.C. 8105]	See new section 9007 of FSRIA below.
Rural Energy for America Program	
The Renewable Energy Systems and Energy Efficiency Program (Section 9006), administered by USDA's Rural Development Agency, authorizes direct loans, loan guarantees, and grants to farmers, ranchers, and rural small businesses to purchase and install renewable energy systems and to make energy efficiency improvements. Grant funds may be used to pay up to 25% of project costs; combined grants and loans or loan guarantees may fund up to 50% of project cost. Eligible projects include those that derive energy from wind, solar, biomass, or geothermal sources. Projects using energy from those sources to produce hydrogen from biomass or water are also eligible. Mandatory CCC funding of $23 million is available for each of FY2003 through FY2007. Unspent money lapses at the end of each year. [7 U.S.C. 8106]	New section 9007 of FSRIA. Renamed as the Rural Energy for America Program. Provides grants and loan guarantees to state governments, tribal, or local governments, land-grant institutions, rural electric cooperatives or utilities to provide energy audits and renewable energy assistance, and financial assistance for energy efficiency improvements and renewable energy systems. Grants up to 25% of cost are provided. Loan guarantees up to $25 million. Combined amount of grant and guaranteed loans limited to 75% of cost. 20% of funds made available in this section to be reserved for grants of $20,000 or less until the end of the fiscal year. Mandatory CCC funds of $55 million in FY2009, $60 million in FY2010, $70 million in FY2011, and $70 million in FY2012. Discretionary funding of $25 million annually is authorized to be appropriated for FY2009-FY2012. [Sec. 9001]
Biomass Research and Development Program	
This program—created originally under the Biomass Research and Development Act of 2000 (BRDA, P.L. 106-224)—provides competitive funding for research, development, and demonstration projects on biofuels and bio-based chemicals and products, under the Biomass Research and Development Initiative, administered jointly by USDA and DOE. Creates Biomass research and Development Board to coordinate government activities in biomass research, and the Biomass Research and Development Technical Advisory Committee to advise on proposal direction and evaluation. Authorizes mandatory CCC funding of $5 million in FY2002 and $14 million for each of FY2003 through FY2007 (available until expended). Additional appropriation authority of $200 million for each of FY2006 through FY2015. [7 U.S.C. 8101]	New section 9008 of FSRIA. Moves the Biomass Research and Development Act of 2000 in statute to Title IX of FSRIA of 2002. Defines biobased product. Expands advisory committee. New technical areas for grants include feedstock development, biofuels and biobased products development, and biofuels development analysis with a minimum of 15% of funding going to each area. Minimum cost-share requirement for demonstration projects increased to 50% and research projects to 20%. Provides for coordination of biomass research and development, including life cycle analysis of biofuels, between USDA and DOE. Authorizes mandatory funding of $20 million for FY2009, $28 million for FY2010, $30 million for FY2011, and $40 million for FY2012. Discretionary funding of $35 million is authorized to be appropriated annually for FY2009-FY2012. [Sec. 9001]
Rural Energy Self-Sufficiency Initiative	
No provision.	New section 9009 of FSRIA. Establishes the Rural Energy Self-Sufficiency Initiative to assist rural communities with community-wide energy systems that reduce conventional energy use and increase the use of energy from renewable sources. Grants are made available to assess energy use in a rural community, evaluate ideas for reducing energy use, and develop and install integrated renewable energy systems. Grants are not to exceed 50% of the total cost of the activity. Appropriations of $5 million annually are authorized for FY2009-FY2012. [Sec. 9001]

2002 Farm Bill (FSRIA, P.L. 107-171) or Other Law (as indicated)	Enacted Farm Bill (P.L. 110-246)
Feedstock Flexibility Program	
No provision.	New section 9010 of FSRIA. Requires that USDA establish (in FY2008) and administer a sugar-for-ethanol program using sugar intended for food use but deemed to be in surplus. USDA would implement the program only in those years where purchases are determined to be necessary to ensure that the sugar program operates at no cost. The use of such sums as necessary is authorized to carry out the program. [Sec. 9001]
Biomass Crop Assistance Program	
No provision.	New section 9011 of FSRIA. Establishes the Biomass Crop Assistance Program (BCAP) to provide producers committing to biomass production or a biomass conversion facility with contracts, which will enable producers in a BCAP project area to receive financial assistance for crop establishment costs and annual payments for biomass production. Producers must be within an economically practical distance from a biomass facility and adhere to resource conservation requirements. Cost-share payments cover costs of establishing crops and for collection, harvest, storage, and transportation to a biomass conversion facility. Annual payments authorized to producers to support biomass production. A report is required no later than 4 years after enactment. Mandatory CCC funds of such sums as necessary are made available for each of FY2008-F2012. [Sec. 9001]
Forest Biomass for Energy Program	
No provision.	New section 9012 of FSRIA. Forest Service competitive research and development program to encourage use of forest biomass for energy. Priority is given to projects that use low-value forest byproduct biomass for the production of energy; develop processes to integrate bioenergy from forest biomass into existing manufacturing streams; and develop new transportation fuels and improve the production of trees for renewable energy. Authorized appropriations of $15 annually for FY2009-FY2012. [Sec. 9001]
Community Wood Energy Program	
No provision.	New section 9013 of FSRIA. Establishes Community Wood Energy Program to provide matching grants to state and local governments to acquire community wood energy systems for public buildings. Participants must also implement a community wood energy plan to meet energy needs with reduced carbon intensity through conservation, reduced costs, utilizing low-value wood sources, and increased awareness of energy consumption. Authorizes discretionary funding of $5 million annually for FY2009-2012. [Sec. 9001]
Biofuels Infrastructure Study	
No provision.	Requests USDA, DOE, EPA, and DOT to jointly report on the infrastructure needs, requirements, and development approaches for expanding the domestic production, transport, and distribution of biofuels. Mandatory funding of $1 million for FY2008, and $2 million annually for each of FY2009-FY2012. Discretionary appropriations of $2 million annually are authorized for FY2009-FY2012. [Sec. 9002]

2002 Farm Bill (FSRIA, P.L. 107-171) or Other Law (as indicated)	Enacted Farm Bill (P.L. 110-246)
Renewable Fertilizer Study	
No provision.	Requires a report within 1 year of appropriations on the production of fertilizer from renewable energy sources in rural areas. Report must identify challenges to commercialization of rural fertilizer production, processes and technologies and the potential impacts of renewable fertilizer on fossil fuel use and the environment. Appropriation of $1 million is authorized for FY2009. [Sec. 9003]
Title XI: Livestock	
Study on Bioenergy Operations	
No provision.	Requires a USDA study on the use of animal manure as a fertilizer; the impact of limitations placed on the use of animal manure on consumers and agricultural operations; and the effects of increased competition for manure due to biofuel uses. [Sec. 11014]
Title XV: Trade and Tax Provisions; Subtitle C Part II – Energy Provisions	
Credit for Production of Cellulosic Biofuel	
Under the American Jobs Creation Act (AJCA) of 2004, (P.L. 108-357), cellulosic ethanol, once developed, would receive the current tax credit of $0.51 per gallon available to any ethanol blended into gasoline as provided through Dec. 31, 2010. [26 U.S.C. 40]	Provides a fourth tax credit under 26 U.S.C. 40, the Cellulosic Biofuel Producer Credit. The credit is $1.01 per gallon less the amount of small-producer ethanol credit claimed and the alcohol mixture credit claimed for ethanol. [Sec. 15321]
Comprehensive Study of Biofuels	
No provision.	The Secretary of Treasury, with USDA, DOE, and EPA shall commission the National Academy of Sciences to produce a report on biofuels, including current and projected production, economic and environmental impacts, government program impacts, and the relative impacts of different types of biofuels. [Sec. 15322]
Alcohol Fuel: Modification of Alcohol Credit	
Any ethanol blended into gasoline is eligible for a tax credit of $0.51 per gallon as provided under current law (AJCA of 2004, P.L. 108-357) through Dec. 31, 2010. [26 U.S.C. 40]	Reduces the ethanol tax credit of $0.51 per gallon to $0.45 per gallon beginning in the first calendar year after the year in which 7.5 billion gallons of ethanol is produced. [Sec 15331]
Alcohol Fuel: Calculation of Volume of Alcohol for Fuel Credits	
Under current law (AJCA of 2004, P.L. 108-357) the volume of bio-alcohol counted as fuel eligible for the tax credit may include up to 5% of the volume as denaturant. [26 U.S.C. 40]	Reduces the permissible volume of denaturant to 2% for purposes of calculating the volume of alcohol eligible for the tax credit. [Sec. 15332]
Alcohol Fuel: Ethanol Tariff Extension	
Under current law (Heading 9901.00.50 of the Harmonized Tariff Schedule (HTS)), imports of ethyl alcohol are subject to a duty of 14.27¢ per liter ($0.54 per gallon) and a duty of 5.99¢ per liter (Heading 9901.00.52; HTS) on imports of ethyl tertiary-butyl ether through Dec. 31, 2008. [19 U.S.C. Chapter 18]	Extends the tariff of $0.54 per gallon for imported ethanol or mixtures of ethanol (headings 9901.00.50 and 9901.00.52 of the HTS) through Dec. 31, 2010. [Sec 15333]

2002 Farm Bill (FSRIA, P.L. 107-171) or Other Law (as indicated)	Enacted Farm Bill (P.L. 110-246)
Alcohol Fuel: Limitations on, and Reductions of, Duty Drawback on Certain Imported Ethanol	
Section 1313 of the Tariff Act of 1930, as amended, permits the refund of duty if the duty-paid good is re-exported or used to make a good that is exported. A person who manufactures gasoline with ethanol subject to the duty imposed under HTS 9901.00.50 (see previous description), can export jet fuel (does not contain ethanol) and still obtain a refund of the duty paid. [19 U.S.C. Chapter 18]	Eliminates the ability to obtain a refund of duty imposed under HTS 9901.00.50, when imported ethanol is re-exported by substituting either ethanol not subject to the duty, or another petroleum product (e.g., jet fuel) that is exported to obtain the refund. [Sec. 15334]

Author Contact Information

Megan Stubbs
Analyst in Agricultural Conservation and Natural Resources Policy
mstubbs@crs.loc.gov, 7-8707

Renewable Energy: Background and Issues for the 110th Congress

Fred Sissine
Specialist in Energy Policy

December 10, 2008

Congressional Research Service
7-5700
www.crs.gov
RL34162

CRS Report for Congress
Prepared for Members and Committees of Congress

Summary

Renewable energy can be used to produce liquid fuels and electricity. A variety of funding, tax incentives, and regulatory policies have been enacted to support renewables as a means for addressing concerns about energy security, air pollution, international competitiveness, and climate change. This report reviews the background for renewables and describes the current congressional debate.

Budget and funding issues are key concerns. The Energy Policy Act of 2005 authorized several new renewable energy demonstration and deployment programs, but most of them have not been funded. Further, the Energy Independence and Security Act of 2007 (P.L. 110-140) authorized several new renewable energy programs that have not yet received appropriations. The Consolidated Appropriations Act for 2008 (P.L. 110-161) increased Department of Energy (DOE) renewable energy funding by $31.4 million (7%). The Continuing Appropriations Resolution for FY2009 (P.L. 110-329, H.R. 2638) continues DOE funding at the FY2008 level through March 6, 2009.

Tax policies are also at issue. The interaction of the federal renewable energy electricity production tax credit (PTC) with state renewable portfolio standard (RPS) policies has forged a strong incentive for wind energy development. The Emergency Economic Stabilization Act of 2008 (P.L. 110-343 [Division B], H.R. 1424) extends the PTC for wind farms for one year (three years for most other renewables) through the end of 2009, provides $800 million for a new category of clean renewable energy (tax credit) bonds, and extends for eight years the 30% level for the business solar tax credit and the 30% residential solar tax credit. Further, the law repeals nearly $17.7 billion in tax subsidies for oil and natural gas and reduces certain other financial incentives that will be used to offset the cost of the tax incentives for renewable energy ($9.1 billion) and energy efficiency ($3.6 billion).

The ethanol fuel issue intensified for much of the 110th Congress. Corn ethanol production climbed rapidly, but appeared to be causing food price increases. Concerns about rising food prices and apparent limits to the long-term potential for corn ethanol have brought a focus on cellulosic ethanol. Cellulosic sources avoid many limits on corn and appear to have much lower net CO_2 emissions, but they require an extensive and costly conversion process. P.L. 110-140 set a new renewable fuels standard (RFS), which starts at 9.0 billion gallons in 2008 and rises to 36 billion gallons in 2022. P.L. 110-343 (H.R. 1424) and the farm bill (P.L. 110-246, H.R. 6124) contain several tax incentives and other provisions for biofuels.

Contents

Tables

Contacts

Renewable energy is derived from resources that are generally not depleted by human use, such as the sun, wind, and water movement. These primary sources of energy can be converted into heat, electricity, and mechanical energy in several ways. There are some mature technologies for conversion of renewable energy such as hydropower, biomass, and waste combustion. Other conversion technologies, such as wind turbines and photovoltaics, are already well developed, but they have not achieved the technological efficiency and market penetration that many expect they will ultimately reach. Although geothermal energy is produced from geological rather than solar sources, it is often included as a renewable energy resource (and is treated as such in this report). Commercial nuclear power is not generally considered to be a renewable energy resource.[1]

Despite fluctuating government policies since the 1970s, a combination of incentives and high energy prices has enabled wind energy to gain a toe-hold in electric power markets and allowed ethanol to secure a modest, but growing, presence in motor fuels markets. Congress is now debating whether to provide additional subsidies, incentives, and mandates to further expand renewable energy use. This report describes the background and primary policy issue areas affecting renewable energy, including budget and funding, tax incentives, electricity regulatory initiatives, renewable fuels, and climate change.

History and Background

The energy crises of the 1970s spurred the federal government, and some state governments, to mount a variety of renewable energy policies. These policies included support for research and development (R&D), technology demonstration projects, and commercial deployment of equipment. For renewable energy, these policies included a focus on the production of both liquid fuels and electricity.

Fuels Production

The Energy Tax Act of 1978 established a 4 cents per gallon excise tax exemption for ethanol blended into gasoline. This incentive expired, and was extended, several times during the 1980s and 1990s. In some cases, the incentive was modified at the same time that it was extended.[2] The Energy Policy Act of 1992 extended the excise tax exemption and created a tax deduction for clean-fuel vehicles that included those using 85% ethanol (E85). It also established a requirement that federal, state, and other vehicle fleets include a growing percentage of alternative-fueled vehicles, including those using ethanol. In 2000, the General Accounting Office (GAO)[3] reported that the excise tax exemption and the alcohol fuel tax credits had been the most important incentives for renewable fuels.[4] By the time that the Energy Policy Act of 2005 (EPACT) was enacted, a variety of tax, grant, loan, and regulatory provisions had been established for renewable fuels. This included some 17 programs spanning five agencies. At present, the major

[1] For further definitions of renewable energy, see the National Renewable Energy Laboratory's website information on "*Clean Energy 101*" at http://www.nrel.gov/learning/.

[2] *A History of Ethanol*. http://e85.whipnet.net/index.html.

[3] This is now the Government Accountability Office.

[4] GAO. *Petroleum and Ethanol Fuels: Tax Incentives and Related GAO Work*. Letter to Senator Tom Harkin. September 25, 2000. (B-286311) 3 p. http://www.gao.gov/new.items/rc00301r.pdf.

tax incentives are a 51 cents per gallon excise tax exemption for ethanol blends, a $1 dollar per gallon tax credit for agri-biodiesel (50 cents per gallon for recycled biodiesel), and the alternative motor vehicle tax credit.[5] However, some believe that the Renewable Fuel Standard (RFS) set by EPACT Section 1501—which requires that motor fuels contain increasing amounts of renewable fuel each year through 2012—may now be the most important policy supporting renewable biofuels.[6]

Electricity Production

The Public Utility Regulatory Policies Act (PURPA, Section 210) created a policy framework that required electric utilities to purchase electricity produced from renewable energy sources. PURPA also empowered the states to set the price for such purchases. PURPA aimed to reduce oil use for power production, encourage the use of renewable energy for power production, and to structure a new dimension of competition to help keep electricity prices down. In the early 1980s, under the influence of PURPA regulation, a convergence of federal and state policies launched commercial deployment of wind and solar energy in California. In particular, the development of early wind farms was driven mainly by a combination of federal and state investment tax credits for wind energy.

As the new wind industry developed, two emerging aspects stimulated further policy changes. First, some firms took advantage of the investment tax credits by capturing the tax benefits at the front end and leaving wind machines that operated poorly or not at all. Recognition of this problem eventually led to the creation of a production-oriented tax credit. Second, in order to obtain third party financing, wind farm developers needed to secure agreements for power purchases that fixed the price for a long-term (10 years or more) period. This led the California Public Utility Commission to promote the development of "standard offer" contracts. These contracts reduced investment risk, established stable revenue streams, and helped launch early wind farm developments.

Oil and natural gas prices slumped during the mid-1980s, and declined more steeply in the late 1980s. Meanwhile, Congress let the residential solar investment tax credit expire in 1985. Funding for Department of Energy (DOE) renewable energy R&D programs also declined, reaching a low point in 1990.

In late 1990 and early 1991, the Persian Gulf War re-ignited interest in renewable energy. Other nations, notably Japan and Germany, began to undertake more aggressive policies to subsidize renewables, especially wind and solar technologies. In the United States, Congress began to increase funding for the Department of Energy (DOE) renewable energy R&D program. In 1992, the United States became a signatory of the United Nations Framework Convention on Climate Change (UNFCCC). This action forged a new environmental motive for support of renewable energy. These national interests were reflected in the Energy Policy Act of 1992 (P.L. 102-486). For electricity, this law made permanent the 10% business investment tax credit for solar and

[5] The 2004 Jobs Bill (P.L. 108-311) revised and extended the excise tax exemption for ethanol, and created the incentives for biodiesel fuel. EPACT extended the ethanol and biodiesel incentives. It also sunset the deduction for clean-fuel vehicles and created a new credit for alternative motor vehicles. For more details see CRS Report RL33572, *Biofuels Incentives: A Summary of Federal Programs*, by Brent D. Yacobucci.

[6] For more about ethanol fuels, see CRS Report RL33290, *Fuel Ethanol: Background and Public Policy Issues*, by Brent D. Yacobucci.

geothermal equipment. It also created a new renewable energy electricity production tax credit of 1.5 cents per kilowatt-hour (kwh) for wind farms and closed-loop (energy crop) biomass.

Climate change concerns spurred other industrialized nations to strengthen renewable energy policies and programs. Through the 1990s, concern about global climate change became an increasingly important motive in the European Union (EU), Japan, and other countries for raising renewable energy production goals and providing incentives to support commercial deployment. The Kyoto Protocol set emission reduction targets for carbon dioxide (CO_2) and other greenhouse gases (GHG). After signing the Protocol, these nations intensified their efforts for commercial deployment of renewable energy. In the United States, concern about climate change was largely offset by a concern about the potential effect of the Kyoto CO_2 emission reduction targets on economic growth and competitiveness. As a result of this economic concern, the United States has taken a more limited effort than many other industrialized nations to support renewable energy as a strategy for addressing climate change. The federal government has continued support for existing funding and subsidies. However, aside from the previously mentioned policies, it has not established major new policies and programs like the feed-in tariff in Germany or the European Union's target for producing 20% of its energy from renewables.[7]

State action on renewable energy has often supplanted federal action or created models for new federal policies. As one example, California has implemented very aggressive programs for renewable energy. In the mid-1990s, the advent of electric industry restructuring led California state policymakers to create a public goods charge on ratepayer electricity use. Part of the resulting revenue was used to fund renewable energy development and deployment programs. Also, California's electricity shortages in 2000 and 2001 prompted the state to expand its renewable energy programs. Motivated by concern over climate change, California has recently adopted more aggressive actions for renewables. This includes a $3 billion solar deployment initiative, and an increase of its renewable portfolio standard to 33% of total electricity production by 2020.

Action in the 110th Congress

Economic and environmental concerns—namely energy security, international competitiveness, high energy prices, air pollution, and climate change—are now driving policy proposals to support renewable energy R&D and market deployment. In the 110th Congress, more than 200 bills were introduced that would support renewable energy.[8] In the first session, the Energy Independence Act (P.L. 110-140) and the Consolidated Appropriations Act (P.L. 110-161) increased support for renewable energy.[9] In the second session, the Emergency Economic

[7] A feed-in tariff directs a utility to purchase electricity generated by renewable energy producers in its service area at a tariff determined by public authorities and guaranteed for a specific period of time. The price and term can vary by technology and over time. For more details, see California Energy Commission, *Notice of IEPR Committee Workshop on "Feed-In" Tariffs*, May 21, 2007. On the Commission's website at http://www.energy.ca.gov/2007_energypolicy/notices/2007-05-21_committee_workshop.html.

[8] For a comprehensive list of renewable energy bills, see CRS Report RL33831, *Energy Efficiency and Renewable Energy Legislation in the 110th Congress*, by Fred Sissine, Lynn J. Cunningham, and Mark Gurevitz.

[9] For a side-by-side comparison of the omnibus bills, see CRS Report RL34135, *Omnibus Energy Efficiency and Renewable Energy Legislation: A Side-by-Side Comparison of Major Provisions in House-Passed H.R. 3221 with Senate-Passed H.R. 6*, coordinated by Fred Sissine.

Stabilization Act (P.L. 110-343, Division B) provided several billion in tax incentives for renewables.

(For more details on the Energy Independence Act, see CRS Report RL34294, *Energy Independence and Security Act of 2007: A Summary of Major Provisions*, by Fred Sissine; for more details on FY2008 appropriations for DOE's renewable energy programs, see CRS Report RL34009, *Energy and Water Development: FY2008 Appropriations*, by Carl E. Behrens et al. and CRS Report RL34417, *Energy and Water Development: FY2009 Appropriations*, by Carl E. Behrens et al.; for more information about renewable energy laws and bills, see CRS Report RL33831, *Energy Efficiency and Renewable Energy Legislation in the 110th Congress*, by Fred Sissine, Lynn J. Cunningham, and Mark Gurevitz.)

Budget and Funding Issues and Action

EPACT Implementation (P.L. 109-58)

As part of the strategy to address energy security, climate change, and other national interests, the Energy Policy Act of 2005 (EPACT, P.L. 109-58) contained several provisions that authorized new programs and spending for renewable energy. Many of those provisions have either gone unfunded or have been funded below the authorized level.

Loan Guarantee Program

Title 17 of EPACT created a DOE loan guarantee program for certain energy technologies that could improve energy security, curb air pollution, and reduce greenhouse gas emissions.[10] Innovative renewable energy power plants and fuel production facilities would be eligible for a federal loan guarantee covering up to 80% of construction costs.[11]

EPACT Framework for Loan Guarantee Program

Many view this program as a key element of EPACT that addresses climate change and supports the commercial development of biofuels, such as cellulosic ethanol. The law authorizes DOE to issue loan guarantees to eligible projects that:

> ... avoid, reduce, or sequester air pollutants or anthropogenic emissions of greenhouse gases ... [and] ... employ new or significantly improved technologies as compared to technologies in service in the United States at the time the guarantee is issued.[12]

Title 17 provides broad authority for DOE to guarantee loans that support early commercial use of advanced technologies, if "there is reasonable prospect of repayment of the principal and interest on the obligation by the borrower."[13] The emphasis on "early commercial use only"

[10] Information about the DOE Loan Guarantee Program is available at http://www.lgprogram.energy.gov/index.html.

[11] The program authorization applies to other types of innovative energy-related technologies, including nuclear, coal, energy efficiency, vehicles, carbon sequestration, and pollution control equipment.

[12] EPACT (P.L. 109-58), Section 1703(a).

[13] EPACT (P.L. 109-58), Section 1702(d).

distinguishes the program from other DOE activities that are focused on research, development, and demonstration. Further, DOE states that the program will support the goals of the President's Advanced Energy Initiative.[14]

Loan Guarantee Program Regulations

In October 2007, DOE issued final loan guarantee regulations.[15] The regulations provide that DOE may issue guarantees for up to 100% of the amount of the loan, subject to the EPACT limitation that DOE may not guarantee more than 80% of the total cost for an eligible project. Under the final rule, if DOE issues a guarantee for 100% of a debt instrument, the loan must be issued and funded by the Treasury Department's Federal Financing Bank. DOE says that it intends to issue loan guarantees only if borrowers and project sponsors pay the "credit subsidy cost" for any loan guarantee they receive.[16]

Subsidy Cost

The subsidy cost is the expected long-term liability to the federal government in issuing the loan guarantee, excluding the administrative cost.[17] Title 17 specifies that DOE must receive either an appropriation for the subsidy cost or payment of that cost by the borrower. No funds have been appropriated for the subsidy cost of loan guarantees. DOE anticipates that the project borrower (sponsor) will pay this cost. Thus, DOE says it does not plan to use taxpayer funds to pay for the credit subsidy cost of the loan guarantees.[18]

Energy Independence Act Provisions

Two provisions of the law (P.L. 110-140) expand the range of facilities eligible for loan guarantees. Section 134 amended EPACT Title 17 to direct that DOE establish a loan guarantee program for facilities that manufacture "fuel efficient vehicles or parts of those vehicles, including electric drive vehicles and advanced diesel vehicles." Section 135 allows DOE, under certain conditions, to establish a loan guarantee program for the construction of facilities that manufacture advanced vehicle batteries and battery systems. Eligible parties would include manufacturers of advanced lithium ion batteries, manufacturers of hybrid electrical systems and components, and software designers.

[14] DOE. *FY2009 Congressional Budget Request*. Vol. 2. February 2008. p. 329.

[15] The process began with a proposed rule on May 16, 2007, which was followed by a comment period. The final rule is at http://www.lgprogram.energy.gov/lgfinalrule.pdf.

[16] DOE. *FY2009 Request*, p. 330.

[17] The Federal Credit Reform Act [Section 502(5A)] defines the subsidy cost as "the estimated long-term cost to the government of a direct loan or a loan guarantee, calculated on net present value basis, excluding administrative costs." The Director of the Office of Management and Budget is responsible for coordinating the estimation of subsidy costs. For more discussion of subsidy costs, see CRS Report RL30346, *Federal Credit Reform: Implementation of the Changed Budgetary Treatment of Direct Loans and Loan Guarantees*, by James M. Bickley.

[18] *DOE FY2009 Request*, p. 329-330.

Program Funding

DOE Loan Guarantee Program funding is shown in **Table 1.** In FY2006, DOE used about $500,000 from three separate appropriation accounts to fund start-up activities for $2 billion in loan guarantee authority.[19] The FY2007 continuing appropriations bill (P.L. 110-5, H.J.Res. 20) provided $7 million from DOE's Departmental Administration Account for program operating costs. Also, P.L. 110-5 raised the loan guarantee program authority to $4 billion, and required that DOE prepare a rulemaking to implement the program.[20]

Table 1. DOE Loan Guarantee Program Funding

($ millions)

FY2006 Apprn.	FY2007 Apprn.	FY2008 Apprn.	FY2009 Request
$0.5	$7.0	$4.5	$19.9

Source: GAO; and *DOE FY2009 Congressional Budget Request*, vol. 2, p. 329.

At both House and Senate energy committee hearings on the DOE FY2008 budget request, concerns were raised that the Loan Guarantee Program had not been implemented. DOE stated that, beginning in FY2008, the administrative activities for the Loan Guarantee Program Office would be funded in a separate discrete appropriation account entitled "Innovative Technology Loan Guarantee Program."[21]

The FY2008 Consolidated Appropriations Act (P.L. 110-161) directed DOE to issue $38.5 billion in new loan guarantee authority through the end of FY2009.[22] The law calls for $10.0 billion of the $38.5 billion to be designated to support renewables, energy efficiency, distributed energy, and transmission and distribution projects.

For FY2009, DOE requests $19.9 million for the Innovative Technology Loan Guarantee Program. This funding would cover administrative and operational expenses to support personnel and associated costs. DOE expects that the amount requested will be offset by collections authorized by EPACT (§1702[h]).[23] The FY2009 DOE request seeks to extend that authority through the end of FY2011. Specifically, DOE's request calls for $20.0 billion of the $38.5 billion to be available through FY2010 to support renewables and certain other projects.[24] The remaining $18.5 billion would be available through FY2011 to support nuclear power facilities.[25]

---

[19] Government Accountability Office (GAO). *Observations on Actions to Implement the New Loan Guarantee Program for Innovative Technologies*. (GAO-07-798T) p. 2.

[20] DOE issued the proposed rule on May 16, 2007. http://www.lgprogram.energy.gov/NOPR-fr-5-16-07.pdf.

[21] DOE. *FY2009 Congressional Budget Request, Budget Highlights*. February 2008. p. 52.

[22] The $38.5 billion of new authority is provided in addition to the $4.0 billion in authority set by the FY2007 appropriations bill. Thus, the two years of appropriations provide for a combined total of $42.5 billion in loan guarantee authority.

[23] DOE. *FY2009 Congressional Budget Request, Budget Highlights*. February 2008. p. 329. Section 1702(h) states that "DOE shall charge and collect fees for guarantees in amounts the Secretary determines are sufficient to cover applicable administrative expenses."

[24] The other projects include uranium enrichment, coal-based power, advanced coal gasification, and electricity delivery.

[25] DOE. *FY2009 Congressional Budget Request*, Vol. 2. February 2008. p. 330.

The Continuing Appropriations Resolution for FY2009 (P.L. 110-329, H.R. 2638) provides for continued funding at the FY2008 level through March 6, 2009.

First Round of Project Solicitations

In February 2007, the FY2007 Continuing Appropriations Resolution (P.L. 110-5) provided $4 billion in authority for loan guarantees. In May 2007, DOE announced a solicitation for the first round of projects. Eligible categories of renewable energy projects included biomass, solar, wind, and hydropower.[26] In October 2007, DOE announced that it was inviting 16 pre-applicants to submit full loan guarantee applications.[27] Among the 16 pre-applicants, eight proposed renewable energy projects. There are six biofuels projects, of which four involve cellulosic ethanol fuel production facilities and two involve biodiesel fuel production facilities. Also, there are two solar projects. One involves concentrated solar-thermal technology, and the other involves the manufacture of thin-film solar photovoltaic equipment.

Second Round of Project Solicitations

On June 30, 2008, DOE announced three solicitations for the 2nd round of solicitations. A total of up to $30.5 billion of loan guarantee authority was established for "advanced energy technologies" that would reduce air pollutants or greenhouse gases. One of the three solicitations was for "renewable energy and advanced transmission and distribution technologies."[28] On October 29, 2008, DOE announced that it was extending the "Renewables Solicitation" application due dates for stand-alone and manufacturing projects and for Part I large-scale integration projects, from December 31, 2008, to February 26, 2009. The deadline for Part II applications for large-scale integration projects will remain set at April 30, 2009.[29]

Biofuels and Other New Program Authorizations

Several biofuels programs authorized by EPACT have not been funded, including sugar cane ethanol (§208), biodiesel (§757), advanced biofuels (§1514), and cellulosic ethanol (§942, §1511, §1512). Unfunded biomass provisions include forest biomass (§210), biomass research and development (§941g), and bioenergy (§971d). Additionally, residential and small business renewable rebates (§206c) and insular areas (§251, §252) have not been funded. Provisions for technologies that would address climate change by reducing greenhouse gas emissions (§1601, §1602) also remain unfunded. Distributed energy (§921) and renewable energy (§931) are funded below authorized levels.

---

[26] The other eligible categories were hydrogen, advanced fossil energy (coal), carbon sequestration, electricity delivery and energy reliability, alternative fuel vehicles, industrial energy efficiency, and pollution control equipment.

[27] DOE. *DOE Announces Final Rule for Loan Guarantee Program.* (Press Release) October 4, 2007. http://www.lgprogram.energy.gov/press/100407.html.

[28] The other solicitations were for energy efficiency and nuclear power facilities. See DOE loan guarantee program chronology at http://www.lgprogram.energy.gov/.

[29] DOE. *DOE Extends Application Deadline for Renewable Energy Loan Guarantee Solicitation.* (press release) http://www.lgprogram.energy.gov/press/102908.pdf.

Energy Independence Act Implementation (P.L. 110-140)

As part of the strategy to address energy security, climate change, and other national interests, the Energy Independence and Security Act of 2008 (EISA, P.L. 110-140) contained several provisions that authorized new programs and spending for renewable energy. Some of those provisions have either gone unfunded or have been funded below the authorized level.

Accelerated Research and Development

Title VI directs DOE to conduct several new programs to accelerate the development of renewable energy and hydrogen technologies. For example, $90 million was authorized for new geothermal programs, but the FY2009 continuing appropriations bill (P.L. 110-329) does not include new funding to cover the cost of the programs. Also, $1 billion was authorized over a 10-year period to support the establishment of a hydrogen prize (H-prize) to accelerate technology development. DOE has issued a solicitation for an organizational host, pending congressional appropriations.[30] P.L. 110-329 does not provide new funding for this program.

International Energy Programs

Title IX, Subtitle A, calls for a new program of "Assistance to Promote Clean and Efficient Energy Technologies in Foreign Countries." The U.S. Agency for International Development (USAID) is directed to report to Congress on efforts to support policies for clean and efficient energy technologies. The Department of Commerce is directed to increase efforts to export such technologies and report to Congress on the results. Other U.S. agencies with export promotion responsibilities are required to increase efforts to support these technologies. Also, a multi-agency Task Force on International Cooperation for Clean and Efficient Energy Technologies is created to support the implementation of clean energy markets in key developing countries. The Senate Appropriations Committee recommended $100 million for FY2009 USAID programs that "directly support zero-carbon" efficiency and renewables programs.[31] However, P.L. 110-329 does not provide appropriations at that level, and it does not call for the new program developments set out in the Energy Independence Act.

Subtitle B directs that an "International Clean Energy Foundation" be established, with the long-term goal of reducing greenhouse emissions. Authorized funds would be used to make grants to promote projects outside the United States that serve as models of how to reduce emissions. The Senate Appropriations Committee recommended $200 million for a U.S. contribution to establish an "International Clean Energy Fund" at the World Bank or other entity, and requires that the contribution "be matched by other sources."[32] However, P.L. 110-329 does not provide appropriations for such a fund.

[30] See http://www.hydrogen.energy.gov/news_h-prize_administrator.html.

[31] S.Rept. 110-425, p. 43.

[32] S.Rept. 110-425, p. 67.

Green Jobs

Title X authorizes up to $125 million in funding to establish national and state job training programs, administered by the Department of Labor, to help address job shortages that impair growth in green industries, such as energy efficient buildings and construction, renewable electric power, energy efficient vehicles, and biofuels development. No funding was proposed for such programs in FY2009 budget requests nor in congressional committee appropriations recommendations.

FY2009 DOE Budget

Energy Efficiency and Renewable Energy

The President's 2008 State of the Union address set out goals to strengthen energy security and confront global climate change, and stated that "... the best way to meet these goals is for America to continue leading the way toward the development of cleaner and more energy-efficient technology."[33] As part of that effort, the Administration proposes to continue its support for the Advanced Energy Initiative (AEI, an element of the American Competitiveness Initiative), which "aims to reduce America's dependence on imported energy sources." The AEI includes hydrogen, biofuels, and solar energy initiatives that are supported by programs in EERE.[34]

According to the FY2009 budget document, the Hydrogen Initiative has a "long-term aim" of developing hydrogen technology that will help the Nation achieve a "cleaner, more secure energy future."[35] Further, current research aims to "enable industry to commercialize a hydrogen infrastructure and fuel cell vehicles by 2020." The Biofuels Initiative seeks to make cellulosic ethanol cost competitive by 2012 using a wide array of regionally available biomass sources. The Solar America Initiative aims to "... accelerate the market competitiveness of photovoltaic systems using several industry-led consortia which are focused on lowering the cost of solar energy through manufacturing and efficiency improvements."[36] Further, the *Budget* states that there is a goal to make solar power "cost-competitive with conventional electricity by 2015."[37]

As **Table 4** shows, DOE's FY2009 request seeks $1,255.4 million for the EERE programs. Compared to the FY2008 appropriation, the FY2009 request would reduce EERE funding by $467.0 million, or 27.1%. The request would eliminate $186.7 million in Congressionally-Directed Assistance and it would reduce Facilities construction spending by $57.3 million.[38] For renewable energy technologies, **Table 4** shows that—compared to the FY2008 appropriation—the key increases are for Biomass Energy ($26.8 million) and Geothermal Energy ($10.2 million). The key decreases are for Water/Hydrokinetic Power (-$6.9 million) and Solar Energy (-$12.3

[33] The White House. State of the Union 2008. http://www.whitehouse.gov/news/releases/2008/01/print/20080128-13.html.

[34] U.S. Executive Office of the President, *Budget of the United States Government, Fiscal Year 2007*. Appendix, p. 390. Also see DOE, *FY2007 Congressional Budget Request: Budget Highlights*, p. 41.

[35] U.S. Executive Office of the President, *Budget of the United States Government, Fiscal Year 2009*, Appendix, p. 393.

[36] U.S. Executive Office of the President, *Budget of the United States Government, Fiscal Year 2009*, Appendix, p. 393.

[37] *U.S. Budget*, p. 59.

[38] Facilities funding for construction tends to be provided in a lump sum. No major construction projects would be cancelled as a result of this proposed reduction.

million). Overall, funding for renewable energy technologies would increase by $20.7 million (4.6%). For deployment programs, the main increase is for the Asia Pacific Partnership ($7.5 million).[39] Also, the request would terminate the Renewable Energy Production Incentive (-$5.0 million).[40]

In contrast to the Administration's request, the House Appropriations Committee recommended $2,531.1 million for DOE's EERE programs in FY2009.[41] This would be a $808.7 million (47%) increase over the FY2008 appropriation and a $1,275.7 million (102%) increase over the DOE request. Compared with the request, the Committee recommendation would embrace a $381.5 million increase for R&D programs. Further, the Committee-approved bill would provide $259.2 million more for energy assistance programs, of which $250.0 million would go to the Weatherization Program—in sharp contrast to DOE's proposal to eliminate it. Also, the Committee recommended $500.0 million for new assistance programs authorized by the Energy Independence and Security Act (EISA, P.L. 110-140).

As a major initiative, the Committee recommended $500.0 million as "initial program investment" for several new programs authorized by EISA. The Renewable Fuel Infrastructure Program (EISA §244) would get $25.0 million to begin grant-giving operations. Aside from the $500.0 million initiative, some additional EISA-related funding would be provided under the technology programs. The most notable examples are $25 million for the production of advanced biofuels (EISA §207) under the Biomass and Biorefinery Program and $33 million for zero net energy commercial buildings (EISA §422) under the Buildings Program.

The committee recommended $134.7 million for Congressionally-Directed Assistance.

In addition to funding recommendations, the House Appropriations Committee report includes three policy directives for DOE. First, DOE would be required to report annually on the return on investment for each of the major EERE program funding accounts. Second, DOE would be directed to make up to $20 million of EERE funds available for "projects at the local level capable of reducing electricity demand." Each project would involve multiple technologies and public-private partnerships. Priority would go to projects that have a substantial local cost-share, help reduce water use, or curb greenhouse gas emissions. Third, DOE would be required to implement "an aggressive program" of minority outreach at Historically Black Colleges and Universities and at Hispanic Serving Institutions to deepen the recruiting pool of scientific and technical persons available to support the growing renewable energy marketplace.

The Senate Appropriations Committee recommended (S. 3258) $1,928.3 million for EERE,[42] which is $205.9 million (12.0%) more than the FY2008 appropriation and $672.9 million (53.6%) more than the request.

---

[39] *DOE Request*, p. 482-483. The Asia Pacific Partnership (APP) is a multinational undertaking that the federal government supports through several agencies. The Department of State is the lead agency for APP. DOE's request for APP in FY2009 would support new renewable power generating capacity, best manufacturing practices for targeted industries, and best design and construction practices for buildings and efficient appliance standards.

[40] For a brief discussion of the Renewable Energy Production Incentive, see the section on "Clean Renewable Energy (Tax Credit) Bonds", below.

[41] See the House report on H.R. 7324.

[42] The Senate Appropriations Committee report directs that $59.5 million of a proposed $72.9 million increase for the Solar Energy Program, will be provided by a transfer from the Basic Energy Sciences Program under the Office of Science.

Compared with the House Appropriations Committee report, the Senate Appropriations Committee recommended $602.8 million, or 23.8%, less for EERE programs. The main difference ($450.0 million) is that the House Appropriations Committee proposes an increase of $500.0 million for a new EISA Federal Assistance Program, while the Senate Appropriations Committee proposes an increase of $50.0 million for a new Local Government/Tribal Technology Demonstration Program. Further, the Senate report recommended less funding than the House report for several technology programs. Relative to the House Committee report figures, the Senate Committee report's proposed decreases for renewable energy R&D include Geothermal (-$20.0 million), Bioenergy (-$15.0 million), and Water Energy (-$10.0 million). The major decreases for energy efficiency include Weatherization (-$48.8 million), Industrial Technologies (-$34.9) million, and Vehicle Technologies (-$24.5 million).

The Senate Appropriations Committee recommended $124.2 million for Congressionally-Directed Projects.

In general, both committee reports recommended higher funding levels than the request. In particular, each included more than $200 million for the Weatherization Program. Both committees disagreed with the DOE request to fund the Asia Pacific Partnership,[43] and neither committee recommended funding it. Both committees called for the Biomass program to emphasize the use of non-food sources for the development of biofuels. The Senate Committee report further stressed R&D efforts to focus on algae as a biofuels source.

Electricity Delivery and Energy Reliability

The FY2009 request includes $134.0 million for the Office of Electricity Delivery and Energy Reliability (OE). The House Appropriations Committee recommended $149.3 million, which is $15.3 million more than the request. The Senate Appropriations Committee recommended $166.9 million, which is $17.7 million more than the House Appropriations Committee recommended. For OE congressionally directed projects, the House Committee report called for $5.3 million, while the Senate Committee report sought $12.9 million.

FY2009 Department of Agriculture (USDA) Request

The FY2009 budget document states that the Administration's 2007 farm bill proposal "... provides more than $1.6 billion in new renewable energy funding and targets programs to cellulosic ethanol projects."[44] In its FY2009 request document, the USDA states that, "While discretionary funding is not being requested, the Administration's farm bill proposal includes funding for renewable energy/energy efficiency loans and grants, and biomass research and

[43] *DOE Request*, p. 482-483. The Asia Pacific Partnership (APP) is a multinational undertaking that the federal government supports through several agencies. The Department of State is the lead agency for APP. DOE's request for APP in FY2009 would support new renewable power generating capacity, best manufacturing practices for targeted industries, and best design and construction practices for buildings and efficient appliance standards. During debate over the FY2008 request for EERE, the Administration threatened to veto the appropriations bill, in part, due to the lack of funding for APP.

[44] *FY2009 Budget of the U.S. Government*. Appendix. p. 120.

development grants.[45] (For more details, see CRS Report RL34130, *Renewable Energy Policy in the 2008 Farm Bill*, by Tom Capehart.)

Congressional Action on FY2009 Appropriations (P.L. 110-329)

House Passes H.R. 2638, Continuing Appropriations Resolution

On September 24, 2008, the House substitute to the Senate substitute to the proposed Department of Homeland Security Appropriations Act, 2008 (H.R. 2638) was brought to the House floor. The substitute was adopted by vote of 370 to 58. Division A—the Continuing Appropriations Resolution, 2009—would continue federal funding at FY2008 levels through March 6, 2009. Two provisions of the resolution would provide additional funding for energy efficiency. Section 129 would provide $7.51 billion for a DOE Advanced Technology Vehicles Manufacturing Loan Program authorized by the Energy Independence Act (P.L. 110-140, §136[d]). The Program would support $25 billion in loans to domestic automobile manufacturers and automobile part manufacturers to cover up to 30% of the costs of re-equipping, expanding, or establishing a manufacturing facility in the United States to produce advanced technology vehicles or components (automobiles and parts that exceed fuel-efficiency standards). Recipients would be required to pay employees and contractors prevailing wage rates, and the program would be scheduled to expire in 2017. Section 130 would provide an additional $250 million for the DOE Weatherization Assistance Program in FY2009. Those additional funds would remain available until expended.

House Passes H.R. 7110, Supplemental Appropriations

On September 26, 2008, the House passed the Supplemental Appropriations Bill for Fiscal Year 2009 (H.R. 7110) by a vote of 264 to 158. The bill would fund a green schools initiative at the Department of Education (DOED) and provide additional funding, above that in the Continuing Resolution (H.R. 2638), for efficiency and renewables programs at DOE.

Chapter 6 (Energy Development) would provide an additional $500 million in FY2009 appropriations for DOE's Office of Energy Efficiency and Renewable Energy (EERE). The purpose of the additional funding is to accelerate the development of technologies that would "diversify the nation's energy portfolio and contribute to a reliable, domestic energy supply." An additional $100 million would be provided to DOE's Office of Electricity Delivery and Energy Reliability (OE) to "modernize the electric grid, enhance security and reliability of the energy infrastructure, and facilitate recovery from disruptions to the energy supply." For the cost of loans authorized by the Energy Independence Act (P.L. 110-140, §135) the bill would provide $1 billion to remain available until expended. Of that amount, $5 million could be used only for administrative expenses to conduct the loan program. The leveraged loan guarantee commitments would be capped at a total of $3.3 billion in loan principal.

---

[45] USDA. FY2009 Budget Summary and Annual Performance Plan. February 2008. p. 44. http://www.obpa.usda.gov/budsum/fy09budsum.pdf.

Senate Adopts H.R. 2638

On September 27, 2008, the Senate adopted the House-passed version of H.R. 2638 by vote of 78 to 12. The Senate did not act on the supplement bill, H.R. 7110.

Tax Credit Issues and Action (P.L. 110-343)

On October 3, 2008, the proposed Emergency Economic Stabilization Act of 2008 (H.R. 1424) was signed into law as P.L. 110-343. Division B contains the Energy Improvement and Extension Act (EIEA), which extends and establishes several tax incentives for renewable energy.[46]

P.L. 110-343 extends or re-establishes several tax incentives that support renewable electricity production, biofuels production, transportation efficiency and conservation, buildings efficiency, and equipment efficiency. The law has four incentives for electricity production: the production tax credit, two solar investment tax credits, and new clean energy (tax credit) bonds. Also, the law has several incentives for biofuels. P.L. 110-343 includes $9.1 billion in renewable energy production (electricity and fuels) tax incentives and $3.6 billion in energy efficiency (transportation and buildings/equipment) tax incentives.[47] The renewable energy incentives include $5.8 billion for the renewable energy electricity production tax credit (PTC), $1.9 billion for business solar (and fuel cell) credits, $1.3 billion for residential solar tax credits, and $267 million for clean renewable energy (tax credit) bonds.[48] (For more about the background and debate on the renewable energy incentives, see the discussion below. For more details about the energy efficiency incentives, see CRS Report RL33831, *Energy Efficiency and Renewable Energy Legislation in the 110th Congress*, by Fred Sissine, Lynn J. Cunningham, and Mark Gurevitz.)

EIEA offsets the cost of the incentives by reducing several existing subsidies, with the effect of generating nearly $17 billion in revenue over 10 years. The largest offset, $6.67 billion, is provided by a modification of the requirements imposed on brokers for the reporting of their customers' basis in securities transactions. A freeze on a deduction for certain types of oil and natural gas production will provide $4.91 billion. Additional revenue will be derived from changes in foreign income taxes under Foreign Oil and Gas Extraction Income (FOGEI) and Foreign Oil Related Income (FORI) rules for the production and sale of oil and gas products ($2.23 billion), a modification of the excise tax for the Oil Spill Liability Trust Fund ($1.72 billion), and an extension of the Federal Unemployment Tax Act (FUTA) surtax ($1.47 billion).[49]

---

[46] EIEA was first introduced in the text of the Senate-adopted substitute to H.R. 6049. A brief legislative history of action on the tax incentives in the first session of the 110th Congress is provided in CRS Report RL34294, *Energy Independence and Security Act of 2007: A Summary of Major Provisions*, by Fred Sissine. Coverage of the more intense action in the second session is provided in CRS Report RL33831, *Energy Efficiency and Renewable Energy Legislation in the 110th Congress*, by Fred Sissine, Lynn J. Cunningham, and Mark Gurevitz.

[47] The law also provides $4.1billion for carbon mitigation and coal provisions.

[48] The Joint Committee on Taxation scores the estimated costs of the tax provisions at http://www.house.gov/jct/x-78-08.pdf. The Congressional Budget Office provides a brief summary of the scored costs at http://www.cbo.gov/ftpdocs/98xx/doc9852/hr1424Dodd.pdf.

[49] For additional discussion of the offset provisions, see CRS Report RL33578, *Energy Tax Policy: History and Current Issues*, by Salvatore Lazzari.

Debate over Proposed Incentives

Aside from several differences over the amount and duration of some incentives, the primary issue was focused on House proposals to fully offset the estimated cost of the incentives by reducing other tax subsidies. For example, during the House floor debate over H.R. 5351,[50] opponents argued that the proposed repeal of oil and natural gas subsidies (§301 and §302) would raise gasoline prices and lead to higher energy costs generally. Further, they contended that such a repeal would cause a decline in oil industry jobs. Also, some opponents argued that the proposed 35% cap on the renewable energy production tax credit (PTC) would severely impair the ability of the credit to stimulate the development of new wind farms.[51]

Proponents argued that the repeal would focus mainly on the five largest oil companies, which have recently made historical record-breaking profits and, thus, do not need the subsidies. Further, they contended that the subsidies currently favor conventional fuels and that the bill would help to bring support into a more equal balance. Proponents also argued that the incentives would spur the development of greater numbers of "green jobs" and help reduce greenhouse gas emissions.[52] (For more details about the proposed revenue offsets, see CRS Report RL33578, *Energy Tax Policy: History and Current Issues*, by Salvatore Lazzari.)

Renewable Energy Electricity Production Tax Credit (PTC)

Electricity produced by certain renewable energy facilities is eligible for an income tax credit based on production. Eligible facilities include those that produce electricity from wind, closed-loop biomass, open-loop biomass (including agricultural livestock waste nutrients), geothermal energy, solar energy, small irrigation power, landfill gas, and trash combustion. The credit's expiration date refers to the deadline for a facility to be placed into initial operation. Once a facility is qualified, a taxpayer may claim the credit annually over a 10-year period that commences on the facility's placed-in-service date.[53]

Background and History

The PTC was established by federal law (P.L. 102-486) in 1992.[54] The credit was originally set at 1.5 cents/kwh and is adjusted annually for the previous year's inflation rate.[55] Since 1992, it has expired and been reinstated three times, and it has been extended two other times.[56] In August

[50] Congressional Record. February 27, 2008. p. H1091-H1131.

[51] The Administration has threatened to veto the bill, stating its opposition to repeal of the oil industry subsidies and to proposals for clean renewable energy (tax credit) bonds and qualified energy conservation bonds. Executive Office of the President. *Statement of Administration Policy on H.R. 5351.* February 26, 2008. 2 p. http://www.whitehouse.gov/omb/legislative/sap/110-2/saphr5351-r.pdf.

[52] Many of these points were also stated in a letter from the Speaker of the House to the President. Office of the Speaker. *Pelosi, Hoyer, Clyburn and Emanuel Send Letter to White House on House-Passed Energy Legislation.* Press Release. February 28, 2008. 2 p. http://speaker.house.gov/newsroom/pressreleases?id=0544.

[53] U.S. Joint Committee on Taxation. *Description and Technical Explanation of the Conference Agreement of H.R. 6, Title XIII, "The Energy Tax Incentives Act of 2005."* July 28, 2005. p. 16. http://www.house.gov/jct/x-60-05.pdf.

[54] Section 1914 of the Energy Policy Act of 1992 (EPACT92, P.L. 102-486).

[55] The adjustment is set retrospectively, after inflation data is available for the previous calendar year.

[56] The most recent expiration occurred during 2004.

2005, the Energy Policy Act of 2005 (P.L. 109-58, §1301) extended the PTC for two years, through the end of calendar year 2007.[57] Also, the credit was expanded to include incremental hydropower and to increase the credit duration to 10 years for open-loop biomass, geothermal, solar, small irrigation power, and municipal solid waste. The Tax Relief Act of 2006 (P.L. 109-432, §201) extended the PTC for one additional year, through the end of 2008. The Emergency Economic Stabilization Act (P.L. 110-343, Division B, §101 and§102) extended the PTC for windfarms for one year, through the end of 2009. The PTC for biomass, geothermal, solar, and some other sources was extended for two years, through the end of 2010. Also, the PTC was expanded to include marine (ocean, tidal, wave) and hydrokinetic (river current) power sources, with eligibility established for three years, through the end of 2011.

Current Status and Past Significance

For claims against 2008 taxes, the credit stands at 2.1 cents/kwh for wind, closed-loop biomass, geothermal, and solar facilities. The credit stands at 1.0 cents/kwh for marine/hydrokinetic, open-loop biomass, small irrigation power, incremental hydropower, and municipal solid waste (including landfill gas and trash combustion facilities).

In 2007, the credit stood at 2.0 cents/kwh for claims against 2006 taxes. To illustrate the credit's significance, this 2.0 cents/kwh represented about one-third of wind production costs in 2006. As **Table 5** shows, half credit (valued at 1.0 cents/kwh in 2006) was provided for electricity produced by facilities that used open-loop biomass, small irrigation water flows, incremental hydropower, or landfill gas from municipal solid waste. In application, the credit may be reduced for facilities that receive certain other federal credits, grants, tax-exempt bonds, or subsidized energy financing. The amount of credit that may be claimed is phased out as the market price of electricity exceeds certain threshold levels.[58]

Revenue Effects

Claims for the PTC were less than $1 million in 1993 and 1994. **Table 6** shows that credit claims started growing more rapidly in 1995 and increased sharply, though erratically, from 1999 through 2005. Wind farm developments accounted for more than 90% of the dollar value of PTC claims through 2005.[59] Given the credit's availability for new projects through 2008, the table shows that the claims for 2006 through 2010 are estimated to increase substantially (in current year dollars) relative to past levels.

Impact on Resource Development

The PTC, combined with other policies, has had a positive though erratic effect on the growth of the wind energy industry. In contrast, it has had very little effect on baseload renewables, such as geothermal and biomass energy, and it has had virtually no effect on solar energy development. The following sections discuss PTC impacts in more detail.

---

[57] A detailed description of the PTC appears in the report *Description and Analysis of Certain Federal Tax Provisions Expiring in 2005 and 2006*, by the Joint Tax Committee, at http://www.house.gov/jct/x-12-05.pdf.

[58] The reductions and phase-out are described in IRS Form 8835. *Renewable Electricity, Refined Coal, and Indian Coal Production Credit*. 2006. p. 2. http://www.irs.gov/pub/irs-pdf/f8835.pdf.

[59] Personal communication with Curtis Carlson, Office of Tax Policy, Department of the Treasury. November 2008.

Impact of Boom-Bust Cycle on Wind Energy Industry

Coupled with rising energy costs, R&D advances, and a variety of state policies, the PTC has stimulated significant growth in wind capacity over the past 10 years.[60] However, the PTC expirations in 2000, 2002, and 2004 caused annual capacity growth to fall sharply in those years, by as much as 80% relative to the previous year. After each expiration, the PTC was reinstated for one- to two-year periods.[61] In 2005, one wind industry representative testified:

> Unfortunately ... two plus one plus one plus one does not necessarily equal five predictable years. Instead, it represents not the sum total of years the credit has been in place, but rather periods of uncertainty, when new wind construction stopped, jobs were eliminated, and costs were driven up. Business thrives on the known and fails on the unknown. The unpredictable nature of the credit has prevented the needed investment in U.S.-based facilities that will drive economies of scale and efficiencies.[62]

In 2007, one renewable energy analyst echoed this observation, testifying that the frequent credit expiration, and short-term nature of reinstatements and extensions, have led to several adverse impacts on wind industry growth. The variability of the credit has caused the growing demand for wind power to be "compressed into tight and frenzied windows of development. This cycle of boom-and-bust has resulted in under-investment in manufacturing capacity in the United States and variability in equipment and supply costs." It may also have caused under-investment in transmission planning and development, further restricting growth.[63]

The American Wind Energy Association has noted that the cycle of decline in wind industry activity actually starts about eight months before a PTC expiration date.[64] Representatives of the wind industry have testified that the cycle of peak manufacturing production demands followed by cutbacks "would be eliminated if a long-term PTC extension was in effect."[65] Opponents of the PTC say that the credit was created to provide temporary economic assistance to help the renewable electricity production industry get started. Further, they say that the PTC was not intended to be a permanent subsidy. Despite 15 years of subsidies, wind still apparently cannot compete without the PTC, opponents note.

Very Limited Impact on Other Renewables

Geothermal power facilities are physically and operationally more like conventional coal-fired power plants than wind machines. There is usually one large, highly capital-intensive plant that uses heat to produce base-load power.[66] However, industry testimony suggests that identifying a

[60] U.S. Congress. Senate. Committee on Finance. *Clean Energy: From the Margins to the Mainstream.* Hearing held March 29, 2007. Testimony of Ryan Wiser, p. 5. http://finance.senate.gov/sitepages/hearing032907.htm.

[61] Senate Finance Committee, *Clean Energy*, Testimony of Ryan Wiser p. 5.

[62] U.S. Congress. House. Committee on Ways and Means. *Tax Credits for Electricity Production from Renewable Sources.* Hearing held May 24, 2005. Testimony of Dean Gosselin, FPL Energy. p. 25-26. http://waysandmeans.house.gov/hearings.asp?formmode=detail&hearing=411.

[63] Senate Finance Committee, *Clean Energy*, Testimony of Ryan Wiser, p. 7.

[64] American Wind Energy Association (AWEA). *Legislative Priorities: Production Tax Credit Extension.* http://www.awea.org/legislative/.

[65] House Ways and Means Committee, *Tax Credits for Renewables*, Testimony of Dean Gosselin, p. 25.

[66] These facilities are often 10 megawatt (mw) to 100 mw in capacity, compared with wind machines that usually range from 2 mw to 5 mw.

suitable geothermal resource is similar to prospecting for oil or natural gas. The costs and risks of exploration for geothermal are as high or higher than those for the oil and gas industry, and the ability to attract financing is far more difficult. Once a resource is verified, permitting and construction can take three to five years or more. Since 1992, there has been very limited development of new geothermal facilities.[67]

In 2005, EPACT increased the amount of the PTC available to geothermal facilities from half to full credit. However, the PTC's short windows of availability have made the credit largely ineffective as an incentive for the geothermal industry. Industry representatives have noted that the largest projects "may not go forward because they face unacceptable risks trying to meet the rigid deadline ... [or to avoid] taking an all-or-nothing gamble on future extensions of the credit."[68] The geothermal industry says a PTC extension of 10 years or more could be sufficient to stimulate a higher level of sustained industry growth.[69]

Representatives of biomass, hydropower, and landfill gas industries say their facilities are more like geothermal facilities than wind machines and, thus, also require a longer-term PTC period. In 2005 testimony, EIA offered a similar observation:

> Short-term extensions of the PTC are likely to have limited impact on qualifying technologies like biomass and geothermal, which have relatively long development periods, even if the credit were large enough to make them economical.[70]

The PTC has been even less valuable for solar energy equipment. Most solar electricity equipment comes as small, widely distributed units that are designed mainly for on-site use, not for power sales to the grid.[71] These aspects make the PTC less valuable for solar than the business and residential investment tax credits (ITC).[72] Due to rules against multiple tax credit use, solar equipment cannot qualify for both the PTC and ITC, and so owners must choose one or the other. Representatives of the solar industry have indicated a clear preference for ITC over PTC.[73] Even with the PTC, solar is too expensive for utility-scale application.

Combined Impact with State Renewable Portfolio Standards

After its creation in 1992, the PTC was virtually unused until states began to establish renewable portfolio standard (RPS) policies.[74] State RPS action began in the mid-1990s.[75] Since then, an

[67] U.S. Congress. Senate. Committee on Energy and Natural Resources. *Implementation of Provisions of the Energy Policy Act of 2005.* Hearing held July 11, 2006. Testimony of Karl Gawell, Geothermal Energy Association (GEA). p. 95. http://frwebgate.access.gpo.gov/cgi-bin/getdoc.cgi?dbname=109_senate_hearings&docid=f:30004.pdf.

[68] Senate Energy Committee, *Implementation of EPACT*, Testimony of GEA, p. 92-93.

[69] Personal communication with Karl Gawell, Geothermal Energy Association, April 6, 2007.

[70] House Ways and Means Committee, *Tax Credits for Renewables,* Testimony of Dr. Howard Gruenspect for the Energy Information Administration (EIA), p. 10.

[71] Also, solar energy equipment has high capital costs and low capacity factors.

[72] House Ways and Means Committee, *Tax Credits for Renewables*, EIA Testimony, p. 6-9.

[73] House Ways and Means Committee, *Tax Credits for Renewables*, Testimony of Chris O'Brien for the Solar Energy Industries Association (SEIA), p. 47-49.

[74] EIA, *AEO2005*, p. 58.

[75] Iowa first established a renewable energy requirement in 1983. However, most states did not consider an RPS until after electricity restructuring policies appeared in the mid-1990s. The following section of this report discusses state RPS activity in greater detail.

increasing number of states have implemented an RPS. **Table 5** shows the trend depicting the close correlation between rising PTC claims and the growing number of states with an RPS. Since the late 1990s, many have noted that the combined effect of the PTC with state RPS policies has been a major spur to wind energy growth.[76]

Credit Design Issues

The variability in tax credit availability has led to erratic growth in energy production, and it has caused the U.S. wind industry to become more dependent on European equipment due to stronger European requirements for renewables.[77] Despite these problems, wind has been the main beneficiary of the credit. A related issue is that the PTC has not been effective at stimulating the development of other renewable energy facilities, which generally need a longer period of credit availability. The main proposal to address the variable impact on wind and the lack of impact on other renewables is the enactment of a longer-term PTC extension. The wind industry prefers an extension of five years or more.

On occasion, the PTC has been expanded to include a broader range of renewable energy resources. This credit design issue surfaced in the 110th Congress, as it addressed the question of whether the credit should be expanded to include production from equipment that uses marine energy (tidal, wave, and ocean thermal) resources and hydrokinetic (river current) resources. P.L. 110-343 (Division B, §102) did expand the credit to cover those resources, establishing a 1.0 cent/kwh credit with a three-year eligibility window.

Extend the Credit to Achieve a Five-Year Period or More

At least two studies have attempted to assess the potential results of a longer-term PTC extension. In one study, EIA examined a 10-year extension and found that wind power would continue to show the largest projected gains.[78] Landfill gas, geothermal, and biomass were also projected to experience some capacity expansion. EIA estimated a 7-fold increase for wind, a 50% increase for biomass, and a 20% increase for geothermal facilities.[79]

In 2007, DOE's Lawrence Berkeley National Laboratory (Berkeley Lab) reported the results of a study that examined the potential benefits of extending the PTC for 5 to 10 years. Relative to a projection with continued cycles of one-year to two-year extensions, it found that the installed cost of wind could be reduced by 5% to 15%. Additional benefits could include better transmission planning and enhanced private R&D spending. Also, Berkeley Lab estimated that a 10-year extension could increase the domestic share of manufactured wind equipment from the current level of 30% to about 70%.[80] The Joint Committee on Taxation has estimated that the one-

[76] DOE. Energy Information Administration (EIA). *Annual Energy Outlook 2006.* (Section on "State Renewable Energy Requirements and Goals: Update Through 2005.) p. 27. Further discussion of the importance of the PTC to RPS is presented in the section under Renewable Portfolio Standard entitled "Federal Tax Credit (PTC) Supports State RPS Policies."

[77] Senate Finance Committee, *Clean Energy*, Testimony of Ryan Wiser, p. 7-9.

[78] Prior to the PTC extension in EPACT05, EIA examined an extension from the end of 2005 through the end of 2015. The extension included all resources covered by the PTC at that time at the values that were in place then. EIA. *Annual Energy Outlook 2005 (AEO2005)*. p. 60.

[79] House Ways and Means Committee, *Tax Credits for Renewables*, EIA Testimony, p. 10.

[80] Senate Finance Committee, *Clean Energy*, Testimony of Ryan Wiser, p. 8-10.

year[81] extension of the wind credit's placed-in-service deadline in P.L. 110-343 would reduce tax revenue to the U.S. Treasury by about $5.8 billion over the 10-year duration of credit claims.[82]

In 2007 testimony, MidAmerican Energy Company suggested that a 5-to-10 year PTC extension would also be the best way to encourage baseload renewables, such as geothermal and biomass. Such an extension, it said, would provide long-term certainty to utilities, independent project developers, and manufacturers. To address budget-related cost concerns for a PTC extension, Mid-American suggested that a long-term extension could be coupled with a gradual phase-down of the credit to 1.5 cents/kwh. Alternatively, if the credit extension were set at something less than five years, Mid-American proposed that a conditional second deadline could be set up that would extend the placed-in-service eligibility period. That extension would require an offsetting reduction in the credit period, the length of time over which credit claims could be filed. The conditions required for an extension to a secondary placed-in-service deadline are that the project must be under construction and have signed power sales contracts before the initial credit expiration date and it must bring the project online before the secondary placed-in-service deadline. For example, if the secondary deadline were set as one year past the initial placed-in-service deadline, a project that met those conditions would be eligible to receive the credit, but only for nine years instead of ten.[83]

Debate Over PTC Extension

Because the PTC was set to expire at the end of 2008, proposals to extend it began early in the first session of the 110th Congress. **Table 2** shows that several proposals—with different time periods—were considered. Although PTC cost was a concern, growing higher in relation to the time period, the debate focused mainly on House paygo requirements that drew Administration veto threats. The House consistently proposed that the cost of the incentives be fully offset. Proposed offsets stressed reduced tax subsidies for oil and natural gas, which the Administration found unacceptable.[84] In the first session, H.R. 2776 proposed a four-year extension that was incorporated into H.R. 3221 and, later, into a House-passed version of H.R. 6.[85] The Senate was unable to pass any tax incentive package in the first session, and the Energy Independence and Security Act of 2007 (P.L. 110-140) was enacted without tax provisions. In the second session, the key bills were H.R. 5351, H.R. 6049, and H.R. 1424.[86]

---

[81] The law provides a two-year extension for production from most other sources and sets a three-year eligibility window for production from marine and hydrokinetic sources.

[82] Joint Committee on Taxation. *Estimated Budget Effects of the Tax Provisions Contained in an Amendment in the Nature of s Substitute to H.R. 1424, Scheduled for Consideration on the Senate Floor on October 1, 2008*. October 1, 2008. http://www.house.gov/jct/x-78-08.pdf.

[83] Senate Finance Committee. *Clean Energy*. Testimony of Todd Raba of MidAmerican Energy Company, p. 3.

[84] Executive Office of the President. Office of Management and Budget. *Statement of Administration Policy on H.R. 6*. December 7, 2007. http://www.whitehouse.gov/omb/legislative/sap/110-1/hr6sap-h_2.pdf.

[85] In Senate floor action on an amendment to H.R. 6 in June 2007, S.Amdt. 1704 (§801) would have extended the PTC for five years, but a cloture motion was defeated (57-36). Also, in August 2007, the House approved H.R. 3221 with a four-year PTC extension (§11001).

[86] Action is discussed in CRS Report RL34294, *Energy Independence and Security Act of 2007: A Summary of Major Provisions*, by Fred Sissine (1st session) and in CRS Report RL33831, *Energy Efficiency and Renewable Energy Legislation in the 110th Congress*, by Fred Sissine, Lynn J. Cunningham, and Mark Gurevitz (2nd session).

Table 2. Selected Wind Production Tax Credit (PTC) Extension Proposals

Bill	House		Senate	
	Extension Period	Final Action	Extension Period	Final Action
First Session				
H.R. 6	4 years (§1501)	adopted (235-181)	2 years (S.Amdt. 3841)	cloture motion defeated (59-40)
Second Session				
H.R. 5351	3 years	adopted (236-182)	—	no action
H.R. 6049	1 year	adopted (263-160)	1 year	adopted (93-2)
P.L. 110-343 (H.R. 1424)	1 year	adopted (263-171)	1 year	adopted (74-25)

Proponents of extending the credit past 2008 argued that the PTC is merited because it corrects a market failure by providing economic value for the environmental benefits of "clean" energy sources that emit less (in many cases, far less) air pollutants and CO_2 than conventional energy equipment. Also, they contended that it helps "level the playing field," noting that there is an even longer history of federal subsidies for conventional energy.[87] For example, they point to the permanent depletion allowance for oil and natural gas that has been in place for many decades.[88]

Opponents of extending the PTC beyond the end of 2008 argued that generally there are no market failures that warrant special tax subsidies for particular types of renewable energy technologies. They argued further that subsidies generally distort the free market and that renewables should not get special treatment that exempts them from this principle. Also, regarding the concern about the environmental problems of "dirty" conventional energy sources, they contended that the most cost-effective economic policy is to put a tax on the pollution from energy sources and let the free market make the necessary adjustments. Another argument against the PTC was that much renewable energy production, particularly from wind and solar equipment, has a fluctuating nature that makes it less valuable than energy produced by conventional facilities.[89]

At a Senate hearing in February 2007, Energy Secretary Bodman testified that the Administration was unlikely to support a five-year or 10-year PTC extension because it would not be consistent with free markets.[90] Consistent with that stance, the Administration's FY2008 budget request did

[87] Federal subsidies for conventional energy resources and technologies and for electric power facilities (including large hydroelectric power plants) have been traced back as far as the 1920s and 1930s. See DOE (Pacific Northwest Laboratory). *An Analysis of Federal Incentives Used to Stimulate Energy Production*, 1980. 300 p.

[88] GAO. *Petroleum and Ethanol Fuels: Tax Incentives and Related GAO Work*. (GAO/RCED-00-301R) September 25, 2000. The report notes that from 1968 through 2000, about $150 billion (constant 2000 dollars) worth of tax incentives were provided to support the oil and natural gas industries.

[89] Some argue further that as the contributions from wind and solar power production rise, their intermittent nature may create grid management problems for electric utilities.

[90] U.S. Congress. Senate. Committee on Energy and Natural Resources. *Proposed Budget for FY 2008 for the* (continued...)

not include a provision to cover a PTC extension beyond 2008. Similarly, the Administration's FY2009 budget request did not include such a provision. However, Section 304 of the Senate version (S.Con.Res. 70) of the budget resolution proposed the creation a deficit-neutral reserve that could be used to support a five-year PTC extension. Further, Section 305 of the House budget resolution (H.Con.Res. 312) also allowed for support of renewable energy tax incentives.

Solar Investment Tax Credits

Residential Credit

The Energy Tax Act of 1978 (P.L. 95-618) established a residential energy investment tax credit (ITC) for solar and wind energy equipment.[91] As energy prices declined, Congress allowed the credit to expire at the end of 1985. In 2005, EPACT (P.L. 109-58, §1335) established a 30% residential solar credit with a cap at $2,000, through the end of 2007.[92] The Tax Relief Act of 2006 (P.L. 109-432, §206) extended the credit through the end of 2008.

The Emergency Economic Stabilization Act of 2008 (P.L. 110-343, Div. B, §106) extends the residential solar tax credit at the 30% level for eight years, through the end of 2016. Further, the annual cap on the credit is increased from $2,000 to $4,000. Also, residential wind equipment and ground source heat pumps are eligible for a 30% credit.

Business Credit

The Energy Tax Act also established a 10% business investment tax credit for solar, wind, geothermal, and ocean energy equipment.[93] The Energy Policy Act of 1992 made permanent the 10% business credit for solar and geothermal equipment. In 2005, EPACT (§1337) increased the solar business credit to 30% through the end of 2007.[94] The Tax Relief Act of 2006 extended the 30% rate through the end of 2008. Without an extension, the credit would have dropped back to 10% after 2008.

P.L. 110-343 (Div. B, §103) extends the business solar tax credit at the 30% level for eight years, through the end of 2016.[95] Further, the credit would be allowed to offset the alternative minimum tax. Also, public utilities would become eligible for the credit.

(...continued)

Department of Energy. Hearing held February 7, 2007. http://energy.senate.gov/public/index.cfm?FuseAction=Hearings.Hearing&Hearing_ID=1601.

[91] The claim against income was set at 30% of the first $2,000 and 20% of the next $8,000. The Crude Oil Windfall Profits Tax Act of 1980 (P.L. 96-223) increased the credit from 30% to 40% of the first $10,000.

[92] Joint Tax Committee, *Description of H.R. 6*, p. 49.

[93] The Windfall Profits Act increased the credit to 15% and extended it through the end of 1985. The Tax Reform Act of 1986 (P.L. 99-514) extended the credit through 1988.

[94] Joint Tax Committee, *Description of H.R. 6*, p. 52-53.

[95] This is the same provision that the House approved in H.R. 6 (§1503) on December 6, 2007. S.Amdt. 3841 included the same provision in section 1502, except that it would have also made certain combined heat and power equipment eligible for the credit. However, the amendment was defeated on a cloture vote and it was not further considered. The Solar Energy Industry Association had endorsed H.R. 550/S. 590, which would have expanded the business credit to include certain solar storage and lighting equipment, and it would have extended the credit at the 30% level for eight (continued...)

The debate over extending these credits was similar to that for the PTC. Opponents argued that subsidies distort the operation of the free market. They also contended that the most effective policy is to impose a tax on energy equipment that causes pollution. The solar industry has testified that the business ITC is the most important tax incentive for solar equipment. Proponents of the credit counter-argued that the credits correct a market failure and help establish equality with subsidies that exist for conventional energy equipment. They also asserted that the subsidy-induced increase in demand helps manufacturers establish economies of scale that will broaden the use of solar equipment and make it more competitive in the long term.

Other Business Tax Credits

P.L. 110-343 established other new business tax incentives. Division B (§104) establishes a 30% credit (capped at $4,000) over an eight-year period (end of 2016) for wind machines with a capacity of 100 kilowatts or less. Also, the permanent 10% credit is expanded to include geothermal (ground source) heat pumps (§105).[96]

Clean Renewable Energy (Tax Credit) Bonds

Non-profit electric utilities provide about 25% of the nation's electricity.[97] Due to their tax-exempt status, they are not eligible for the PTC. To address the cost and risk barriers for developing renewable energy facilities, these organizations have sought incentives comparable to the PTC. Using a design that parallels the PTC, the Energy Policy Act of 1992 (EPACT92) established a renewable energy production incentive (REPI) that provided 1.5 cents/kwh, adjusted for inflation.[98] REPI typically receives about $5 million per year, through DOE appropriations. This limited funding and annual uncertainty may have severely limited REPI's potential. DOE data for 2004 shows, for example, that funding covered only about 10% of requests for REPI payments.[99]

In 2005 testimony, the American Public Power Association (APPA) stated that REPI was "woefully underfunded," and the National Rural Electric Cooperative Association (NRECA) proposed that a "clean energy bond" be created to establish an incentive for non-profit electric utilities that would be more comparable in scope to the PTC.[100] Subsequently, EPACT (§1303) established clean renewable energy bonds (CREBs), a tax credit bond that allowed the bond holder to receive a federal tax credit in lieu of interest paid by the issuer.[101] EPACT authorized

(...continued)

years.

[96] Additional energy-efficient equipment was made eligible for the 10% credit. For more details, see the online databases at http://www.energytaxincentives.org/ and at http://www.dsireusa.org/.

[97] These non-profit organizations include public power utilities, cooperative electric utilities, and federally owned power utilities.

[98] For background on REPI, see the Database of State Incentives for Renewable Energy, http://www.dsireusa.org/library/includes/incentive2.cfm?Incentive_Code=US33F&State=federal¤tpageid=1&ee=0&re=1.

[99] For historical details of REPI's use, see the table entitled "REPI Appropriation Summary," on DOE's website at http://www.eere.energy.gov/wip/repi.cfm.

[100] U.S. Congress. House. Committee on Ways and Means. *Tax Credits for Electricity Production from Renewable Sources.* Hearing held May 24, 2005. Testimony of APPA (p. 61-63) and NRECA (p. 67-69).

[101] Thus, CREBs allow a bond issuer to borrow at a zero percent interest rate. Eligible bond issuers include state and local governments, cooperative electric companies, and certain other non-profit organizations. For the bondholder, the (continued...)

$800 million in CREBs for 2006 and 2007.[102] In late 2006, the Internal Revenue Service (IRS) reported requests totaling $2.6 billion in bond authority.[103] The Tax Relief Act of 2006 (§202) authorized a second round of CREBs through the end of 2008, adding $400 million more in total bond authority.[104]

P.L. 110-343 (Div. B, §107) established a new category of CREBs (New CREBs) for state/local/tribal governments, public power providers (utilities), and cooperative electric companies.[105] The "New CREBs" differ from the previously issued CREBs in four aspects. First, issuers of New CREBs will be subjected to a shorter three-year period for use of the bond proceeds, two years less than the previous five-year period for CREBs. Second, the tax credit rate will be lower, set at 70% of the previous rate for CREBs.[106] Third, taxpayers can carry forward unused credits into future years. Fourth, the tax credit benefits can be separated from bond ownership.[107]

A national limit of $800 million was set for New CREBs, of which one-third will be available for state, local and tribal governments; one-third for public power providers; and one-third for cooperative electric companies. The revenue drain on the U.S. Treasury is estimated at a total of $267 million over the period from 2009 through 2018.[108] The Administration repeatedly stated its opposition to the New CREBs that the House approved in the first session (H.R. 3221 and the House version of H.R. 6) and in the second session (H.R. 5351 and H.R. 6049).[109] For example, it contended that the CREBs are "expensive and highly inefficient," and that New CREBs would be "inconsistent with the Federal Credit Reform Act of 1990 and/or unduly constrain the Administration's ability to effectively manage Federal credit programs.[110] Proponents of the New CREBs counter-argue that the New CREBs would "help limit the environmental consequences of continued reliance on power generated using fossil fuels." The tax-credit bonds, they argue, can attract investment from taxpayers that are unable to benefit from tax credits.[111]

(...continued)

tax credit is also treated as taxable interest. For example, a bondholder in a 30% tax bracket who receives a $100 tax credit from the bond purchase would also have $30 treated as taxable interest income, leaving a net tax credit of 70%. See https://www.appanet.org/files/PDFs/CREB.pdf.

[102] This included $500 million for governmental borrowers.

[103] DOE's description of the requests and $800 million IRS allocation is provided at http://apps1.eere.energy.gov/news/news_detail.cfm/news_id=10423.

[104] DOE's description of the requests and $400 million IRS allocation is provided at http://apps1.eere.energy.gov/news/news_detail.cfm/news_id=11575.

[105] This provision is identical to section 1506 of H.R. 6 passed by the House on December 6, 2007. S.Amdt. 3841 included an identical provision (§1505). However, the amendment was defeated on a cloture vote and it was not further considered.

[106] The previous tax credit rate for CREBs was set as the rate that would permit issuance of CREBs without discount and interest cost to the issuer.

[107] H.Rept. 110-214. *Renewable Energy and Energy Conservation Tax Act of 2007.* June 27, 2007. p. 40.

[108] Joint Committee on Taxation. *Estimated Revenue Effects of the Tax Provisions Contained in H.R. 5351.* February 27, 2008. http://www.house.gov/jct/x-20-08.pdf.

[109] Executive Office of the President. Office of Management and Budget. Statement of Administration Policy on H.R. 5351. February 26, 2008. p. 2. http://www.whitehouse.gov/omb/legislative/sap/110-2/saphr5351-r.pdf.

[110] Executive Office of the President. Office of Management and Budget. Statement of Administration Policy on H.R. 2776 and H.R. 3221. August 3, 2007. p. 2. http://www.energy.gov/media/SAP_on_HR2776_and_HR3221.pdf

[111] U.S. Congress. House. Committee on Ways and Means. *Renewable Energy and Energy Conservation Tax Act of 2007.* (H.Rept. 110-214) p. 39.

Table 3. Clean Renewable Energy Bonds History

Public Law	Bond Authorization	Requests for Bond Authority	Allocation Date
P.L. 109-58 (§1303)	$800 million	$2.6 billion	Nov. 20, 2006
P.L. 109-432 (§202)	$400 million	$898 million	Feb. 8, 2008
P.L. 110-343 (Div. B, §107)	$800 million	—	—

Revenue Offsets Debate

The Emergency Economic Stabilization Act (P.L. 110-343, Div. B, Title III) requires nearly $17.0 billion in revenue offsets, primarily to support incentives for renewables (Title I) and efficiency (Title II).[112] Debate over the revenue offset provisions in H.R. 1424 directly paralleled the House and Senate floor debates over similar proposals in other key bills during the first session (H.R. 3221 and H.R. 6) and second session (H.R. 5351 and H.R. 6049). In those debates, opponents argued that the reduction in oil and natural gas incentives would dampen production, cause job losses, and lead to higher prices for gasoline and other fuels. Proponents counter-argued that record profits show that the oil and natural gas incentives were not needed and that the new incentives would help spur the development of "green" jobs.

Regulatory Issues and Action

Renewable Portfolio Standard (RPS)

Under a renewable energy portfolio standard (RPS), retail electricity suppliers (electric utilities) must provide a minimum amount of electricity from renewable energy resources or purchase tradable credits that represent an equivalent amount of renewable energy production. The minimum requirement is often set as a percentage of retail electricity sales. More than 28 states have established an RPS, with most targets ranging from 10% to 20% and most target deadlines ranging from 2010 to 2025.[113] Most states have established tradable credits as a way to lower costs and facilitate compliance. State RPS action has provided an experience base for the design of a possible national requirement.

[112] H.R. 5351 also includes $1.83 billion for "New York Liberty Zone" tax credits for transportation infrastructure projects proposed in the Administration's FY2009 budget. For more discussion of the revenue offset provisions, see CRS Report RL33578, *Energy Tax Policy: History and Current Issues*, by Salvatore Lazzari.

[113] The Federal Energy Regulatory Commission (FERC) posts a tally of state RPS action that is updated regularly. http://www.ferc.gov/market-oversight/mkt-electric/overview/elec-ovr-rps.pdf.

State RPS Debate

Opponents often contend that state RPS policies are not worth implementing because the incremental costs of renewable energy may lead to substantial increases in electricity prices. RPS proponents often counter by presenting evidence that renewable energy costs would be modest and arguing that RPS creates employment, reduces natural gas prices, and produces environmental benefits.[114]

Federal Tax Credit (PTC) Supports State RPS Policies

The renewable energy electricity production tax credit (PTC) is the single most important form of federal support for state RPS policies. The PTC can "buy-down" the cost of renewable energy by about $20/mwh on a long-term levelized cost basis. Thus, assumptions about the future availability and level of the PTC can have a major impact on planning for state RPS policies.[115] Otherwise, federal agency involvement with state RPS programs has primarily involved support for planning and analysis.[116]

Federal RPS Debate

RPS proponents contend that a national system of tradable credits would enable retail suppliers in states with fewer resources to comply at the least cost by purchasing credits from organizations in states with a surplus of low-cost production. Opponents counter that regional differences in availability, amount, and types of renewable energy resources would make a federal RPS unfair and costly.

During the first session of the 110th Congress, RPS action began with Senate floor consideration of S.Amdt. 1537 to H.R. 6. The amendment proposed a 15% RPS target. The proposal triggered a lively debate, but was ultimately ruled non-germane. In that debate, opponents argued that a national RPS would disadvantage certain regions of the country, particularly the Southeastern states. They contended that the South lacks a sufficient amount of renewable energy resources to meet a 15% renewables requirement. They further concluded that an RPS would cause retail electricity prices to rise for many consumers.

---

[114] DOE. Lawrence Berkeley National Laboratory. Weighing the Costs and Benefits of State Renewables Portfolio Standards: A Comparative Analysis of State-Level Policy Impact Projections. March 2007. p. 58. http://eetd.lbl.gov/ea/ems/reports/61580.pdf. This survey of 28 state RPS cost projection studies found two that estimated rate increases greater than 5% and 19 that estimated rate increases less than 1%. Of the latter 19 studies, six estimated rate decreases. The study concludes that "when combined with possible natural gas price reductions and corresponding gas bill savings, the overall cost impacts are even more modest."

[115] DOE. Lawrence Berkeley National Laboratory. *Weighing the Costs and Benefits of State Renewables Portfolio Standards: A Comparative Analysis of State-Level Policy Impact Projections.* March 2007. p. 50. http://eetd.lbl.gov/ea/ems/reports/61580.pdf.

[116] Under its State and Local Program, the Environmental Protection Agency (EPA) has provided online workshops (conference calls) that have promoted collaboration between various states with an RPS in place. FERC has prepared studies and rulemakings related to transmission, grid interconnection, and other RPS-related policies. NREL has prepared various studies of state RPS programs and activities. EIA has prepared studies projecting impacts of RPS proposals on electricity and natural gas prices. Some of these EIA studies are cited under the below section on "Federal RPS Debate."

RPS proponents countered by citing an EIA study that examined the potential impacts of the 15% RPS proposed in S.Amdt. 1537. It indicated that the South has sufficient biomass generation, both from dedicated biomass plants and existing coal plants co-firing with biomass fuel, to meet a 15% RPS. EIA noted further that the estimated net RPS requirement for the South would not make it "unusually dependent" on other regions and was in fact "below the national average requirement.... " Regarding electricity prices, EIA estimated that the 15% RPS would likely raise retail prices by slightly less than 1% over the 2005 to 2030 period. Further, the RPS would likely cause retail natural gas prices to fall slightly over that period.

In House floor action on H.R. 3221, an RPS amendment (H.Amdt. 748) was added by a vote of 220 to 190. The bill subsequently passed the House by a vote of 241 to 172. The RPS amendment would set a 15% target for 2020, of which up to four percentage points of the requirement could be met with energy efficiency measures. Key points and counterpoints of the Senate debate were repeated. On the House floor, RPS opponents also contended that biomass power technologies were not yet ready for commercial use and that certain usable forms of biomass were excluded. Proponents acknowledged that there is a need to expand the definition of biomass resources, and offered to do so in conference committee.

On December 6, 2007, the House approved the same RPS provision as section 1401 of the omnibus energy bill, H.R. 6. However, the Senate passed H.R. 6 without an RPS provision. Thus, the Energy Independence and Security Act (P.L. 110-140) did not contain an RPS. (For more details see CRS Report RL34116, *Renewable Energy Portfolio Standard (RPS): Background and Debate Over a National Requirement*, by Fred Sissine.)

Other Regulatory Issues

Wind Energy

Major wind developments in Europe have expanded from land-based operations to include some offshore coastal areas. Proposals to develop offshore wind have emerged in the United States as well. During the 109th Congress, a major debate erupted over safety, economic, and environmental aspects of a proposal by Cape Wind Associates to develop a 420-megawatt offshore wind farm in Nantucket Sound, south of Cape Cod, Massachusetts. Cape Wind and other proponents say the project is a safe, clean way to develop renewable energy and create jobs. Opponents of the project have collaborated to create the Alliance to Protect Nantucket Sound. The Alliance says that the project poses threats to the area's ecosystem, maritime navigation, and the Cape Cod tourism-based economy.

EPACT (§388) placed regulatory responsibility for offshore wind developments with the Minerals Management Service (MMS) of the Department of the Interior. In 2006, MMS announced that an environmental impact statement (EIS) would be prepared for the project. In February 2007, Cape Wind submitted its draft EIS to MMS.[117] MMS released its Draft Environmental Impact Statement in January 2008.[118] The study found that environmental, fishery, and marine transportation impacts would range from negligible to minor. On-shore visual impacts would be

[117] Cape Wind has posted its draft EIS at http://www.capewind.org/article137.htm.

[118] MMS. *Draft Environmental Impact Statement* Available on the MMS website at http://www.mms.gov/offshore/RenewableEnergy/DEIS/Volume%201%20-%20Cape%20Wind%20DEIS/Cape%20Wind%20DEIS.pdf.

moderate. After the report was released, MMS began a two-month review and comment period. Also, the Coast Guard Act of 2006 (P.L. 109-241, §414) directs the Coast Guard to determine the status of navigational safety aspects for the Cape Wind Project. The parties to the debate are waiting for the final results of the EIS and Coast Guard study.

There is also a concern that tall wind turbines create false radar signals that may disrupt civilian and military radar equipment.[119] This led to federal actions to temporarily halt several wind farm developments. The Defense Authorization Act for FY2006 directed the Department of Defense (DOD) to study the issue and report to Congress. In 2006, the Sierra Club filed suit to compel DOD to complete the radar study. DOD released the report in late 2006,[120] and allowed most of the delayed projects to resume action. However, the report concluded that some mitigation strategies would have to be conducted on a case-by-case basis and that the development of additional mitigation measures would require further research and validation.

The impact of wind turbines on wildlife has also become a focus of concern. H.R. 3221 (§7231-7234) would have required the Department of the Interior to form a committee to recommend guidance to minimize and assess impacts of land-based wind turbines on wildlife and wildlife habitats. State and federal laws (and regulations) would not be preempted. However, this provision was not included in the final version of H.R. 6 that was enacted as the Energy Independence and Security Act (P.L. 110-140).

Marine (Tidal, Wave, and Ocean) Energy and Hydrokinetic (River Current) Energy

Technology that generates electricity from marine sources—including ocean waves, tides, and river currents—has reached the pre-commercial stage. Tax incentives and other programs have been established in Florida, Maine, and New Jersey to encourage commercial development. MMS has authority under EPACT (§388) to regulate development of ocean energy resources on the outer continental shelf (OCS). The Federal Energy Regulatory Commission (FERC) has asserted its authority to regulate these technologies, which it considers to be forms of hydropower. As these technologies develop to commercial scale, environmental issues are likely to arise, over which several other agencies appear to have regulatory jurisdiction. As technologies advance and new incentives become available, the regulatory struggle between MMS and FERC, and the potential regulatory roles of other agencies, may grow in importance.[121]

The 110th Congress,[122] took two major actions to promote marine and hydrokinetic power technologies. First, the Energy Independence and Security Act (P.L. 110-140) directs DOE to create an R&D program focused on technology that produces electricity from waves, tides, currents, and ocean thermal differences (§633). A report to Congress is required. Further, DOE is

---

[119] More information on this issue is available on DOE's website at http://www.eere.energy.gov/windandhydro/windpoweringamerica/ne_issues_interference.asp.

[120] The report is available at http://www.defenselink.mil/pubs/pdfs/WindFarmReport.pdf.

[121] For more information, see CRS Report RL33883, *Issues Affecting Tidal, Wave, and In-Stream Generation Projects*, by Nic Lane.

[122] The 109th Congress considered, but did not enact, legislation for these technologies that would have authorized guaranteed loans and direct revenues from Outer Continental Shelf (OCS) leases to fund ocean energy development. Also, a proposal to expand the renewable energy production tax credit (PTC) to include these technologies was approved by the Senate, but it was dropped in conference committee.

instructed to award grants to institutions of higher education (or consortia thereof) to establish National Marine Renewable Energy Research, Development, and Demonstration Centers (§634). The FY2008 Consolidated Appropriations Act (P.L. 110-161) provided $9.9 million for DOE's Water/Marine Energy Technology Program. The FY2009 Continuing Appropriations Act (P.L. 110-329) provides funding through March 9, 2009, at the same level as FY2008.

Second, P.L. 110-343 (Div. B, §102) expanded the renewable energy electricity production tax credit to include production from marine and hydrokinetic sources. The credit is set at 1 cent/kwh for the 2008 tax year. The window of eligibility will be open for three years, through the end of 2011.

Renewable Fuels and Energy Security

Types of Renewable "Biofuels"

Renewable fuel is defined to include ethanol, biodiesel, and certain other sources. Ethanol is the only one produced in large quantity.

Corn Ethanol

In the United States, ethanol is produced mainly from corn grown on farms.[123] It is most often used as a 10% blend with gasoline. Ethanol's high cost has been a key barrier to increased commercial use. This barrier has been addressed mainly by a 51-cent per gallon tax credit for fuel use. Also, there has been a debate over the net energy benefit of using corn ethanol.[124] National ethanol production was estimated at 6.48 billion gallons in 2007.[125] However, due to ethanol's lower heat content,[126] this is equivalent to about 4.34 billion gallons of gasoline, or about 285,000 barrels of oil per day (b/d).

Corn Ethanol Impacts and Debate

The U.S. Department of Agriculture (USDA) estimates that 20% of the 2006 corn crop was used to produce ethanol. The rapid growth in agriculture-based biofuel production generated a sharp upturn in corn, grain, and oilseed prices in late 2006. At the end of 2006, corn ethanol plant capacity expansion was on record pace. The rapid growth in production and plant capacity has

[123] Ethanol is the major farm-based renewable fuel. Corn provides 98% of ethanol production. Biodiesel is another important farm-based fuel, produced mainly from soybean oil. However, annual production is nearly 99% less than that for corn ethanol. For more information on farm-based renewable fuels, see CRS Report RL32712, *Agriculture-Based Renewable Energy Production*, by Randy Schnepf.

[124] For more information about ethanol developments and issues, see CRS Report RL33564, *Alternative Fuels and Advanced Technology Vehicles: Issues in Congress*, by Brent D. Yacobucci, and CRS Report RL33290, *Fuel Ethanol: Background and Public Policy Issues*, by Brent D. Yacobucci.

[125] DOE. *U.S. Ethanol Production Totaled 6.48 Billion Gallons in 2007*. March 12, 2008. http://www.ethanolrfa.org/industry/statistics/, http://apps1.eere.energy.gov/news/news_detail.cfm/news_id=11633.

[126] DOE, EIA, *Ethanol*. EIA reports that the heat content of ethanol is about 3.5 million Btu per barrel (42 gallons); see http://www.eia.doe.gov/oiaf/ethanol3.html. Also, EIA's *Monthly Energy Review*, at http://www.eia.doe.gov/emeu/mer/append_a.html, reports that the heat content of motor gasoline is 5.25 million Btu per barrel. Thus, on a per volume basis, ethanol has about 67% of the heat content of gasoline.

raised concerns that further acceleration of ethanol production may pose more challenges, including the development of pipeline capacity and the potential for more food price increases.[127]

Supporters argue that ethanol displaces petroleum imports, thus improving energy security. They further contend that its use can lead to lower emissions of air pollutants and greenhouse gases, especially if higher-percentage blends are used. Opponents argue that various federal and state incentives for ethanol distort the market and provide "corporate welfare" for corn growers and ethanol producers. Further, they assert that the energy and chemical inputs that fertilize corn and convert it into ethanol actually increase energy use and emissions. However, proponents counter-argue that ethanol provides modest energy and emissions benefits relative to gasoline.

Cellulosic Ethanol

Cellulosic ethanol can be produced from dedicated fuel crops, such as fast-growing trees and switchgrass. Switchgrass grows well on marginal lands, needing little water and no fertilizer. This allows its growing area to be much larger than that for corn.[128] Cellulosic feedstocks may be cheaper and more plentiful than corn, but they require more extensive and costly conversion to ethanol. Both DOE and USDA are conducting research to improve technology and reduce costs. The United States and Canada have pilot production facilities. Canada has one commercial-scale plant in operation, and the first U.S. commercial plants are expected to start operating in 2009.

Renewable Fuel Standard (RFS)

New Goals Set By the Energy Independence Act

Section 202 of the Energy Independence and Security Act of 2007 (P.L. 110-140) extends and increases the RFS. The standard requires minimum annual levels of renewable fuel in U.S. transportation fuel. The previous standard was 5.4 billion gallons for 2008, rising to 7.5 billion by 2012.[129] The new standard starts at 9.0 billion gallons in 2008 and rises to 36 billion gallons in 2022. Starting in 2016, all of the increase in the RFS target must be met with advanced biofuels, defined as cellulosic ethanol and other biofuels derived from feedstock other than corn starch—with explicit carve-outs for cellulosic biofuels and biomass-based diesel.[130]

The law gives the EPA Administrator authority to temporarily waive part of the biofuels mandate, if it were determined that a significant renewable feedstock disruption or other market circumstance might occur. Renewable fuels produced from new biorefineries will be required to reduce by at least 20% the life cycle greenhouse gas (GHG) emissions relative to life cycle emissions from gasoline and diesel. Fuels produced from biorefineries that displace more than

[127] For more information on renewable energy initiatives in the 2008 farm bill proposals, see CRS Report RL34130, *Renewable Energy Policy in the 2008 Farm Bill*, by Tom Capehart.

[128] For more information about using cellulosic biomass for ethanol production, see CRS Report RL32712, *Agriculture-Based Renewable Energy Production*, by Randy Schnepf.

[129] The previous standard was set by section 1501 of the Energy Policy Act of 2005 (EPACT, P.L. 109-58). Actual production had been exceeding EPACT targets.

[130] The RFS includes an "advanced biofuels mandate," which begins with 600 million gallons in 2009 and rises to 21 billion gallons in 2022. The cellulosic ethanol portion of the advanced biofuels mandate starts with 100 million gallons in 2010 and rises to 16 billion gallons in 2022.

80% of the fossil-derived processing fuels used to operate a biofuel production facility will qualify for cash awards. Several studies are required on the potential impacts of the RFS expansion on various sectors of the economy.

Implementation Concerns

In February 2008, the Senate Committee on Energy and Natural Resources held an oversight hearing on the new RFS.[131] Both leaders of the Committee, the Chairman[132] and the Ranking Member,[133] expressed concern that the RFS set by the Energy Independence Act may need changes in order to be implemented effectively. One major focus of concern is that the law may unintentionally preclude new technologies and feedstock sources, such as woody biomass from federal lands, urban and commercial waste, and biocrude from algae. (For more details on issues related to the RFS, see CRS Report RL34265, *Selected Issues Related to an Expansion of the Renewable Fuel Standard (RFS)*, by Brent D. Yacobucci and Tom Capehart.)

Potential to Reduce Oil Imports

Table 7 shows baseline EIA data for U.S. oil use and Persian Gulf Imports in 2006 and EIA projections for selected future years through 2030.[134] The table also shows ethanol production estimates for the current RFS of 36 billion gallons by 2022.[135] At its peak in 2022, the current RFS would displace an estimated 1.57 million barrels per day (mbd), or about 59% of projected Persian Gulf imports for that year.

Biofuels Funding and Tax Issues

Biofuels Technology Funding Initiative

The Bush Administration's Biofuels Initiative, part of the Advanced Energy Initiative (AEI), was designed to increase funding for cellulosic ethanol development with the goal of accelerating its commercial use.[136] In 2006, DOE formed a joint research effort between its Office of Energy Efficiency and Renewable Energy (EERE) and the Office of Science to develop cellulosic biotechnology that would enable the production of 60 billion gallons per year.[137] The research

---

[131] U.S. Senate. Committee on Energy and Natural Resources. *The Energy Market Effects of the Recently-Passed Renewable Fuel Standard.* Hearing held February 7, 2008. http://energy.senate.gov/public/index.cfm?FuseAction=Hearings.Hearing&Hearing_ID=1676.

[132] The Chairman's statement is available on the Committee's website, at http://energy.senate.gov/public/index.cfm?FuseAction=PressReleases.Detail&PressRelease_id=235445&Month=2&Year=2008&Party=0.

[133] The Ranking Member's statement is available on the Committee's website, at http://energy.senate.gov/public/index.cfm?FuseAction=PressReleases.Detail&PressRelease_id=235447&Month=2&Year=2008.

[134] To facilitate comparison, all figures in the table are shown in terms of millions of barrels per day, mbd.

[135] The RFS scenario is identified by its ultimate target, expressed in billions of gallons per year of ethanol production in a certain future year. The ethanol figures in **Table 7** were converted from billions of gallons per year to millions of barrels per day. They assume 100% corn ethanol, with 67% of the heat content of gasoline by volume.

[136] The White House, *Fact Sheet: President Bush's Four-Part Plan to Confront High Gasoline Prices*, April 26, 2005, at http://www.whitehouse.gov/news/releases/2006/04/20060425-2.html.

[137] DOE, *Factsheet on a Scientific Roadmap for Cellulosic Ethanol*, p. 1. Assuming that the 60 billion gallons per year is provided by ethanol, that would be equal to 3.9 million barrels per day of ethanol. Using the fact that ethanol has (continued...)

plan aims for biotechnology breakthroughs to increase the quantity of biomass (e.g., switchgrass) per acre and to breed the plants to have more cellulose. The plan would cut costs through biorefinery breakthroughs that reduce the number of conversion steps and shift the process from chemical steps to biological steps.[138]

As **Table 4** shows, DOE's FY2009 budget request would have provided $225.0 million for DOE's Biomass Program that supports the Biofuels Initiative and the RFS goals. This would have been a $26.8 million increase from the $198.2 million appropriated for FY2008. However, the Consolidated Security, Disaster Assistance, and Continuing Appropriations Act (P.L. 110-329) continues FY2009 funding at the FY2008 level through March 6, 2009. It does not provide for a funding increase.

Tax Incentives Provided in P.L. 110-343 (Division B) [139]

The law has four key tax incentive provisions for biofuels. Section 201 makes a 50% tax deduction available for the cost of building facilities that produce cellulosic biofuels, with the incentive available for four years (end of 2012). Section 202 extends for one year (end of 2009) the $1.00 per gallon production credit for biodiesel and the 10 cents per gallon credit for small biodiesel producers. Also, it extends the $1.00 per gallon production credit for biomass-derived diesel fuel.[140] Section 207 extends the alternative refueling stations credit for one year, through the end of 2009. The credit value is set at 30%, with a cap at $30,000. Section 203 clarifies that the production incentives in sections 201, 202, and 207 are available only for fuels produced in the United States.

Climate Change

This section discusses the potential for renewable energy to reduce carbon dioxide (CO_2) emissions by displacing fossil fuel use.

CO_2 Emissions Reduction Estimates

In most cases renewable energy appears to release less carbon dioxide (CO_2) than fossil fuels.[141] Thus, renewables are seen as a key long-term resource that could substitute for significant amounts of fossil energy that would otherwise be used to produce vehicle fuels and electricity.

(...continued)

about 67% of the heat content of gasoline by volume yields an estimate of 2.6 million barrels of oil equivalent per day. See http://www.er.doe.gov/News_Information/News_Room/2006/Biofuels/factsheet.htm.

[138] DOE, *Factsheet on a Scientific Roadmap for Cellulosic Ethanol*, p. 2.

[139] A summary of the provisions is available on the Senate Finance Committee website at http://finance.senate.gov/sitepages/leg/LEG%202008/100208%20Economic%20Stabilization%20Summary.pdf.

[140] Section 202 eliminates a previous gap in the credit for biodiesel and agri-biodiesel and it broadens the range of eligible processes that can be used to produce renewable diesel fuel.

[141] Because renewable energy is often developed for energy security, air pollution reduction, or other purposes, it is an example of a "no-regrets" strategy for CO_2 emission reductions. Wind and solar energy have zero CO_2 emissions in operation but may need an energy storage back-up system (such as batteries or fuel cells) that do require fossil fuel use. When biomass is developed as an energy crop, the CO_2 emissions are near zero because each new crop absorbs the same amount of emissions as are released by combusting the previous crop—unless fertilizer is used.

The potential percentage of renewable energy substitution can depend on many factors, including energy prices, energy demand growth,[142] technology cost, and market penetration. As renewable energy production displaces fossil fuel use, it would also reduce CO_2 emissions in direct proportion, except perhaps for biofuels and biopower.[143]

In general, the combustion of biomass for fuel and power production releases CO_2 at an intensity that may be close to that for natural gas. However, the re-growth of biomass material, which absorbs CO_2, often offsets this release. Hence, net emissions occur only when combustion is based on deforestation. In a "closed loop" system, biomass combustion is based on rotating energy crops, there is no net CO_2 release unless fertilizer is used, and any fossil fuel displacement, including decreased natural gas use, would tend to reduce CO_2 emissions.

Support for Renewables to Curb CO_2

Since 1988, the federal government has initiated programs to support renewable energy as a CO_2 mitigation measure at DOE, USDA, EPA, the Agency for International Development (AID), and the World Bank. AID and the World Bank have received funding for renewable energy-related climate actions through foreign operations appropriations bills.

States have undertaken a variety of programs that support renewables to curb CO_2. These programs often have reasons other than climate change for supporting renewables. California and New York are notable examples that have sizable programs for R&D and market deployment.[144] These programs are funded in large part by a surcharge on electricity use, often identified as a public goods charge.[145] As noted in a previous section of this report, many states have enacted a renewable portfolio standard. However, a growing number of states have also undertaken climate programs that specifically include renewables as one mitigation measure.[146] Many local governments have also undertaken climate programs that include renewables as a component.[147]

Climate Security Act (S. 3036)

The proposed Lieberman-Warner Climate Security Act (S. 3036) was introduced during the second session.[148] It would have established a cap-and-trade system to reduce greenhouse gas

---

[142] The use of energy efficiency measures can have a significant effect on energy prices and demand growth.

[143] Non-biomass renewables also tend to reduce emissions of other air-borne pollutants that cause urban smog, acid rain, and water pollution.

[144] California's renewable energy program is at http://www.energy.ca.gov/renewables/, and its climate program is at http://www.climatechange.ca.gov/; for more about New York's renewable energy program go to http://www.powernaturally.org/.

[145] The Database of State Incentives for Renewable Energy (DSIRE) has information about virtually all state renewable energy programs at http://www.dsireusa.org/.

[146] For more information see CRS Report RL33812, *Climate Change: Action by States To Address Greenhouse Gas Emissions*, by Jonathan L. Ramseur.

[147] Information about local government programs is available from the EPA website at http://www.epa.gov/climatechange/wycd/stateandlocalgov/local.html and from Cities for Climate Protection Campaign of the International Council for Local Environmental Initiatives at http://www.iclei.org/index.php?id=391.

[148] S. 3036 was introduced to replace S. 2191. An attempt to take up S. 3036 failed on a cloture vote. Additional description of the renewable energy provisions and the Senate floor process is provided in CRS Report RL33831, *Energy Efficiency and Renewable Energy Legislation in the 110th Congress*, by Fred Sissine, Lynn J. Cunningham, and (continued...)

emissions. The bill contained several provisions for energy efficiency and renewable energy. Revenue from the auctioned allowances could be used for multiple purposes, including accelerated deployment of renewable energy, energy efficiency, and other new energy technologies. Each year, nearly 40% of the revenue from auctions would go to efficiency and renewables. Compared with S. 3036, the proposed Boxer substitute to the bill would have established an even broader array of incentives for the deployment of energy efficiency and renewable energy measures. Eight of the 17 titles in the substitute contained such measures, including grants, worker training, incentives to facility developers, and leverage for private financing to support international partnerships.

Legislation

Major Laws Enacted in the First Session

FY2008 Appropriations (P.L. 110-161)

DOE's FY2008 budget request sought $1,236.2 million for DOE's Energy Efficiency and Renewable Energy (EERE) programs. In H.R. 2641, the House approved $1,873.8 million for EERE and the Senate Appropriations Committee recommended $1,715.6 million for EERE. The Consolidated Appropriations Act of 2007 (H.R. 2764) subsumed H.R. 2641, and the enacted law included $1,723.7 million for EERE. (Details of the FY2008 appropriations are available in the "Key Policy Issues—Department of Energy" section of CRS Report RL34009, *Energy and Water Development: FY2008 Appropriations*, by Carl E. Behrens et al.)

Energy Independence and Security Act (P.L. 110-140)

At the end of its first session, the 110th Congress enacted a major omnibus energy bill focused on improving energy efficiency and increasing the availability of renewable energy. Highlights of the major provisions enacted are:

- *Corporate Average Fuel Economy (CAFE).* Title I sets a target of 35 miles per gallon for the combined fleet of cars and light trucks by model year 2020.
- *Renewable Fuels Standard (RFS).* Title II sets a modified standard that starts at 8.5 billion gallons in 2008 and rises to 36 billion gallons by 2022.
- *Appliance and Lighting Standards.* Title III legislates new standards for broad categories of incandescent lamps (light bulbs), incandescent reflector lamps, and fluorescent lamps. Further, a required target is set for lighting efficiency, and energy efficiency labeling is required for consumer electronic products. Efficiency standards are set by law for external power supplies, residential clothes washers, dishwashers, dehumidifiers, refrigerators, refrigerator freezers, freezers, electric motors, residential boilers, commercial walk-in coolers, and

(...continued)
Mark Gurevitz.

commercial walk-in freezers. Further, DOE is directed to set standards by rulemaking for furnace fans and battery chargers.

(For more details about the provisions in P.L. 110-140, see CRS Report RL34294, *Energy Independence and Security Act of 2007: A Summary of Major Provisions*, by Fred Sissine).

Major Laws Enacted in Second Session

Farm Bill (P.L. 110-246) Provisions

The enacted version of the farm bill (H.R. 6124; "Food, Conservation, and Energy Act of 2008") became law on June 18, 2008.[149] The law contains provisions that extend and/or expand upon renewable energy (and energy efficiency) provisions of the Farm Security Act of 2002 (P.L. 107-171). Several programs for grants, loans, and tax incentives were put in place to support renewable energy (and/or energy efficiency). (For more details, see CRS Report RL34130, *Renewable Energy Policy in the 2008 Farm Bill*, by Tom Capehart.)

FY2009 Appropriations (P.L. 110-329)

As **Table 4** shows, DOE's FY2009 budget request sought $1.255 billion for DOE's Energy Efficiency and Renewable Energy (EERE) programs. P.L. 110-329 (H.R. 2638) provides continuing appropriations through March 6, 2009, at the FY2008 level. The law provides an additional $250 million to the DOE Weatherization Assistance Program and provides $7.5 billion for a $25 billion loan to help U.S. automakers retool facilities to produce advanced technology energy-efficient vehicles. The law covers appropriations through March 6, 2009.

Emergency Economic Stabilization Act (P.L. 110-343)

Division B of the Emergency Economic Stabilization Act (P.L. 110-343) contains the text of the Senate-passed version of H.R. 6049, which provides several tax incentives for efficiency and renewables. The Senate crafted its substitute to H.R. 6049 as a response to the House-passed version of the bill and to Administration-expressed concerns about House provisions for renewable energy bonds and revenue offsets. The highlights of key provisions enacted into law are as follows:

- *Renewable Electricity Production Tax Credit (PTC).* The credit for wind farms is extended for one year, through the end of 2009. Other equipment are eligible for two years, through 2010. Newly eligible marine technologies are eligible for three years, through 2011.
- *Solar Investment Tax Credits (ITC) for Residential and Commercial Sectors.* The law extends the existing 30% credit for each sector for eight years, through the end of 2016.

[149] The House and Senate overrode an Administration veto to enact the Food, Conservation, and Energy Act of 2008 (P.L. 110-234, H.R. 2419) on May 22, 2008. Due to a technical error that left one title out of the copy vetoed by the President, a second identical bill, H.R. 6124 was passed by both chambers. Upon the President's veto of H.R. 6124, the House overrode the veto by a vote of 317 to 109 and the Senate overrode the veto by a vote of 80 to 14. H.R. 6124 was enacted as P.L. 110-246.

- *Clean Renewable Energy Bonds (CREBs).* The law authorizes a new round of state and local bond issuances, with a total national value of $800 million.
- *Energy Conservation Bonds.* The law authorizes a new state and local program, with a total national value of $800 million.
- *Revenue Offsets.* The cost of incentives are offset by a freeze in certain oil and natural gas deductions, a reduced foreign tax credit for certain foreign oil and gas income, reduced deductions for certain securities transactions, a change in the Federal Unemployment Tax Act (FUTA) surtax, and an increase of the Oil Spill Liability Trust Fund tax.

Other Laws and Bills

In the 110th Congress, more than 460 bills with provisions for energy efficiency or renewable energy were introduced. A general description of the renewable energy provisions in those bills, including those enacted into law, is available in CRS Report RL33831, *Energy Efficiency and Renewable Energy Legislation in the 110th Congress*, by Fred Sissine, Lynn J. Cunningham, and Mark Gurevitz. The report also groups the bills by policy and issue areas, provides a table that identifies recent action on the bills, and discusses recent action.

Table 4. DOE Renewable Energy Budget for FY2006-FY2009

(selected programs, $ millions)

Program	FY2006	FY2007	FY2008	FY2009 Request	FY2009 House	FY2009 Senate
Biomass & Biorefinery Systems	$89.8	$199.7	$198.2	$225.0	$250.0	$235.0
—Cellulosic Ethanol Auction	10.4	0.0	5.0	0.0	—	—
Solar Energy Technology	81.8	157.0	168.5	156.1	220.0	229.0
—Photovoltaics	58.8	138.4	136.7	137.1	—	156.8
—Concentrating Solar	7.3	15.7	29.7	19.0	—	50.0
—Solar Heating & Lighting	1.4	3.0	2.0	0.0	—	22.2
Wind Energy Technology	38.3	48.7	49.5	52.5	53.0	62.5
Geothermal Technology	22.8	5.0	19.8	30.0	50.0	30.0
Water/Marine Technology	0.5	0.0	9.9	3.0	40.0	30.0
Subtotal, Renew. Technologies	233.2	407.0	445.9	466.6	613.0	586.5
International Renewables	3.9	9.5	0.0	0.0	7.0	0.0
Tribal Energy	4.0	4.0	5.9	1.0	6.0	6.0
Renewables Prod'n Incentive	5.0	4.9	5.0	0.0	5.0	5.0
Asia Pacific Partner. (Renew.)	0.0	—	0.0	7.5	0.0	0.0
Subtotal, Renew. Deployment	12.9	18.4	10.9	8.5	18.0	11.0
Subtotal, Renewables	246.1	425.4	456.8	475.1	631.0	597.5
Hydrogen Technologies	153.5	189.5	211.1	146.2	170.0	175.0
Vehicle Technologies	178.4	183.6	213.0	221.1	317.5	293.0

Program	FY2006	FY2007	FY2008	FY2009 Request	FY2009 House	FY2009 Senate
Building Technologies	68.2	103.0	109.0	123.8	168.0	176.5
Industrial Technologies	55.9	55.8	64.4	62.1	100.0	65.1
Federal Energy Management	19.0	19.5	19.8	22.0	30.0	22.0
Subtotal, Efficiency R&D	475.0	551.4	617.3	575.2	785.5	731.6
Facilities (Nat. Renew. Lab)	26.1	107.0	76.2	14.0	33.0	37.0
Program Management	115.2	110.2	114.9	141.8	147.6	136.8
—Weatherization/State Grants	278.7	263.5	271.3	50.0	300.0	251.2
—Renewables Deployment	12.9	18.4	10.9	8.5	18.0	11.0
—Cong.-Directed Assistance[a]	—	0.0	186.7	0.0	134.7	124.2
—Prior Year Balances	—	—	-0.7	-0.7	-0.7	0.0
Federal Assistance Subtotal	316.9	281.7	468.1	57.8	452.0	386.4
Total Appropriation, EE & RE	1,166.1	1,457.2	1,722.4	1,255.4	2531.1	1,928.3
Office of Electricity Delivery & Energy Reliability (OE)[b]	158.2	134.4	138.6	134.0	149.3	166.9

Sources: DOE *FY2009 Congressional Budget Request*, vol. 3, February 2008; DOE FY2007 Operating Plan; Congressional Record, December 17, 2007 (Book II), H.R. 2764, Division C. For more details, see CRS Report RL34009, *Energy and Water Development: FY2008 Appropriations*, by Carl E. Behrens et al.

a. In FY2006, there was $159.0 million in congressionally-directed funds spread over EERE accounts.

b. The Distributed Energy Program was moved from EERE to OE in FY2006.

Table 5. Production Tax Credit Value and Duration by Resource

Energy Resource	Credit Amount for 2008 (cents/kwh)	Credit Period for Facilities Placed in Service after August 8, 2005 (years)
Wind	2.1	10
Closed-Loop Biomass	2.1	10
Open-Loop Biomass (includes agricultural livestock waste nutrient facilities)	1.0	10
Geothermal	2.1	10
Solar (pre-2006 facilities only)	2.1	10
Small Irrigation Power	1.0	10
Incremental Hydropower	1.0	10
Municipal Solid Waste (includes landfill gas and trash combustion facilities)	1.0	10

Source: Joint Committee on Taxation. Technical Explanation of H.R. 7060, The "Renewable Energy and Job Creation Tax Act of 2008,"a Scheduled for Consideration by the House of Representatives on September 25, 2008." (JCX-75-08) September 25, 2008. p. 7. http://www.house.gov/jct/x-75-08.pdf.

Table 6. Production Tax Credit Claims, History and Projections

($ millions)

Year	Public Law	Credit Lapse (months)	PTC Claims ($ current)	Deflator ($ 2007)	PTC Claims ($ 2007)	Number of States with RPS
History						
1995	P.L. 102-486		3.2	0.7711	4.2	2
1996	P.L. 102-486		9.3	0.7859	11.9	3
1997	P.L. 102-486		9.4	0.7996	11.7	6
1998	P.L. 102-486		13.9	0.8093	17.2	9
1999	P.L. 102-486, P.L. 106-170	6 months	28.9	0.8199	35.3	11
2000	P.L. 106-170		50.1	0.8365	59.9	12
2001	P.L. 106-170		70.6	0.8562	82.5	12
2002	P.L. 107-147	2 months	131.6	0.8726	150.8	13
2003	P.L. 107-147		142.8	0.8903	160.4	13
2004	P.L. 108-311	9 months	207.3	0.9134	227.0	18
2005	P.L. 108-311		333.2	0.9427	353.5	21
Total, History			1,000.40		1,114.3	
JCT Future Estimates						
2006	P.L. 109-58		900	0.9739	924	23
2007	P.L. 109-58		900	1.0000	900	24
2008	P.L. 109-432		1,000	1.0193	981	28
2009			1,600	1.0399	1,539	
2010			1,200	1.0606	1,131	
Total, Future Estimates			5,600		5,475	

Source: Historical data on PTC claims for 1995 through 2005 were obtained from Mr. Curtis Carlson, Office of Tax Analysis, Department of the Treasury. Estimates of PTC claims for 2005 through 2010 were obtained by combining estimates from the Joint Committee on Taxation for the PTC provisions in P.L. 109-58, P.L. 109-432, and P.L. 110-343.

Table 7. Renewable Fuels Compared with Persian Gulf Imports

(millions of barrels per day, mbd)

Year	Oil Use or Oil Use Equivalent (mbd)[a]				As a Percent of Persian Gulf Imports	
	Total Oil Use	Persian Gulf Imports	7.5-in 2012 (EPACT)	36-in-2022 (P.L. 110-140)	7.5-in 2012	36-in-2022
2005 Actual	20.66	2.59	0.17	0.17	6.6%	6.6%
2006 Actual	20.45	2.67	0.24	0.24	9.0%	9.0%
2007	20.52	2.72	0.20	0.20	7.4%	7.4%
2008	20.57	2.76	0.24	0.37	8.7%	13.4%
2009	20.40	2.57	0.27	0.46	10.5%	17.9%
2010	20.66	2.63	0.30	0.52	11.4%	19.8%
2011	20.87	2.69	0.32	0.55	11.9%	20.4%
2012	21.04	2.75	**0.33**	0.57	**12.0%**	20.7%
2017	21.30	2.70	—	0.91	—	33.7%
2022	20.83	2.65	—	**1.57**	—	**59.2%**

Sources: For Total Oil Use and Persian Gulf Imports, see EIA, Energy Information Administration. *Annual Energy Outlook 2008.* Oil use data were obtained from Supplemental Table 10. Those data were converted with the equivalence that one quad per year is approximately equal to 0.5106 million barrels per day, from EIA Monthly Energy Review, December 2006, Table A3: Approximate Heat Content of Petroleum Consumption. Persian Gulf oil import data were obtained from Supplemental Table 117. For the 7.5-in-2012 renewable fuel standard (RFS), see P.L. 109-58 (EPACT), §1501. For new RFS "36-in-2022" standard, see P.L. 110-140 (Energy Independence and Security Act of 2007). Note that all displacements assume 100% ethanol, with 67% of the heat content of gasoline by volume. The ethanol figures also reflect the conversion that 42 gallons equal one barrel.

a. The ethanol figures for 7.5-in-2012 (EPACT), and 36-in-2022 (P.L. 110-140) assume 100% corn ethanol, with 67% of the heat content of gasoline by volume. The ethanol figures also reflect the conversion that 42 gallons equal one barrel.

Author Contact Information

Fred Sissine
Specialist in Energy Policy
fsissine@crs.loc.gov, 7-7039

Order Code RL33541

CRS Report for Congress

Background on Sugar Policy Issues

Updated July 26, 2007

Remy Jurenas
Specialist in Agricultural Policy
Resources, Science, and Industry Division

Prepared for Members and Committees of Congress

Background on Sugar Policy Issues

Summary

The sugar program, authorized by the 2002 farm bill (P.L. 107-171), is designed to protect the price received by growers of sugarcane and sugar beets, and by firms that process these crops into sugar. To accomplish this, the U.S. Department of Agriculture (USDA) makes loans available at mandated price levels to processors, limits the amount of sugar that processors can sell domestically, and restricts imports. In support of the program, sugar crop growers and processors stress the industry's importance in providing jobs and income in rural areas. Food and beverage firms that use sugar argue that U.S. sugar policy imposes costs on consumers, and has led some food manufacturers to move jobs overseas where sugar is cheaper.

In a major policy change, the 2002 farm bill reactivated sugar marketing allotments that limit the amount of domestically produced sugar that processors can sell. The level at which USDA sets the national sugar allotment quantity, in turn, has implications for sugar prices. Accordingly, sugar crop producers and processors on one side, and sugar users on the other, have sought to advance their interests by influencing the decisions that USDA makes on allotment and import quota levels.

The issue of additional sugar imports "crowding out" domestic production was divisive when Congress debated the Dominican Republic-Central American Free Trade Agreement (DR-CAFTA) in 2005. Since then, attention on sugar trade issues has turned to the potential impact of free trade in sugar and high-fructose corn syrup (HFCS) — a substitute and cheaper sweetener — between the United States and Mexico, which takes effect on January 1, 2008. Unrestricted sugar imports from Mexico are projected to result in budget outlays (estimated at $1.4 billion over 10 years) as U.S. processors default on price support loans. This outlook conflicts with the current objective that the program operate at no cost and has become a key issue in crafting sugar provisions for the 2007 farm bill.

The House Agriculture Committee-reported farm bill (H.R. 2419) would mandate a sugar-for-ethanol program intended to address any sugar surplus that arises as a result of imports. USDA would be required to purchase as much U.S.-produced sugar as necessary to maintain market prices above support levels. Purchased sugar would then be sold to bioenergy producers for processing into ethanol. The CCC would provide open-ended funding for this program. Other provisions would increase minimum guaranteed prices for raw cane and refined beet sugar by almost 3%, and tighten the rules that USDA must follow to implement marketing allotments and administer import quotas (i.e., remove discretionary authority). These provisions reflect recommendations made by sugar crop producers and processors.

Food and beverage manufacturers that use sugar oppose these provisions, arguing that they "would take a sugar program from bad to worse," would increase costs by $100 million annually to consumers, and would restrict the availability of sugar for food use in the domestic market. They have signaled their intent to offer amendments during House floor debate to strike some of the committee-reported provisions and/or to extend the current program. This report will be updated.

Contents

List of Figures

For additional information, see CRS Report RL34103, *Sugar Policy and the 2007 Farm Bill*, by Remy Jurenas.

Sugar Policy Issues

Recent Developments

On July 26, 2007, the House Rules Committee reported out a rule (H.Res. 574; H.Rept. 110-261) that will be followed in floor debate on the 2007 farm bill (H.R. 2419). One amendment that will be permitted to be offered would strike all of the House farm bill's sugar provisions (including the sugar-for-ethanol program) and extend current program authority through 2012.

On July 19, 2007, the House Agriculture Committee completed consideration of its farm bill. The sugar provisions (reflecting recommendations made by the domestic sugar producers and processors) call for increasing sugar price support levels by almost 3%, revising marketing allotment authority to guarantee the domestic sector a minimum 85% share of the U.S. marketplace, and mandating that surplus sugar be purchased for resale for processing into ethanol as one way to meet the program's no-cost objective. Sugar producers and processors support the measure, appreciative that current sugar policy was not "weakened." Domestic manufacturers of food and beverage products that use sugar, represented by the Sweetener Users Association (SUA), responded that the proposed program "would take the U.S. sugar program from bad to worse," increase costs to consumers, and result in sugar program costs of almost $2 billion during the farm bill's five years.

History of and Background on the Sugar Program

Governments of every sugar-producing nation intervene to protect their domestic industry from fluctuating world market prices. Such intervention is necessary, it is argued, because both sugar cane and sugar beets must be processed soon after harvest using costly processing machinery. When farmers significantly reduce production because of low prices, a cane or beet processing plant typically shuts down, usually never to reopen. This close link between production and capital-intensive processing makes price stability important to industry survival.

The United States has a long history of protection and support for its sugar industry. The Sugar Acts of 1934, 1937, and 1948 required the U.S. Department of Agriculture (USDA) to estimate domestic sugar consumption and divide this market by assigning quotas to U.S. growers and foreign countries. These acts also authorized payments to growers when needed as an incentive to limit production and levied excise taxes on sugar processed and refined in the United States. This type of sugar program expired in 1974. For the next seven years, the U.S. market was relatively open to foreign sugar imports, with mandatory price support provided only in 1977 and 1978, and discretionary support in 1979. Congress reinstated mandatory price support for sugar in the Agriculture and Food Act of 1981 and the Food Security Act of 1985. Subsequently, the 1990 farm bill, the 1993 budget reconciliation bill, and the 1996 and 2002 farm bills extended sugar program authority. The last bill extends it up through the 2007 crop year (i.e., most of FY2008).

Even with price protection available to producers, the United States historically has not produced enough sugar to satisfy domestic demand and thus continues to be a net sugar importer. Prior to the early 1980s, domestic sugar growers supplied roughly 55% of the U.S. sugar market. This share has grown over the last 25 years, reflecting the price protection provided by the sugar program. In FY2006, domestic production filled 73% of U.S. sugar demand for food and beverage use. As high-fructose corn syrup (HFCS) displaced sugar in the United States during the early 1980s, and domestic sugar production increased later in that decade, foreign suppliers absorbed the entire adjustment and saw their share of the U.S. market decline significantly. The import share of U.S. sugar food use in FY2006 was 27%.

Current U.S. sugar policy maintains domestic sugar prices above the world market price, and is structured to protect the domestic sugar producing sector (sugar beet and sugarcane producers, and the processors of their crops) and to ensure a sufficient supply. During 2006, the U.S. raw sugar domestic futures price averaged 22.1¢/lb., compared to the world raw sugar futures price of 15.5¢/lb. Because of the price differential, U.S. consumers and manufacturers of foods and beverages pay more for sugar than they would if imports were allowed to enter without any restriction. Various studies show that over the last 15 years, U.S. sugar users paid between $400 million and $1.9 billion more for sugar annually. These "cost to user" estimates vary widely, and largely reflect the extent of the difference between the higher U.S. price and the lower world price for sugar in the time period examined, and the differing assumptions and methodology analysts use to develop such estimates.

The sugar program differs from grains, rice, peanut, and cotton programs in that USDA makes no direct payments to beet and cane growers and processors. Structured this way, taxpayers do not directly support the program through federal government outlays. This fact is highlighted as a positive feature by the sugar production sector and program supporters. The program's support level and import protection, though, keep the U.S. sugar price above the price of sugar traded internationally, and constitute an indirect subsidy to the production sector by way of higher costs paid by U.S. sugar users and consumers. Program opponents frequently refer to this subsidy component to argue for changes to U.S. sugar policy.

Main Features of U.S. Sugar Policy

U.S. sugar policy uses three tools to ensure that domestic growers and sugar processors receive a minimum price for their sugar. This price is largely determined by the statutorily-set loan rate. "Marketing allotments" limit the amount of domestically-produced sugar that can be sold when imports are estimated below a specified level. Import quotas restrict the amount of foreign sugar allowed to enter the U.S. market. USDA decisions in administering these tools are intended to balance available sugar supply (i.e., domestic output plus imports) with U.S. food demand for sugar so that market prices do not fall below effective support levels. The 2002 farm bill further requires USDA to operate the sugar program on a "no-cost" basis (i.e., to result in no federal government outlays).

Price Support Loans

USDA extends "non-recourse" price support loans to processors of sugarcane and sugar beets rather than directly to the farmers who harvest these crops. Loans are available only to processors who agree to pay growers for deliveries of sugar beets and sugarcane at USDA-set minimum payment levels. Their "non-recourse" feature means a processor can exercise the legal right to hand over sugar it initially offered USDA as collateral for the loan to meet its repayment obligation, if the market price is below the effective support level when the loan comes due. These loans at times can be attractive to sugar processors as a source of short-term credit at below-prime interest rates.

Loan Rates. The 2002 farm bill freezes loan rates through the 2007 crop year, at 18¢/lb. for raw cane sugar and 22.9¢/lb. for refined beet sugar. These rates have not changed since 1995. The loan support for beet sugar is set higher than for raw sugar because it is available immediately after processing in refined form, ready for industrial food and beverage use and for human consumption. By contrast, raw cane sugar must go through a second stage of processing at a cane refinery to be converted into white refined sugar.

Effective Support Levels. The above loan rates do not serve as the intended price floor for sugar. In practice, USDA's aim is to support the raw cane sugar price at not less than 20.72¢ to 21.46¢/lb. (i.e., the state's price support level *plus* an amount that covers a processor's cost to transport raw cane sugar to a cane refinery *plus* the interest paid on any price support loan taken out *plus* location discounts). Similarly, USDA seeks to support the refined beet sugar price at not less than 23.5¢ to 27.13¢/lb. (i.e., the regional loan rate *plus* specified marketing costs *plus* interest paid on a price support loan *plus* a cash discount). To ensure that market prices do not fall below these "loan forfeiture," or "effective" price support, levels, USDA administers sugar marketing allotments and import quotas. A loan forfeiture (turning over sugar pledged as loan collateral to USDA) occurs if a processor concludes that the domestic market price when the loan comes due is below the "effective" sugar support level for its state or region. It is this level — not the loan rate — that represents the minimum level of program benefits intended for sugar crop growers and processors.

Marketing Allotments

The 2002 addition of marketing allotments to support domestic sugar prices reflected the sugar production sector's willingness to accept reduced sales in return for the assurance of price protection. Allotments serve as a tool to ensure that any growth in U.S. sugar demand is first met by the U.S. sugar sector, and to guarantee both the beet and cane sectors a specific share of the U.S. market. By regulating the amount of sugar that processors can sell, USDA is expected to meet the program's no-cost objective by keeping market prices above effective support levels, and thus not acquiring sugar as a result of any loan forfeitures.

Allotments Required When Sugar Imports Are Below 'Trigger' Level. USDA is required to announce marketing allotments when it projects annual sugar imports will be *below* 1.532 million short tons (ST) — referred to as the "trigger level." By limiting the amount of sugar that beet sugar refiners and raw cane mills can sell, this mechanism ensures that the United States meets its market access commitments for sugar imports under the World Trade Organization (WTO) and NAFTA agreements (see "Import Quotas").

USDA must weigh several factors to calculate the amount of domestic sugar that can be sold (the "overall allotment quantity" or OAQ) — (1) estimated consumption, (2) "reasonable" ending stocks, (3) beginning stocks, and (4) imports for human consumption. The formula that USDA uses to determine the OAQ must accommodate imports under both trade agreements up to the 1.532 million ST level. USDA can further adjust the OAQ if necessary to avoid loan forfeitures. Second, the OAQ must be split between the beet and cane sectors using 54.35% and 45.65% shares, respectively. Separate rules specify how each sector's allotment is to be allocated (i.e., distributed) to each processing firm. Once detailed calculations are made, each firm can sell only as much sugar as stated in its allotment notification received from USDA. Sugar produced in excess of a firm's allotment must be held off the market (referred to as "blocked stocks"). **Figure 1** illustrates how USDA implemented marketing allotments for FY2006, the second year it activated some of the less-known features of this authority.

Allotments Suspended When Imports Exceed Trigger Level. If USDA estimates sugar imports will be *above* 1.532 million ST, USDA must suspend marketing allotments (with the one exception noted below). If allotments are triggered, USDA is still required to make available price support loans to raw cane sugar processors and beet sugar refiners. Suspending allotments, though, would raise considerable uncertainty for domestic sugar prices. Depending upon the U.S. sugar production and food use outlook at that point in time, and estimated imports of sugar for food consumption, price scenarios could vary considerably. If sugar processors use the suspension to release blocked stocks of sugar, sugar prices (depending upon the additional import amount and the amount of stocks released) could decline to below loan forfeiture levels. This would likely result in USDA acquiring sugar from processors that decide not to repay their loans. If U.S. demand for sugar is higher than can be met by the domestic sugar sector (either from projected output and/or blocked stocks), prices as additional imports enter under a suspension could very well rise, and likely stay above loan forfeiture levels.

Figure 1. Implementation of Sugar Marketing Allotments, FY2006

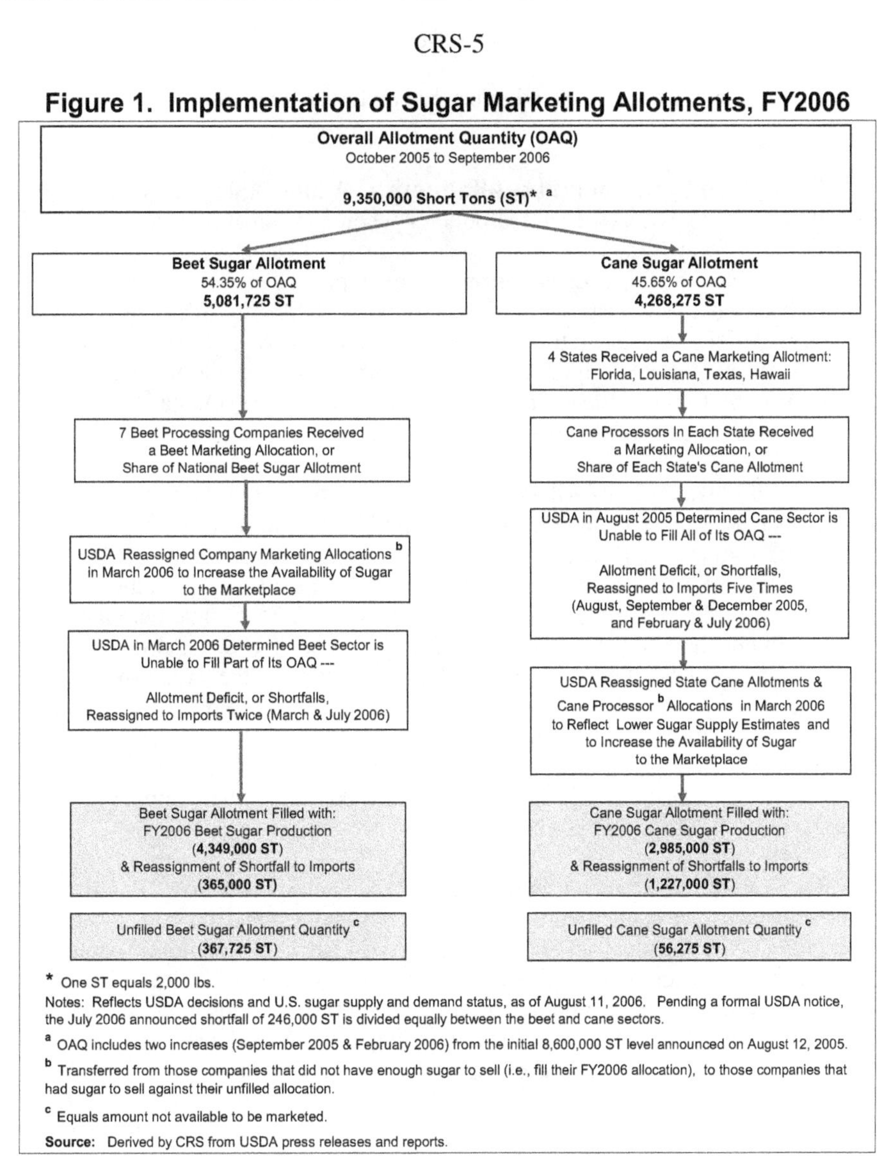

* One ST equals 2,000 lbs.

Notes: Reflects USDA decisions and U.S. sugar supply and demand status, as of August 11, 2006. Pending a formal USDA notice, the July 2006 announced shortfall of 246,000 ST is divided equally between the beet and cane sectors.

[a] OAQ includes two increases (September 2005 & February 2006) from the initial 8,600,000 ST level announced on August 12, 2005.

[b] Transferred from those companies that did not have enough sugar to sell (i.e., fill their FY2006 allocation), to those companies that had sugar to sell against their unfilled allocation.

[c] Equals amount not available to be marketed.

Source: Derived by CRS from USDA press releases and reports.

The USDA initial determination of the OAQ by August 1 for the marketing year that begins October 1, and any subsequent adjustments, are the most significant decisions made in implementing marketing allotment authority. Accordingly, sugar processors and food manufacturers (e.g., users of sugar) weigh in to influence USDA's decision-making process on this issue. Each group differs in how USDA should define "reasonable" ending stocks — a key determinant of the level of domestic sugar prices in the last quarter of the marketing year (July to September) when price support loans come due. However, USDA must estimate a year in advance (even though market conditions can change significantly during this period) an OAQ level intended to result in market prices 12 months later that are above effective support levels. Sugar processors favor a smaller OAQ, hoping to benefit from sugar prices well above effective support levels. Food manufacturers advocate a larger OAQ, hoping that year-end prices end up lower, and close to loan forfeiture levels.

Exception to Suspending Allotments. A little-noticed exception in the law allows USDA not to suspend allotments when additional imports exceed the trigger level, because the beet or cane sector is unable to supply sugar against their allotment. Since the law does not allow, for example, a cane sugar "deficit" to be met with any available "surplus" of beet sugar, USDA can "reassign" such an allotment deficit, or "shortfall," to imports. This occurred three times late in FY2005, and five times in FY2006, when USDA reassigned allotment deficits to imports in order to alleviate a tight supply situation (see "Import Quotas" and **Figure 1** for more on the relationship between allotments and imports).

FY2006 and FY2007 Allotment Announcements. On February 2, 2006, USDA increased the **FY2006** OAQ to 9.35 million ST, in order to respond "to a continuing tight sugar market" caused largely by Hurricanes Katrina, Rita, and Wilma, which reduced domestic supplies. This decision allowed both the beet and cane sectors (unlike previous years) to sell all of their FY2006 sugar output against their allotment share of the OAQ. To meet the balance of U.S. sugar demand that could not be met from domestic sources, USDA reassigned the large cane sector's and small beet sector's "allotment shortfall" to imports (**Figure 1**). Earlier, on September 30, 2005, USDA announced the initial details of FY2006 marketing allocations for beet processing companies. Cane state allotments and processor allocations were announced later on March 22, 2006, after the company-specific impacts from the 2005 season's hurricanes had become clear.

On July 27, 2006, USDA set the **FY2007** OAQ at 8.75 million ST, reflecting its assessment of the upcoming year's sugar supply and demand outlook. Projecting that 2006/2007 U.S. sugar production and carry-in stocks would not be enough to meet domestic needs and rebuild ending stocks "to reasonable levels," USDA immediately announced a supply shortfall of 350,000 ST. This entire deficit was reassigned to imports (reflected in the related sugar import quota announcement) to help exporters facilitate shipping arrangements. USDA stated this was intended to ensure that sufficient sugar is available to the U.S. market earlier than would otherwise occur. USDA also signaled that "appropriate adjustments" to the OAQ would be made during FY2007 to ensure the domestic market is adequately supplied, to avoid loan forfeitures, and to prevent market disruptions.

Import Quotas

USDA restricts the quantity of foreign sugar allowed to enter the United States for refining and sale for domestic food and beverage use. By controlling the amount of sugar allowed to enter, USDA seeks to ensure that market prices do not fall below effective price support levels, and thus not acquire sugar due to loan forfeitures (i.e., meet the "no-cost" requirement). Though the sugar import quotas are not directly addressed by farm bill provisions, USDA and the Office of the U.S. Trade Representative (USTR) administer them as an integral part of the sugar program.

Tariff-rate quotas (TRQs) are used to restrict sugar imports to the extent needed to meet U.S. sugar program objectives.[1] The U.S. market access commitment made under WTO rules means that a minimum of 1.256 million ST of foreign sugar must be allowed to enter the domestic market each year. Although the WTO commitment sets a minimum import level, policymakers may allow additional amounts of sugar to enter if needed to meet domestic demand (as occurred in FY2005 and FY2006, and announced for FY2007). In addition, the United States is committed to allow sugar to enter from Mexico under NAFTA provisions. The complex terms are detailed in a schedule and a controversial and disputed side letter, which lay out the rules and the formula to calculate how much sugar Mexico can sell to the U.S. market. Through FY2008, the maximum amount that can enter from Mexico is 276,000 ST (see "Sweetener Disputes with Mexico — Mexico's Terms of Access to the U.S. Sugar Market," below). Under the WTO agreement, foreign sugar enters under two TRQs — one for raw cane, another for a small quantity of refined sugar. Under the NAFTA TRQ and the DR-CAFTA TRQ, sugar can enter either in raw or refined form.

The USTR allocates the WTO TRQs among 41 eligible countries, including Mexico and Canada. The amount entering under the "in-quota" portion is subject to a zero or low duty. Sugar that enters in amounts above the WTO quota is subject to a prohibitive tariff (78% in 2003, according to the International Trade Commission), serving to protect the U.S. sugar-producing sector from the entry of additional foreign sugar. The tariff on above-quota sugar entering from Mexico under NAFTA (equivalent to about 7% in 2007) will fall to zero on January 1, 2008. In addition, other TRQs limit the import of three categories of sugar-containing products (SCPs — products containing more than 10% sugar, other articles containing more than 65% sugar, and blended syrups).

FY2006 Import Quota Decisions. USDA to date has set the FY2006 WTO raw and refined TRQs for sugar imports at 2.515 million ST, a level double the U.S. minimum WTO commitment. On September 29, 2005, USDA announced a separate sugar TRQ of 268,000 ST for Mexico, having determined that it is a net 'surplus producer' under NAFTA's terms. The increases in both the WTO and NAFTA sugar TRQs to levels above what would occur in a typical recent year represented USDA decisions to boost short-term supplies to address a tight supply situation caused by the delayed sugar beet harvest in North Dakota and Minnesota, hurricane-related losses to the sugarcane crops in Louisiana and Florida, the hurricane-related closure of a large sugar cane refinery in New Orleans from late August to December 2005, and to make available high quality refined sugar for immediate use. On July 27, 2006, USDA announced that raw sugar under the FY2007 TRQ will be allowed to enter early (starting August 7) in order to meet refiners' needs for more flexibility in acquiring and processing raw sugar.

---

[1] A TRQ combines two policy instruments used to restrict imports: quotas and tariffs. The quota component works together with a specified tariff level to provide the desired degree of import protection. Imports entering under the quota portion of a TRQ are usually subject to a lower, or sometimes a zero, tariff rate. This "in-quota" amount represents the minimum that a country has committed to allow to enter under multilateral or other trade agreements. Imports above the quota's quantitative threshold (referred to as "above-quota") face a much higher (usually prohibitive) tariff.

FY2007 Import Quota Decisions. On July 27, 2006, USDA announced that the FY2007 raw and refined TRQs for sugar imports will be 1.544 million ST, 23% higher than the U.S. minimum WTO commitment. On that same day, USDA announced a NAFTA TRQ of 268,000 ST for Mexico, stating both countries had jointly determined that Mexico is projected to be a net 'surplus producer' of sugar in FY2007. The first announcement reflects USDA's assessment that imports above the WTO minimum commitment is needed to meet domestic needs and rebuild stocks. The second is part of a U.S.-Mexican agreement announced to resolve outstanding bilateral sweetener disputes (see "Sweetener Disputes with Mexico," below).

Sugar Imports, the Allotment Suspension Trigger Level, and DR-CAFTA. One concern raised during the congressional debate on DR-CAFTA was that the additional amount of sugar that enters, when added to existing U.S. sugar access commitments under the WTO and NAFTA agreements, would exceed 1.532 million ST — the trigger for suspending marketing allotments. Some pointed out that adding the DR-CAFTA's first year access commitment to those in the two existing trade agreement commitments would result in a sugar import level of 1.652 million ST (see **Figure 2**).

Figure 2. U.S. Sugar Imports Compared to Allotment Suspension Trigger: Trade Agreement Commitments; FY2003-FY2006 Actual; and FY2007-FY2008 Estimates

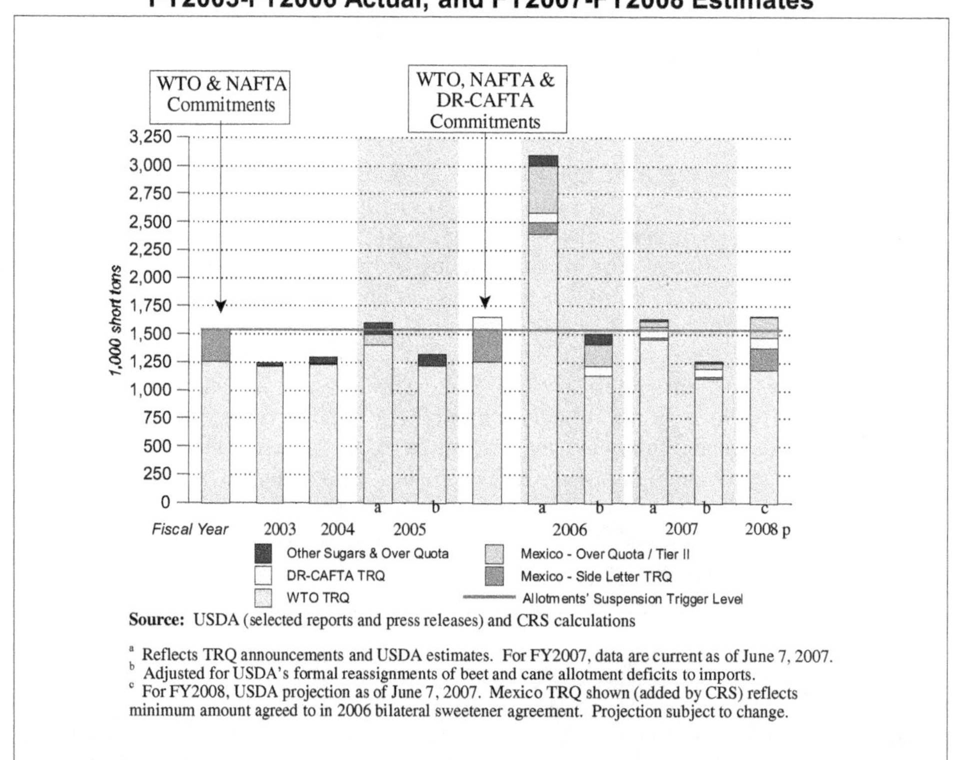

Source: USDA (selected reports and press releases) and CRS calculations

[a] Reflects TRQ announcements and USDA estimates. For FY2007, data are current as of June 7, 2007.
[b] Adjusted for USDA's formal reassignments of beet and cane allotment deficits to imports.
[c] For FY2008, USDA projection as of June 7, 2007. Mexico TRQ shown (added by CRS) reflects minimum amount agreed to in 2006 bilateral sweetener agreement. Projection subject to change.

Sugar imports in the first two years of current sugar program authority (FY2003 and FY2004) were well below the trigger level. With significantly lower domestic sugar supplies caused by weather-related problems in FY2005 and FY2006, permitted sugar imports were above this level. However, above average imports did not result in the suspension of allotments because USDA determined the additional imports were needed to meet U.S. sugar demand (see "Exception to Suspending Allotments," above). To explain further, USDA decided late in FY2005 to allow additional imports to add to domestic supplies tightened by the effects of Hurricane Katrina. This resulted in FY2005 imports of 1.6 million ST — 68,000 ST above the trigger level. However, because much of the late-year increase in imports was due to the inability of the cane sugar sector to fully utilize its increased allotment, these imports (used to cover the cane allotment "shortfall") did not count against the trigger level.

For FY2006, USDA projected lower sugar output and reassigned increases in the cane and beet allotment deficits to imports to boost the availability of both raw and refined sugar. Taken together, these decisions (plus imports under DR-CAFTA) equaled FY2006 sugar imports of 3.095 million ST — 1.563 million ST above the trigger level — but did not result in the suspension of allotments. Similarly, USDA currently projects FY2007 sugar imports at 1.634 million ST — 102,000 ST above the trigger level. Allotments again were not suspended because of the need for additional imports. For FY2008, USDA's import estimates plus the U.S. commitment to allow at least 193,000 ST of Mexican sugar to enter under a NAFTA TRQ could total 1.657 million ST — 125,000 ST above the trigger level (see **Figure 2**).

The issue of the potential impact of sugar imports on the U.S. sugar program under DR-CAFTA has receded for now, because USDA included these imports when setting the FY2006 and FY2007 OAQs. Looking ahead, whether sugar imports in FY2008 — the last year covered by current sugar program authority — activate the trigger to suspend allotments will depend on the extent to which domestic sugar production (particularly in the hurricane-affected cane sector) recovers, and the degree that U.S. commitments under existing trade agreements allow more sugar to enter than the U.S. sugar market can absorb and still allow USDA to administer the program at no-cost. The issue of whether allotments might be suspended if imports result in more sugar than the U.S. market needs (i.e., exceeds the trigger level), though, likely was made moot by a pledge made by the Secretary of Agriculture during congressional debate on DR-CAFTA. In a June 29, 2005, letter to the chairmen of the House and Senate Agriculture Committees, he specified the steps he will take to ensure the sugar program operates as authorized through FY2008 if he estimates imports under DR-CAFTA, NAFTA, and other trade agreements exceed the trigger level (see "Sugar Trade Issues — Sugar in DR-CAFTA — Sugar Deal to Secure Votes," below).

Legislative Activity in the 109th Congress

House Amendment to FY2007 Agriculture Appropriations. During floor debate on the FY2007 agriculture appropriations bill (H.R. 5384) on May 23, 2006, the House rejected (135-281) an amendment offered by Congressman

Blumenauer to effectively reduce the program's loan rates by about 6%, to not more than 17¢/lb. for raw cane sugar and 21.6¢/lb. for refined beet sugar. He argued that the sugar program artificially raises the price paid by consumers and food manufacturing firms, most of the program benefits go to large producers, and that it causes environmental damage in the Everglades. Opponents countered that the program does not cost taxpayers, is supported by two-thirds of those polled in a recent survey, and that any vote on a proposed program change should occur when the 2007 farm bill is debated. In June 2005, an identical amendment offered to the FY2006 agriculture appropriations bill (H.R. 2744) was rejected by the House on a vote of 146 ayes to 280 noes.

Senate Oversight Hearing. On May 10, 2006, the Senate Agriculture Committee held a hearing to review the implementation of the sugar provisions of the 2002 farm bill. The USDA's Undersecretary stated that the sugar program operated relatively smoothly through July 2005, but characterized it as "highly prescriptive, containing many rigid, and sometimes contradictory, rules" that increased the complexity of program administration. He noted that the opening of the U.S. and Mexican sugar sectors after NAFTA takes full effect in January 2008 mean "alternative [sugar policy] approaches will need to be explored." Anticipating upcoming debate on the next farm bill, sugar crop growers, processors, and most cane refiners stated that they will seek an extension of the current program, arguing that existing policy works well and has provided supply stability. Sugar users (food and beverage manufacturers) called for changes that would allow the program to operate more flexibly, and want to "explore common ground" with the sugar producing sector. The president of the only independent cane refinery expressed concern about two issues that in his view current policy does not handle well, which he would like addressed in the 2007 farm bill. A spokesman for the sugar industry of Mauritius, one of the 40 countries with a share of the U.S. import quota, emphasized how important continued access to the U.S. market is for the stable revenue flow and positive developmental impact provided by the U.S. program's price premium (i.e., the difference between the U.S. price and the lower world price).

Administration's FY2007 Budget Proposal. The Bush Administration's FY2007 budget proposal included legislative changes to reduce farm program spending, which included a 1.2% marketing assessment on domestic sugar processed by raw cane mills and sugar beet refiners. This would effectively have lowered loan rates by more than 0.2¢/lb, and generated receipts of an estimated $34 million in FY2007, and $364 million over 10 years. Congress did not address this specific proposal when each chamber passed its FY2007 budget resolution, and no additional legislative action occurred.

In 2005, the Administration proposed an identical marketing assessment in its FY2006 budget package. The Senate Agriculture Committee adopted a modified version, which was dropped by conferees when they completed action on the FY2006 budget reconciliation bill (S. 1932) in December 2005. The Senate-adopted provision would have assessed a penalty on non-recourse sugar loans forfeited by a sugar processor, equal to 1.2% of the loan rates for raw cane and refined beet sugar. This would have represented the sugar sector's contribution toward budget deficit reduction targets.

Sugar Trade Issues

The United States imports sugar to cover the balance of its domestic food needs (16% in FY2005; 27% in FY2006) that are unmet by the U.S. sugar production sector. Sugar imported under market access commitments made by the United States in trade agreements (such as NAFTA) or under prospective free trade agreements (FTAs), together with the growing increase in imports of sugar-containing products not subject to import restrictions, can increase available sugar supplies and push prices down. Increases in sugar imports, though, can also serve to dampen any rise in domestic sugar prices, particularly when domestic sugar output falls noticeably.

The U.S. sugar production sector argues that liberalizing trade in sugar should be addressed in multilateral World Trade Organization negotiations, rather than in hemispheric and bilateral free trade agreements (FTAs). Its concern is that additional market access provided to prospective FTA partners, many of which are major sugar exporters with weak labor and environmental rules, would undermine the U.S. sugar program and threaten the sector's viability. Sugar users advocate including sugar in all trade negotiations, eyeing the possibility that increased imports might contribute to program reform and lower sugar prices.

The sugar provisions in DR-CAFTA, strongly opposed by the U.S. sugar producing sector but favored by most other U.S. commodity groups, drew much attention during congressional debate. The debate drew attention again to NAFTA's sugar provisions. Free trade in sweeteners between Mexico and the United States will take effect in January 2008, after two longstanding trade disputes were resolved in July 2006. How to handle the prospect of unrestricted sugar imports from Mexico starting in 2008, added to sugar imports under other trade agreements, will be a major issue in the debate on the sugar program's future when Congress considers the 2007 farm bill.

Sugar in Trade Agreement Negotiations

Whether, and on what terms, to liberalize trade in sugar and sugar-containing products in prospective trade agreements is a difficult issue that U.S. negotiators face. With the U.S. sugar price higher than the world sugar price, exporting countries want these agreements to provide increased access for their sugar to benefit from the more lucrative U.S. price. The U.S. sugar production sector opposes any additional entry of sugar and products under bilateral and regional trade agreements. It is concerned increased imports would undermine its market share, threaten the viability of the domestic sugar program, and result in significant loan forfeitures. U.S. manufacturers that use sugar in food products and beverages favor opening up the U.S. market to additional imports, anticipating lower sugar prices over time.

Sugar trade has been more of an issue in negotiating bilateral and regional FTAs (witness the debate on DR-CAFTA) than in multilateral negotiations under the WTO. With Brazil, Colombia, Guatemala, and Thailand considered to be major low-cost sugar producing and exporting countries, FTAs with them could allow for additional sales of sugar to the U.S. market above levels now permitted under their shares of the U.S. sugar TRQ. Brazil's negotiators frequently mentioned that increased market access for its sugar in the U.S. market was one of their key agricultural priorities in the currently dormant hemispheric Free Trade Area of the Americas (FTAA). Since the U.S. objective in negotiating an FTA is to eliminate eventually all border protection on all imports, the removal of current U.S. quotas and tariffs on imports of sugar and sugar-containing products would begin to undermine the operation of the domestic sugar program as now structured (e.g., make it impossible for USDA to operate it at "no-cost"). By contrast, any multilateral agreement that might eventually emerge from the WTO negotiations could reduce trade-distorting policies that countries (including the United States) use to support their sugar and other commodity sectors (see "Sugar in WTO Negotiations," below).

Key Interest Group Views. The American Sugar Alliance (ASA) representing sugar crop farmers and processors has argued that the Bush Administration's efforts should be to "reform the world sugar market through comprehensive, sector-specific WTO negotiations" and not through regional or bilateral trade agreements. ASA supports the goal of global free trade (including for sugar) through the WTO, which it views as the best venue for comprehensively addressing "the complex array of government policies that distort the world sugar market." ASA contends that the subsidies used by many countries "encourage the dumping of sugar at a fraction of what it costs to produce it." To support its position, ASA released in 2003 a commissioned report it says documents the non-transparent and indirect subsidies that major sugar producing and exporting countries use to assist their sugar sectors. For this reason, ASA opposes including sugar market access provisions in FTAs, arguing that the most damaging government policies (citing Brazil's sugarcane-ethanol subsidies, and the Mexican government's ownership of sugar mills) will not be addressed by bilateral or regional negotiations. It further argues that U.S. consumers would not benefit in the form of lower prices from increased imports under such agreements. ASA opposed the DR-CAFTA and has argued against including sugar in all other FTAs in hearings before USTR and the ITC.

The Sweetener Users Association (SUA — composed of industrial users of sugar and other caloric sweeteners and the trade associations that represent them) have supported the Bush Administration's proposals tabled at the WTO to further liberalize agricultural trade. SUA expects that under trade liberalization, "world sweetener markets will operate more efficiently and fairly," as EU's export subsidies are phased out and U.S. sugar import quotas become more market oriented. SUA argues that liberalizing trade in sugar would benefit the U.S. economy through lower prices, keep food manufacturing jobs in the United States rather than see them move overseas, and help maintain a viable cane refining industry with its well-paid union jobs. It also contends increased imports would encourage product innovation and stimulate demand, stimulate competition, and thwart excessive industry concentration. The SUA supported DR-CAFTA and favors the Administration's objectives in the other bilateral FTAs, the FTAA, and WTO talks.

Sugar in DR-CAFTA. The sugar provisions in DR-CAFTA were among the most hotly debated issues during congressional consideration in the summer of 2005. The U.S. sugar industry strongly opposed DR-CAFTA, arguing that the amount of increased access offered the six countries "would destroy the U.S. sugar industry." Spokesmen emphasized that the increased imports would depress domestic sugar prices, make it impossible to operate the U.S. sugar program on a no-cost basis, increase government costs as processors forfeit on their price support loans, and "drive efficient American producers out of business." While acknowledging that the Administration understood the consequences of reducing the over-quota tariff, one industry spokesperson pointed out that under the current sugar program, additional imports would act to displace domestic sugar output. The sugar industry also feared that approving DR-CAFTA would set a precedent for U.S. negotiators to include sugar in other FTAs being negotiated with several sugar exporting countries (see "Sugar in Prospective FTAs," below). It pointed out total sugar export availability of actual and potential FTA candidates is 27 million metric tonnes (MT), compared with U.S. sugar output of 8 million MT. The Sweetener Users Association supported the DR-CAFTA, stating it will enhance competition in the U.S. sugar market, increase export opportunities for other U.S. food and commodity sectors in the six countries, and result in increased employment in U.S. confectionery and other sugar-using industries.

The six countries covered by this agreement (Costa Rica, Dominican Republic, El Salvador, Guatemala, Honduras, and Nicaragua), prior to its approval, were already allowed to sell to the United States a combined minimum of 311,700 MT of sugar annually under allocations, or shares, of the TRQ. This amount represented a 28% share of the minimum U.S. raw sugar TRQ (1.12 million MT) established under its WTO commitment, and entered on a duty-free basis. Under DR-CAFTA, these six countries together secured access in year 1 to export to the U.S. market an additional 109,000 MT of sugar, a 35% increase over their current quota. Increasing on average about 3% per year, by year 15, these countries combined will be eligible to sell duty-free an additional 153,140 MT of sugar. Thereafter, the quota would increase by almost 2% (2,640 MT) annually in perpetuity. The over-quota tariff would stay at the current high level (78% in 2003) indefinitely, and not decline. The agreement includes a "compensation" mechanism that the United States can exercise at its sole discretion in order to manage U.S. sugar supplies. If activated, the United States commits to compensate the six countries for sugar they would not be able to ship under the above market access provisions.

In justifying DR-CAFTA's sugar provisions, U.S. Trade Representative officials pointed out that the additional access granted all six countries will equal about 1.2% of current U.S. sugar consumption in year 1, increasing to 1.7% in year 15. They also emphasized that the U.S. sugar sector will be protected by the compensation mechanism, the prohibitive tariff on above-quota imports, and the stipulation that a country can only export its net sugar surplus to the U.S. market. USTR's lead agricultural negotiator also stated that the import increase would not interfere with how sugar marketing allotments function.

Sugar Deal to Secure Votes. Under what circumstances, and how the sugar "compensation" mechanism might operate, drew much questioning when USTR officials testified before the Senate Finance Committee in April 2005. Several senators raised concerns about what would happen to sugar marketing allotments if USDA projects sugar imports, because of additional imports under DR-CAFTA, will be above the trigger level that would require USDA to suspend allotments. To address these concerns, Secretary of Agriculture Johanns, on June 28, 2005, reached an agreement with two Senators that helped sway enough votes the next day in the

Finance Committee for the DR-CAFTA implementing bill. In a letter to the chairmen of the House and Senate Agriculture Committees, the Secretary laid out the steps he would take to ensure the sugar program operates as authorized only through FY2008 if he estimates annual sugar imports under DR-CAFTA, NAFTA, and other trade agreements exceed the 1.532 million ST trigger level. These will include donating surplus commodities in USDA inventories or making cash payments as compensation to sugar exporters in Central America or Mexico to not ship sugar to the U.S. market under DR-CAFTA's and NAFTA's terms. The letter pledged that USDA would also divert (by purchasing) surplus sugar imports for ethanol and other non-food uses, and would complete a study to be submitted to Congress by July 1, 2006, on the feasibility of converting sugar into ethanol.[2] Reactions by Members to the letter were mixed, with some skeptical about the assurance and others remaining opposed to DR-CAFTA, in part because of the costs that USDA would incur in meeting its pledge (as scored by the Congressional Budget Office (CBO)). The U.S. sugar industry rejected USDA's commitment, calling it "a repackaged, short-term offer" that did not address its long term concerns about (1) sugar that could enter in future trade agreements; (2) a resolution of the dispute on Mexico sugar access to the U.S. market; and (3) the continuation of the features of the current sugar program after FY2008.

FTA Negotiations with Australia. U.S. trade negotiators excluded sugar from the trade agreement concluded with Australia in February 2004. The sugar industry "applauded the Administration's decision to exclude market access commitments on sugar," pointing out an FTA can be negotiated without including sugar and that this can "serve as a template for all future FTA negotiations." The National Confectioners Association representing candy manufacturers "condemned" the negotiating results, stating that limiting access to Australian-produced sugar is "damaging" to U.S. candy firms and jobs. Other major commodity groups reacted that excluding sugar in negotiating other FTAs would harm their export interests.

Sugar in the Peru, Colombia, and Panama FTAs. In the FTA with **Peru** announced in December 2005, U.S. negotiators offered access for an additional 11,000 MT of sugar to the U.S. market. This represents an almost one-fifth increase in Peru's minimum share of the U.S. raw cane TRQ (47,674 MT). Peru's preferential raw cane sugar quota would rise 2% each year, and the U.S. protective tariff on over-quota sugar would continue indefinitely. Peru would be permitted to ship sugar only if it has a net sugar surplus (i.e., sugar left over once its domestic market is supplied, and taking imports into account). This is expected to occur infrequently, in light of Peru's sugar trade flows in recent years. The American Sugar Alliance (ASA) signaled it would not oppose this agreement, noting the amount is "more reasonable than the excessive access granted in CAFTA." In the **Colombia** FTA concluded in February 2006, the United States offers access for an additional 50,000 MT of sugar.

---

[2] On July 10, 2006, USDA issued this report examining the economic feasibility of converting sugarcane, sugar beets, molasses, raw cane sugar, and refined sugar into ethanol. The study concluded that while such conversion would be profitable with current high demand for ethanol and record ethanol prices, it would be unprofitable when compared to ethanol prices projected for mid-2007. This report can be accessed at [http://www.usda.gov/oce/reports/energy/EthanolSugarFeasibilityReport3.pdf].

This represents a two-fold increase over Colombia's present minimum share of the U.S. raw cane TRQ (25,273 MT). The new quota would increase about 1.5% annually, and the high U.S. tariff on entries above the quota level would remain in place in perpetuity. Panama, in the FTA concluded in mid-December 2006, received three small preferential TRQs for sugar and sugar-containing products allowed to be sold duty-free in the U.S. market. The largest duty-free TRQ (6,000 MT for raw sugar) will expand 1% annually for only 10 years and then be capped at 6,600 MT indefinitely; all over-quota tariffs on sugar product will remain at current high levels also indefinitely. All three preferential quotas represent a 23% increase over Panama's current minimum 2.7% share (30,540 MT) of the U.S. raw cane TRQ. The new quotas in the aggregate represent most of the sugar surplus that Panama traditionally has available to export each year. All three FTAs include a sugar compensation provision similar to that found in DR-CAFTA.

Sugar in WTO Negotiations. U.S. sugar policy may face change if negotiators reach a multilateral agreement that commits countries to proceed further in significantly liberalizing agricultural trade. For this reason, the U.S. sugar production sector and U.S. food and beverage manufacturers that use sugar are watching carefully to see whether the outcome of these negotiations protects and/or advances their respective economic interests.

Launched in late 2001, WTO member countries agreed that the negotiating objectives for agriculture in the Doha Development Round should be: (1) substantial reductions in trade-distorting domestic support, (2) the phase-out, with a view to total elimination, of all export subsidies, and (3) substantial improvements in market access (i.e., reductions in tariffs and expansion of quotas). In August 2004, negotiators agreed upon a "framework agreement" to be followed to meet these objectives. In advance of the December 2005 Hong Kong ministerial conference, the United States, the EU, and two other country groups presented various agriculture "modalities" proposals (formulas, schedules, end dates) to add specifics to this framework. Trade ministers, though, were not able to finalize a modalities package, and subsequently failed to achieve a breakthrough at a crucial meeting of a core group of countries held in July 2006. Though negotiations were then "indefinitely suspended," WTO's director in late January 2007 declared that the process was now back "to full negotiating mode."[3] Efforts among four key trading countries, including the United States, to develop a compromise package to present to other countries to consider collapsed on June 21, 2007, putting into doubt the possibility of concluding a deal by the end of this year. WTO's director responded that a deal is still possible, and announced that talks will now shift back to Geneva to involve all of WTO's country members.

Interest groups with a stake in U.S. sugar policy have closely monitored the "market access" component of these negotiations. Of most significance is the issue of how much (expressed as a percentage of tariff lines) a country should be allowed to protect its "sensitive" agricultural commodities, particularly imports from the least developed countries (LDCs). The American Sugar Alliance had expressed concern

---

[3] For additional information, see CRS Report RL33144, *WTO Doha Round: The Agricultural Negotiations*, by Charles E. Hanrahan, Randy Schnepf.

that the U.S. October 2005 proposal "could dump up to 750,000 tons of unneeded foreign sugar on a chronically oversupplied U.S. market" but did not seek to discipline indirect sugar subsidies, and would allow developing countries "to escape reforms." Others, though, pointed out that modalities that are closer to the EU proposal would leave the U.S. sugar program unchanged. The Sweetener Users Association as a member of an agribusiness/commodity/farm group coalition had called on other countries to match the U.S. proposal to reduce trade-distorting subsidies "with equally ambitious" tariff reductions and export competition commitments.

The ministerial declaration adopted in Hong Kong called for "duty-free, quota-free" imports from LDCs for at least 97% of all of a country's products. For the United States, this could apply to sugar imported from these countries, unless U.S. negotiators decide to exempt sugar from this obligation (i.e., to place sugar in the 3% category). Also, what duties and quotas would apply to imports of sensitive agricultural products (i.e., applicable to sugar imports) from all other countries could also affect U.S. sugar policy.

Sweetener Disputes with Mexico

Longstanding differences over the level of market access for Mexican sugar to the U.S. market, and over Mexican barriers on imports of U.S. high-fructose corn syrup (HFCS), were resolved in a bilateral agreement reached by both governments in late July 2006. The terms of this agreement will apply until January 1, 2008, when bilateral free trade in sweeteners takes effect under NAFTA's original terms.[4] Until this breakthrough, the almost decade-long impasse had largely reflected divergent views held by the U.S. and Mexican governments of NAFTA's sugar provisions; Mexican government actions on behalf of its powerful sugar industry to force a change in the U.S. position, and to protect the sector's share of the Mexican sweetener market in light of increased imports of HFCS from the United States; each government's filing of trade dispute cases before NAFTA and WTO panels challenging the other's handling of these disputes; and unsuccessful efforts by government and private industry negotiators on both sides to arrive at some compromise. Resolution of both disputes appeared to be prompted by the U.S. government's recognition that sugar imports will be needed anyway to meet domestic sugar demand and rebuild stocks, in return for gaining access to Mexico's market for U.S. HFCS sales. Though the agreement in the short term settles the parameters of U.S.-Mexican sweetener trade, some observe that temporarily resolving the underlying problems has simply postponed dealing with them until 2008 and after. As a result, the development of the sugar title in the 2007 farm bill will involve exploring various options to address imports of Mexican sugar.

2006 Sweetener Agreement. In the July 27, 2006, exchange of letters between U.S. and Mexican agricultural trade negotiators, both governments agreed

[4] A precedent for the outline of this agreement was set in the bilateral agreement announced on September 29, 2005, that applied only for FY2006. Then, the United States announced a NAFTA duty-free TRQ of 250,000 MT for sugar imports from Mexico. Mexico reciprocated by establishing a duty-free 250,000 MT TRQ for imports of U.S. HFCS.

on a number of measures "intended to promote an orderly transition to" free trade in sweeteners on January 1, 2008. In brief, the agreement provides for Mexican sugar access to the U.S. market equal to the amount of access that U.S. HFCS will have to Mexico's market during this period. Specific provisions call for:

- The United States to provide duty-free access for Mexican sugar as follows: 250,000 MT in FY2007, and from 175,000 to 250,000 MT in the first quarter of FY2008 (with the exact amount to be jointly determined by July 1, 2007, based on market conditions at that time).[5]

- Mexico to allow duty-free entry for an equivalent amount of U.S.-produced HFCS in FY2007 (250,000 MT), with the FY2008 quantity determined in the same way that the sugar TRQ amount is determined.[6]

- The United States to allow not less than 21,774 MT of Mexican refined sugar to enter duty-free by September 30, 2006 (i.e., reflecting USDA's objective to provide an adequate supply of high-

---

[5] The 250,000 MT amount reflects U.S. implementation of the NAFTA sugar side letter, which the Clinton Administration negotiated with Mexico in November 1993 in order to secure enough votes for House approval of NAFTA. The Mexican government subsequently contested the validity of this side letter, which effectively placed a lower cap on duty-free imports of Mexican sugar allowed to enter the U.S. market than would have been the case under the original NAFTA agreement. Under the side letter's provisions, Mexico can ship not more than 250,000 MT of sugar duty-free to the U.S. market each year through FY2008, but only if Mexico shows a "net production surplus" (defined as Mexican sugar production, minus Mexican sugar consumption, minus Mexican HFCS consumption). NAFTA committed both countries to allow sugar to enter free (not subject to a tariff nor any quota restriction) from the other in 2008.

[6] Under NAFTA, Mexico committed to reduce its base tariff on HFCS imports (15%) from the United States to zero within 10 years (i.e., 2003). As HFCS sales to Mexico's soft drink sector began to displace bottlers' use of Mexican-produced sugar and Mexico's sugar sector sought to renegotiate NAFTA's sugar provisions, the Mexican government responded in 1998 by imposing high anti-dumping (AD) duties on U.S. HFCS imports. With the AD duties, U.S. HFCS sales leveled off but continued. Subsequent NAFTA and WTO dispute panels ruled that these duties were not consistent with Mexico's trade commitments. To comply with the WTO's final ruling, Mexico created in May 2002 a new TRQ for U.S. HFCS equal to the NAFTA sugar TRQ (under the sugar side letter's provisions) that the United States had announced for Mexico for FY2002. This policy also subjected over-quota HFCS imports to a prohibitive 156% or 210% tariff. Concurrently, Mexico's Congress imposed a 20% "soda tax" in 2002 — applied to soft drinks sweetened with corn syrup but not sugar. Both actions eliminated the Mexican soft drink sector's demand for HFCS, and resulted in negligible U.S. HFCS exports in the 2002-2005 period. In a subsequent case filed by the United States arguing that the tax was discriminatory in its application, the WTO accepted the U.S. position. In its final decision issued in March 2006, the WTO found the tax inconsistent with WTO trade rules and called upon Mexico to bring its laws into compliance. Mexico agreed to eliminate this tax, which its Congress did in December 2006, as part of this sweetener agreement.

quality refined sugar to users before the 2006 domestic crops are available for processing).[7]

- Mexico to meet its NAFTA commitment to allow duty-free entry of not less than 7,258 MT of sugar or syrup goods from the United States in each of FY2005, FY2006, and the first quarter of FY2008 (i.e., totaling 21,774 MT).

- The United States to eliminate its over-quota tariffs on imports of sugar from Mexico, and Mexico to eliminate its over-quota tariffs on imports of HFCS from the United States, effective January 1, 2008.[8]

- Only through December 31, 2007, Mexico to apply import licensing procedures on the permitted in-quota amounts of HFCS imports, and to develop and apply "bilaterally agreed" import licensing procedures on the specified amount of in-quota sugar imports, from the United States.

- Each country not to take any discriminatory action (e.g., imposing a tax or any other internal measure) to limit imports of the sensitive sweetener product from the other (for Mexico, HFCS; for the United States, sugar).

- Both countries to continue to consult and work to resolve other ongoing disputes involving trade in sweeteners, in order to facilitate the transition to free trade on January 1, 2008.

- Both countries to establish a joint industry/government task force to (1) help both governments prepare for the elimination of tariffs on sweeteners in January 2008 and (2) periodically review product shipments against this agreement's TRQs to ensure that they are promptly and fully utilized.

The letter exchange also confirms a separate bilateral agreement reached earlier and submitted to the WTO stating that Mexico will eliminate its "soda tax" no later than January 31, 2007.

Reactions to Agreement. The American Sugar Alliance expressed concern that this agreement "will not accomplish" the objective of an "orderly transition" to bilateral free trade in 2008 for U.S. sugar producers "under current market

---

[7] This represents Mexico's share of the additional FY2006 refined sugar TRQ of almost 91,000 ST, announced by USDA on July 27, 2006. This allocation to Mexico reflects U.S. trade commitments under the WTO.

[8] The U.S. over-quota tariff on raw sugar from Mexico under NAFTA is 1.56¢/lb. (equivalent to an *ad valorem* duty of about 7%) in 2007, and will fall to zero in 2008. Mexico will continue to apply its MFN over-quota tariff on U.S. HFCS products in the interim — ranging from 156% to 210%, depending upon the sweetness intensity, not NAFTA's zero rate as initially set.

conditions." Its letter to Administration officials stated that the amount of Mexican sugar to be imported, together with sugar that will enter from other countries, "will oversupply and disrupt the U.S. sugar market." Particularly noted were the fall in the raw cane sugar futures price immediately after the July 27 announcement and a "more likely surge" in over-quota imports from Mexico in 2007 as the NAFTA tariff is further reduced. ASA also asserted that the agreement "abandons the fundamental principles that have governed NAFTA sweetener trade." It pointed out that USDA's estimates of deficits in Mexican sugar and HFCS supply and demand in FY2006 and FY2007 cannot be reconciled with USDA's determination that Mexico is a "net surplus producer" and granted access to the U.S. market, under the NAFTA side letter formula. The Corn Refiners Association welcomed the agreement, viewing its HFCS provisions as setting into "motion an irreversible path to free trade in January 2008, as the NAFTA intended." Acknowledging that it does not resolve all outstanding issues or compensate U.S. corn refiners for 10 years of losses, CRA stated that the agreement "solidifies the promise for increasing U.S.-owned corn sweetener presence in Mexico." The Sweetener Users Association commended USDA and USTR for a "successful resolution of these issues" and noted that in the long term, the agreement puts both countries "on a path toward an open North American sweetener market in 2008."

Potential Impact. The extent to which increased Mexican consumption of HFCS — a cheaper sweetener than sugar and primarily imported from the United States — displaces Mexican-produced sugar in that market will be the major factor influencing how much Mexican sugar is actually shipped to the U.S. market during the sweetener agreement period and beyond. Estimates vary on how much HFCS could displace the sugar consumed by Mexico's soft drink industry, which would be the primary user of HFCS under free trade. In 2004, the second year that the "soda tax" was in effect, Mexico's soft drink sector used a record 1.6 million MT of sugar — an amount that theoretically could all be displaced by HFCS over time. Analysts, though, estimate that HFCS could displace a sizeable, but smaller, portion of the sugar consumed by Mexico's soft drink sector (from 600,000 MT to 1 million MT).[9] The low end of this range reflects the view that a large portion of the soft drink bottlers that are owned by Mexican sugar processors will continue to use sugar rather than switch to HFCS. The high end assumes that Mexico's soft drink sector will substitute HFCS for sugar to the same extent as occurred in the U.S. soft drink sector.

Other factors will influence the quantity of sugar that Mexican processors could export to the U.S. market beginning in 2008. Some of these could reduce the level of actual shipments; a few factors could raise that level. Also, what transpires during the agreement's implementation period through year-end 2007 will affect the climate in which Congress considers the new sugar program. These factors include:

- how much more U.S.-produced HFCS is actually exported to Mexican soft drink bottlers. Though there currently is excess capacity in U.S. plants, the extent to which HFCS production is

[9] Frank Jenkins, "World and US Sugar and Sweetener Market Outlook," presentation to ASA's 23rd International Sweetener Symposium, August 7, 2006, p. 18; *The Future of U.S. Sugar Policy*, prepared for ASA by McKeany-Flavell Company, Inc., June 14, 2006, p. 8.

increased to take advantage of regained access to the Mexican market will depend to some extent on whether firms find it more profitable to produce HFCS for export rather than produce ethanol to meet growing U.S. demand. With the high price of U.S. corn raising HFCS prices above historical levels, this alternative sweetener could become less competitive pricewise with Mexican sugar, and possibly reduce Mexican bottler demand for HFCS.

- how Mexican wholesale sugar prices compare to comparable U.S. raw sugar prices, and how Mexican sugar mills respond. If they are higher — the trend seen in much of the period since 2000 — Mexican mills may be more inclined to sell into their domestic market rather than export sugar to the U.S. market. Though historically high U.S. sugar prices in 2005/2006 encouraged Mexican firms to export sugar to obtain a better price than in their home market, the U.S. price already is falling back toward its historical range. As the U.S. market returns to a more balanced supply and demand situation, this could reduce the incentive for Mexican sugar mills to export as much to the United States. However, if U.S. HFCS exports to Mexico increase substantially and contribute to a fall in Mexican sugar prices, Mexican mills would find selling sugar to the United States a more lucrative option, if U.S. sugar policy (and in turn, U.S. wholesale sugar prices) stays the same.

- the degree to which the Mexican government intervenes, directly or indirectly, in the marketplace to ensure adequate domestic supplies and reduce sugar price increases. The level of Mexican sugar production, while trending up, does vary from year to year. Since Mexican government policy is to have sugar mills hold in reserve sugar equal to at least three months of domestic consumption,[10] lower output in some years could reduce the amount available for export. For this reason, the government can require mills to obtain permits before they are allowed to export sugar. The government also does at times permit sugar imports to dampen price increases.

- how the Mexican government proceeds to handle a number of seemingly intractable problems in its sugar sector. The sector employs more than 2 million workers, many producing and cutting sugarcane on very small plots, and is a powerful political force in the countryside. In 2005, Mexico's Supreme Court reversed the government's expropriation of about half of the country's sugar mills in 2001. The question that arises is what will happen to many of these operations. Will the government step in to inject financial support or let them fail, or will their private owners be able to return them to profitability? Various initiatives to diversify the sugar

[10] USDA, Foreign Agricultural Service, "Mexico Sugar Semi-Annual Report 2005," September 23, 2005, p. 6.

industry are being explored, with much attention focused on developing an ethanol industry that uses sugarcane as a feedstock. Relatedly, uncertainty remains over the legal status of a 2005 law to reintroduce government involvement in the Mexican sugar producing sector — opposed by the government and sugar mills, but supported by sugarcane associations and their political allies in Congress. Ultimately, the outcome of the debate over whether to introduce more market-oriented policies or to maintain the policy status quo will determine whether Mexico's sugar sector contracts to primarily meet domestic demand, or survives to likely generate a surplus of sugar that will seek an outlet in the U.S. market.

- whether Mexican sugar mills invest in processing equipment to improve the quality of sugar sold to U.S. sugar users. Reports indicate that shipments of Mexican "refined" sugar during FY2006 did not meet the standard that U.S. food manufacturers are accustomed to procuring and/or contained foreign material. This required U.S. importers to further process the sugar to remove impurities, or for example, to remove metal shavings. With the additional costs involved, U.S. sugar users may be less willing in a less turbulent market environment to purchase Mexican refined sugar, unless its quality improves.

2007 Farm Bill Debate on the Sugar Program

Sugar Program Options

In 2007, Congress will consider legislation to replace the expiring 2002 farm bill provisions for farm commodity price and income support programs, including those for sugar. In advance of this, interest groups with a direct stake in future U.S. sugar policy laid out their positions. In addition, others have floated a number of other policy options that might receive consideration. Taken together, these proposals include (1) extending the current program (i.e., keep the loan program with present price support levels, together with marketing allotment authority); (2) eliminating the current domestic program, but retaining authority to impose import quotas reflecting U.S. trade agreement commitments; (3) reducing current sugar price support levels, but retaining the loan features of the current program and repealing all marketing allotment authorities; (4) replacing the existing program with a market-oriented approach used for such program crops as grains, cotton, and oilseeds (e.g., direct payments, counter-cyclical payments, marketing loan gains — with benefits determined relative to specified loan rates and to-be-determined target prices); (5) authorizing a buyout of sugar marketing allotments (possibly similar to the quota buyout packages authorized for peanuts in 2002 and tobacco in 2004); (6) adopting a revenue insurance approach for all producers that bases government payments on producer revenue losses rather than on low commodity prices; and (7) making direct payments to sugar crop producers based upon historical acreage levels, among other possible options and variations. Converting sugar into ethanol is also expected to receive attention, in an expanded energy title in the farm bill, as one way to address any sugar surplus scenario and make the United States less reliant on foreign energy sources.

Factors That Will Affect the Debate. Key factors that will influence the debate are the likelihood of unrestricted sugar imports from Mexico once NAFTA takes full effect in 2008 and the funding level that the Agriculture Committees secured for farm bill programs in the FY2008 budget resolution.

Interest Group Positions. The **American Sugar Alliance** advocates an extension of the current "no-cost" sugar program, arguing that the status quo is the best way to preserve a viable U.S. sugar industry. ASA expressly opposes converting the sugar program into the type now used to support program crops, where USDA makes payments to producers when prices fall below specified levels. ASA emphasizes such a change would be "costly" and "unworkable" for the U.S. sugar producing sector, projecting a cost of $1.3 billion annually.[11] ASA further argues that problems would arise in implementing this type of program because the sugar industry is structured differently than other commodities, large sugar producers would quickly hit current payment limits on farm subsidies, bankruptcies and consolidations in the U.S. sugar sector could occur, and a costly program would be politically difficult to approve and maintain in light of the current federal budget outlook. Given a fixed or smaller budget for farm programs, sugar producers state that other commodity producers would not want to reduce their subsidies to make room for a subsidy program for the sugar producing sector.

Separately, recognizing that sugar imports from Mexico are inevitable under NAFTA, ASA wants sweetener trade with Mexico to be managed once free trade takes effect in 2008 so that the current U.S. sugar program can continue. This would mean that Mexico would control the quantity of sugar its sugar mills are allowed to sell (e.g., by operating a program similar to the U.S. sugar marketing allotment program) or agree to export its excess sugar to other countries besides the United States. A Mexican sugar company executive responded that marketing allotments would be "very unpopular" at this time with Mexican sugar processors, who would view such a move as an effort to limit their access to the U.S. market just as NAFTA's sugar provisions take full effect.[12]

In issuing its 2007 farm bill proposals on April 23, 2007, the **American Farm Bureau Federation** expressed its support for the continuation of the current sugar

---

[11] ASA released its assessment in mid-2006 to counter discussion of the policy option to change the current "no-cost" sugar program to a "standard" payment program. The study also concluded that this approach would cause "a dramatic drop in [sugar] producer price and income," and not benefit consumers with lower sugar prices since "food manufacturers historically have not passed savings from lower input costs" on to them. This study, cited in footnote 9, can be viewed at [http://www.sugaralliance.org/library/resourcedocs/MF-SugarPolicy-Final.pdf]. In response, the Sweetener Users Association issued an analysis critical of the ASA-commissioned study, available at [http://www.sweetenerusers.org/Cover%20Letter%20and%20The%20Future%20of%20U.S.%20Sugar%20Policy1.pdf].

[12] *Inside U.S. Trade*, "U.S., Mexico Sugar Producers At Odds Over Integrating NAFTA Market," August 11, 2006.

production and marketing program. Earlier, at its annual meeting held in early March, the **National Farmers Union** (another general farm organization) called for continuing the current U.S. sugar program.

The **Sweetener Users Association** is calling for changes to the current sugar program, in order to respond to external pressures (primarily increased sugar imports from Mexico) and "to achieve a more workable policy for the entire industry." Reflecting a shift in position from what it advocated in the 2002 farm bill debate, spokesmen emphasize that sugar users "need a steady, reliable supply of sugar and a healthy domestic production and processing sector" that is geographically dispersed enough to handle supply disruptions caused by hurricanes and quality problems associated with some imports. Initially, food manufacturers advocated that the sugar program needs to be reformed to meet this objective, and proposed that budget outlays for a different type of sugar program "can be acceptable and defensible if they help to make policies more market-oriented or help adjust to our trade commitments." A SUA spokesman stated that the existing program cannot be sustained anyway on a no-cost basis, pointing to estimates made by CBO which project that lower U.S. prices due to increased sugar imports from Mexico will result in loan forfeitures by sugar processors and accompanying USDA outlays. Recognizing that its call for a closer look at costly options to change the program would not be seriously considered by policymakers at this time, the SUA in mid-May 2007 instead proposed a scaled-back set of recommendations. It supports continuing the beet and raw cane sugar loan rates at current levels, but wants to impose a forfeiture penalty of about one cent per pound when a processor hands over sugar to USDA instead of paying off a price support loan. It also favors abolishing marketing allotments, recommends changes in how U.S. sugar quotas are administered, and calls for USDA to collect and report prices for raw and refined sugar.[13]

On April 23, 2007, a coalition of sweetener users and public interest, consumer, and taxpayer advocacy groups announced they had formed the **Sugar Policy Alliance** to seek reforms of the sugar program during 2007 farm bill debate.[14]

The **Corn Refiners Association** views sugar marketing allotments as "a barrier to sweetener trade with Mexico" because they cap sugar imports from Mexico at 250,000 MT and, in effect, do not allow U.S. HFCS sales to Mexico to increase above this level. CRA states it will not support the U.S. sugar program in the 2007 farm bill "if marketing allotments or any aspect of the sugar program jeopardizes full implementation of free trade in sweeteners under the NAFTA."[15]

---

[13] Chicago Tribune, "U.S. sugar growers, users work on a deal," August 9, 2006; *Inside U.S. Trade*, "Sugar Producers, Users Remain At Odds Over Future of Program," August 18, 2006; and *The Dyergram*, "Users Unveil Farm Bill Proposal...," June 4, 2007, p. 1. For additional perspective on SUA's views, see "Congress Should Reform the U.S. Sugar Program," at [http://www.sweetenerusers.org/Congress%20Should%20Reform%20Sugar%20Program.pdf].

[14] [http://www.prnewswire.com/cgi-bin/stories.pl?ACCT=104&STORY=/www/story/04-23-2007/0004571442&EDATE=]

[15] CRA 2007 Farm Bill Position, August 2006, available at [http://www.corn.org/
(continued...)

CRS-24

USDA's Farm Bill Proposal. On January 31, 2007, the Secretary of Agriculture released its proposals for congressional consideration in this year's debate. For sugar, USDA proposes that the objectives of operating a no-cost program by managing supplies be continued with two changes. These are (1) the removal of the 1.532 million ton import trigger that automatically suspends allotments and (2) discretionary authority to administer these allotments to reduce the program's future cost exposure (estimated at $1.4 billion over 10 years). In response, the ASA stated that USDA's proposal is a "positive development" but expressed concerns over the discretionary authorities the Secretary would exercise. The SUA did not issue any public statement on USDA's proposal.

Status of Sugar in 2007 Farm Bill Debate to Date

On June 6, 2007, during markup of titles under its jurisdiction, the House Agriculture Subcommittee on Specialty Crops, Rural Development and Foreign Agriculture extended the current sugar program without change through 2012. Loan rates would not change; marketing allotments would continue to be applied as long as imports do not exceed 1.532 million tons. However, the full House Agriculture Committee expanded and revised these provisions in the chairman's mark issued on July 6.

As approved by the committee, the major features of the sugar program in H.R. 2419 would:

- increase loan rates by almost 3% (from 18¢/lb. to 18.5¢/lb. for raw cane sugar; from 22.9¢/lb. to 23.5¢/lb. for refined beet sugar),
- guarantee the domestic sugar producing sector a minimum 85% share of the U.S. sugar market (irrespective of import levels),[16]
- mandate that USDA sell sugar acquired as a result of loan forfeitures and surplus sugar purchased to maintain prices above loan forfeiture levels to bioenergy producers for processing into ethanol (a gasoline additive),
- prescribe USDA's implementation (i.e., limit its discretion) of sugar import quota authorities

The American Sugar Alliance has come out in support of the committee's bill. A coalition of food and beverage firms and trade associations, and public interest groups, oppose the new provisions, and plan to offer amendments to delete some of them during House floor debate. The Bush Administration opposes the increase in sugar loan rates and is concerned about language that would limit USDA's ability to administer import quotas. For more discussion on the House farm bill's sugar

[15] (...continued)
CRAFarmBillTalkingPoints.doc].

[16] This would replace the current import trigger provision, which when activated, requires USDA to suspend marketing allotments. If retained, the suspension of USDA's ability to manage domestic sugar supplies would likely lead to considerable price uncertainty (from the sugar producing sector's perspective) and make it difficult, if not impossible, to operate the sugar program at no cost.

provisions and reactions, see CRS Report RL34103, *Sugar Policy and the 2007 Farm Bill.*

In other legislative activity, two introduced bills would repeal the sugar program and/or related authorities, and likely allow for sugar crop producers to take advantage of a new approach to handle farming risk. Section 106 of **H.R. 2720** would repeal current authorities for both the program and the sugar tariff-rate quota; Section 107 would require USDA to establish a recourse loan program for sugar, meaning that any loans taken out by processors would have to be repaid when due (i.e., they would no longer have the right to hand over, or forfeit, sugar, to USDA if the market price falls below the effective price support level). Section 1006 of **S. 1422** would repeal the sugar program, but would not establish a recourse loan program. Both bills appear to make sugar beet and sugarcane producers eligible to take advantage of proposed farmer-held risk management accounts that could be used to purchase crop insurance, cover income losses, or invest in other on-farm improvements.

The Senate Agriculture Committee plans to hold its markup of farm bill provisions in mid-September, with the prospect that floor debate on the sugar program will surface again in October.

Sugar Policy and the 2008 Farm Bill

Remy Jurenas
Specialist in Agricultural Policy

January 30, 2009

Congressional Research Service
7-5700
www.crs.gov
RL34103

CRS Report for Congress
Prepared for Members and Committees of Congress

Summary

Congress reauthorized the sugar price support program with some changes in the Food, Conservation, and Energy Act of 2008 (P.L. 110-246, the enacted 2008 farm bill). The sugar program is designed to guarantee the price received by sugar crop growers and processors and is intended to operate at "no cost" to the U.S. Treasury. To accomplish this, the U.S. Department of Agriculture (USDA) controls supply by limiting the amount of sugar that processors can sell domestically under "marketing allotments" and restricts imports. At the same time, USDA seeks to ensure that supplies of sugar are adequate to meet domestic demand. "No cost" is achieved if USDA applies these tools in a way that maintains market prices above minimum price support levels.

Since January 1, 2008, sugar imports from Mexico no longer face quotas or duties under the North American Free Trade Agreement. Other imports are allowed entry under quotas found in other free trade agreements (FTAs). To address the potential for a U.S. sugar surplus caused by additional imports under these trade agreements, the enacted farm bill mandates a sugar-for-ethanol program. USDA is now required to purchase as much U.S.-produced sugar as necessary to maintain market prices above support levels, to be sold to bioenergy producers for processing into ethanol. Funding is open-ended for this program. Other provisions increase the minimum guaranteed prices for raw sugar and refined beet sugar by 4% to 5%, mandate an 85% market share for the U.S. sugar production sector, and remove certain discretionary authority that USDA exercises to administer import quotas.

The enacted sugar provisions reflect the proposal presented to the House and Senate Agriculture Committees by producers of sugar beets and sugarcane and the processors of these crops. They favored continuing the structure of the current sugar price support program but sought changes to enhance their position in the U.S. marketplace. Their sugar-for-ethanol provisions ensure that the prospect of imports adding to U.S. sugar supplies under any future trade agreements will not undermine the program's price guarantee and the sugar industry's market share. Food and beverage manufacturers that use sugar opposed the proposed program's provisions, arguing that costs to consumers will increase and that new requirements will restrict the flow of sugar for food use in the domestic market. The Bush Administration opposed these provisions, with the President identifying them as one reason why he vetoed the farm bill.

USDA has continued to estimate a tight domestic sugar supply in FY2009 largely due to reduced beet production. Its import quota decisions made to date and its estimate of sugar expected to enter from Mexico and other FTA partners do not point to a sugar surplus. As a result, USDA announced in September 2008 that the sugar-for-ethanol program will not be implemented this year. Attention now turns to how USDA will implement newly enacted rules dealing with the timing of additional raw cane sugar versus refined sugar imports, because of the implications for market prices.

For background information on the sugar program and a review of more recent developments, please see CRS Report R40995, *Sugar Policy Issues*, by Remy Jurenas.

Contents

Tables

Appendixes

Contacts

Overview of Sugar Program

The sugar program is designed to guarantee the minimum price received by growers of sugarcane and sugar beets, and by the firms (raw sugar mills and beet refiners) that process these crops into sugar. To accomplish this, the USDA limits the amount of sugar that processors can sell domestically under "marketing allotments" and restricts imports. USDA is required to operate the sugar program on a "no-cost" basis. This means USDA must regulate the U.S. sugar supply using allotments, import quotas, and related authorities so that domestic market prices do not fall below minimum price levels. These are based on the loan rates for raw cane sugar and refined beet sugar set out in law, which USDA uses to derive effective support levels (see "Level of Sugar Price Support" below, for explanation). If the market price is below the support level when a sugar price support loan comes due, its "non-recourse" feature means a processor can exercise the legal right to forfeit, or hand over, sugar offered to USDA as collateral for the loan in fulfillment of its repayment obligation. Should this occur, USDA would record a budgetary expense, or outlay, for such a transaction.

This report focuses on the issues raised by the sugar program provisions in the House and Senate farm bills. Also, see the **Appendix** for a side-by-side comparison of the sugar provisions in the enacted 2008 farm bill with previous law and the House and Senate farm bill provisions.

Issues in 2008 Farm Bill Debate

Consideration of future U.S. sugar policy revolved primarily around four issues. These were where to set the level of minimum price guarantees to be made available to processors, how to use two tools to manage U.S. sugar supply, authorizing any sugar surplus to be used as a feedstock for ethanol, and accounting for projected program costs. Though industrial users of sugar in food and beverage products initially explored converting the sugar program to operate similar to the programs in place for the major grains, oilseeds and cotton, this policy option did not receive further attention.

Level of Sugar Price Support

USDA is required to extend price support loans to sugar processors that meet certain conditions on passing program benefits to the farmers that supply them with sugar beets or sugarcane. These loans are made at statutorily set loan rates,[1] and account for most of the effective support level made available to producers and processors. USDA is required to use its other tools to protect this price guarantee.[2] Loan rates for raw cane sugar have not changed since 1985; for refined beet sugar, since 1992. These minimum prices have guaranteed producers of sugar crops and the

[1] For sugar, the loan rate is the price per pound at which the Commodity Credit Corporation (CCC)—USDA's financing arm—extends nonrecourse loans to processors. This short-term financing at below market interest rates (e.g., 0.625% for loans taken out in January 2009) enables processors to hold their commodities for later sale.

[2] The loan rates alone do not serve as the intended price guarantee, or floor price, for sugar. In practice, USDA sets marketing allotments and import quota levels in order to support raw cane sugar and refined beet sugar at slightly higher price levels. Each price level takes into account the loan rate, interest paid on a price support loan, transportation costs (for raw sugar), certain marketing costs (for beet sugar), and discounts. These are frequently referred to as "loan forfeiture levels" or the level of "effective" price support.

processors that convert these crops into sugar, a U.S. price that since the early 1980s has ranged from two to four times the price of sugar traded in the world marketplace.

The enacted 2008 farm bill will increase sugar loan rates by 4% to 5% by FY2012. Conferees split the difference between the House- and Senate-proposed rate increases and adopted the Senate approach that proposed to increase rates in stages each year. The loan rate for raw cane sugar would rise in quarter-cent increments from the current 18.0¢ per pound to 18.75¢/lb., beginning with the 2009 sugarcane crop. The refined beet sugar loan rate, beginning with the 2009 sugar beet crop, would similarly increase in stages, from the current 22.9¢ per pound to 24.1¢/lb in FY2012.[3] The **Appendix** provides loan rates for each of the fiscal years covered by 2008 farm bill authority.

Growers and processors had initially sought a one cent increase in the raw cane sugar loan rate (with a corresponding increase in the refined beet sugar rate), and had acknowledged their satisfaction with receiving half of their request in the House-passed farm bill. They argued that the increase in the loan rate is needed to cover increased production costs, particularly energy inputs. Sugar users countered that the House-proposed higher loan rates would increase costs to taxpayers by an additional $100 million annually. They also noted that while the bill's ethanol provisions (see "Sugar for Ethanol" below) "are supposedly designed to deal with surpluses," the loan rate increase "can only encourage *higher* surplus production."[4] The Bush Administration, in its statement of administration policy on the House and Senate farm bills, opposed the increase in the loan rates for sugar.

Implementation

On September 30, 2008, USDA announced loan rates for 2008-crop sugar as required by the 2008 farm bill. The national average loan rate is 18.0¢/lb. for raw cane sugar, and 22.9¢/lb. for refined beet sugar, the same as for the previous year's crop. In turn, these national loan rates are adjusted to reflect transportation cost differentials. Reflecting this, the raw cane sugar loan rate will range from 16.37¢/lb. in Hawaii (if sugar pledged for a loan is stored on the islands) to 18.22¢/lb. in Louisiana. The refined beet sugar loan rate will range from 21.95¢/lb. in Idaho, Oregon, and Washington to 24.34¢/lb. in Michigan and Ohio.

Controlling Sugar Supply to Protect Sugar Prices

The sugar program uses two tools—import quotas and marketing allotments—to ensure that producers and processors receive price support benefits. By regulating the amount of foreign sugar allowed to enter and the quantity of sugar that processors can sell, USDA can for the most part keep market prices above effective support levels, meet the no-cost objective, and ensure that domestic sugar demand is met. If these tools are implemented as intended, the likelihood that USDA acquires sugar due to loan forfeitures is remote.

---

[3] The loan rate for refined beet sugar reflects the requirement that it be set each year equal to 128.5% of that year's raw cane sugar's loan rate, beginning in FY2010.

[4] Letter to Members of Congress, from food and beverage companies and trade associations, and public interest groups, July 13, 2007.

Import Quotas

The United States must import sugar to cover demand that the U.S. sugar production sector cannot supply. However, USDA restricts the quantity of foreign sugar allowed to enter for refining and/or sale to manufacturers for domestic food and beverage use. Quotas are used to ensure that the quantity that enters does not depress the domestic market price to below support levels. Quota amounts are laid out in U.S. market access commitments made under World Trade Organization (WTO) rules and under bilateral free trade agreements (FTAs).

The sugar program authorized by the 2002 farm bill accommodated, or made room for, imports of up to 1.532 million short tons raw value (STRV) each year.[5] This import level is one of the four factors that USDA used to establish the national sugar allotment (called the "overall allotment quantity"), and reflected U.S. trade commitments under two trade agreements in effect when the 2002 program was authorized (**Table 1**).

Table 1. Annual U.S. Sugar Import Commitments When the 2002 Farm Bill Was Enacted

		short tons
World Trade Organization Quota (minimum)[a]		1,256,000
North American Free Trade Agreement (NAFTA)—Mexico Quota (maximum)[b]		276,000
	Total	**1,532,000**

a. Covers both raw sugar and refined sugar.

b. Applied only through the end of calendar year 2007.

Since January 1, 2008, however, U.S. sugar imports from Mexico are no longer restricted. Under NAFTA, Mexico no longer faces any tariff or quantitative limit on the amount of sugar exported to the U.S. market. With this opening, though, imports could fluctuate from year to year for various reasons. First, the amount of Mexican sugar exported to the U.S. market will depend largely upon the extent that U.S. exports of historically cheaper high-fructose corn syrup (HFCS) displace Mexican consumption of Mexican-produced sugar. Surplus Mexican sugar, in turn, would then likely move north to the United States.[6] Second, Mexico's sugar output, though trending upward, does vary from year to year, depending upon weather and growing conditions. Mexican government policy also is to hold three months worth of sugar stocks in reserve and to allow sugar imports when needed to meet demand and lower prices.[7] Third, Mexican sugar prices in recent years have for the most part been higher than U.S. sugar prices. To the extent that this occurs in the future, the incentive for a Mexican sugar mill to export sugar north in search of a better price is reduced. Fourth, U.S. buyers' concerns about the quality of Mexican sugar may limit the amount that actually flows north in the next few years.

[5] One short ton equals 2,000 pounds.

[6] However, the 2007 to mid-2008 increase in U.S. HFCS prices due to the higher cost of corn—its main input—did reduce its competitiveness against Mexican-priced sugar. To the extent that the HFCS price falls below sugar prices in the future, the incentive increases for Mexican bottlers of soft drinks to shift to HFCS.

[7] U.S. sugar processors also are now free to export sugar to Mexico to take advantage of the occasional higher prices there.

Also, the United States has committed under other existing and pending bilateral FTAs to allow for additional sugar imports.[8] Such imports in 2013, the fifth year of the sugar program authorized by the 2008 farm bill, could total from about 420,000 tons to 1.215 million tons *above* existing WTO and FTA trade commitments and *above* the maximum amount of sugar allowed to enter from Mexico in 2007. The wide range reflects two varying assumptions made to estimate by how much HFCS use in Mexico might displace sugar consumption in Mexico and create a surplus available for export to the U.S. market.[9]

Legislation

The sugar program provisions in the enacted 2008 farm bill did not directly address the issue of additional sugar imports. Instead, a new sugar-for-ethanol program is authorized to handle the price-related impact of such imports (Section 9001 in the energy title; see "Sugar for Ethanol" and "Sugar Program Costs" below). However, other provisions prescribe how USDA must now administer import quotas. To cover shortfalls (because of hurricanes or other disastrous events) in what domestic sugar processors can sell under allotments, USDA is directed to ensure that most imports enter in the form of raw cane sugar rather than refined sugar. While most permitted imports have historically entered in raw form, USDA decisions to allow large quantities of refined sugar to enter after the late 2005 hurricanes significantly affected the competitive position of cane refineries in Louisiana and Florida that process raw sugar. Unlike 2001-2002, when the Congress considered the last farm bill, most cane refineries are now a key part of vertically integrated operations owned by raw sugar processors and/or sugarcane producers. The 2008 farm bill's policy change is intended to ensure that these cane refineries (which process raw sugar into refined sugar) can more fully use their operating capacity. Also, limiting the entry of refined sugar enhances the position of the domestic beet sector to increase their sales of refined sugar.

Conferees, though, did not adopt provisions found only in the House-passed bill that would have directed USDA to regulate when and how much raw cane sugar imports are allowed to be shipped to U.S. cane refineries. While USDA announced shipping patterns in FY2003-FY2005, the impact of the hurricanes led to a decision not to follow this long-standing practice in FY2006-FY2008. USDA justified removing these restrictions because of "changes occurring over time in the domestic marketing of cane sugar." The House-passed provisions could be viewed as intending to increase the transaction costs for countries that export larger amounts of sugar to the U.S. market and to give a slight competitive edge to domestic processors with respect to buyers. Food and beverage firms opposed "micro-managing" the timing of imports, noting that the application of such rules will limit the ability of cane refiners to efficiently use their processing capacity and could lead to serious shortfalls at times in the amount of sugar supplied to the market.[10] In commenting on the House bill, the Bush Administration expressed concern over requiring shipping patterns for quota sugar imports. Also, several countries eligible to ship sugar to the U.S. market expressed concern that the proposed regulation of the flow of imports would run counter to U.S. trade commitments. Because of the concern expressed that prescribing how sugar import shipping patterns should be administered would open up the United States to

[8] All of the sugar access provisions in the Dominican Republic-Central American FTA (DR-CAFTA) already are in effect. Congress has yet to consider the FTAs with Panama and Colombia, which would grant additional access for each country's sugar into the U.S. market.

[9] The assumptions are laid out in an analysis that appeared in USDA's Economic Research Service's *Sugar and Sweeteners Outlook*, January 30, 2007, pp. 21-25.

[10] Letter to Members of Congress, July 13, 2007.

challenges by sugar exporting countries in the WTO, these provisions were dropped in conference.[11]

Implementation

In line with the changes made by the 2008 farm bill, USDA on September 9, 2008, announced that the FY2009 raw sugar tariff-rate quota (TRQ)[12] will be set at 1,231,497 STRV—the U.S. minimum access commitment for raw sugar imports under WTO rules. Relatedly, USDA announced the TRQ for refined and specialty sugars at 104,251 STRV. This amount is 80,000 ST higher than the U.S. minimum refined sugar TRQ (24,251 STRV), increased in order to meet U.S. demand for organic sugar not available from the domestic producing sector. Both announcements reflect the enacted 2008 farm bill's requirement that USDA set both the raw sugar and refined sugar TRQs at the minimum levels required by U.S. WTO trade commitments by October 1, 2008. USDA's accompanying statement acknowledged that it expects the domestic market will require additional supplies of sugar (i.e., imports) during FY2009, and indicated that appropriate adjustments will be made to these TRQ levels and the national marketing allotment level (see below) to ensure the availability of adequate supplies of sugar.[13]

The enacted 2008 farm bill limits USDA's ability to allow additional sugar imports under the TRQs established to meet U.S. WTO trade commitments. However, USDA is given some discretionary authority to exercise on this matter. Any decision to increase the raw sugar TRQ and/or the refined sugar TRQ before April 1, 2009, now requires USDA to first declare that an emergency sugar shortage exists because of "war, flood, hurricane, or other natural disaster" or another similar event as determined by the Secretary of Agriculture. USDA could interpret the law's discretionary language to determine that an emergency sugar shortage exists and allow for additional imports to the extent it determines that market prices will not fall below support levels and result in loan forfeitures. The sugar production sector likely will argue that there is no need for additional imports, pointing out that there is no physical shortage of sugar. Industrial sugar users likely will petition USDA to allow for additional refined sugar imports, pointing out that projected low ending stocks are keeping refined sugar prices considerably above the historical average. To date, USDA has not yet issued regulations detailing how such a determination would be made. In the interim, sugar crop producers/processors and major sugar users are expected to weigh in on what USDA under the Obama Administration develops as rules to detail what constitutes an emergency sugar shortage.

Marketing Allotments

In the 2002 farm bill, the domestic production sector accepted mandatory limits on the amount of sugar that processors can sell—known as marketing allotments—in return for the assurance of price protection. It viewed allotments as a way to try to capture any growth in U.S. sugar demand,

[11] The World Trade Organization administers trade dispute settle procedures whereby a country can file a case against another alleging that the latter operates a program or policy that runs counter to WTO rules. In this context, the prospect arose that a sugar exporting country might allege that the proposed shipping patterns provision were discriminatory or trade distorting.

[12] The quota component of a TRQ provides for duty-free access of a specified quantity of a commodity. Imports above this quota are subject to a prohibitive tariff.

[13] "USDA Announces Fiscal Year 2009 Sugar Program," September 9, 2008, as accessed at http://www.usda.gov/wps/portal/!ut/p/_s.7_0_A/7_0_1OB?contentidonly=true&contentid=2008/09/0226.xml.

and assumed that the then-U.S. sugar import quota commitments would continue without change (see "Import Quotas" above). The statute, however, stipulated that if (1) USDA estimates imports will be above 1.532 million short tons, and (2) that such imports would lead USDA to reduce the amount of domestic sugar that U.S. processors can sell, then USDA must suspend marketing allotments. Suspending allotments because of additional imports raises the prospect of downward pressure on market prices if most U.S. sugar demand is already met. If the additional imports were to cause the price to fall below support levels, forfeitures would occur and USDA would be unable to meet the no-cost requirement. Including the allotment suspension provision in the 2002 farm bill was designed to ensure that USDA not lose control over managing U.S. sugar supplies for fear of the consequences that could be unleashed (i.e., demonstrate its inability to implement congressional policy).

Legislation

Implementation of the 2002 farm bill's marketing allotment authority resulted in the U.S. sugar production sector's share of domestic food consumption ranging from a low of 73% in FY2006 to a high of 89% in FY2004. Concerned that their market share would decline as sugar imports increase under various trade agreements (see "Import Quotas" above), sugar producers and processors decided to pursue a different approach in formulating their proposal for the 2008 farm bill. Adopted by farm bill conferees, an important new provision guarantees that the domestic production sector always benefits from a minimum 85% share of the U.S. sugar-for-food market. USDA is now required to announce an "overall allotment quantity" (OAQ)—the amount of sugar that all processors combined can sell—that represents at least 85% of estimated domestic sugar consumption. This is intended to address the sector's objective that imports not displace the ability of U.S. sugar processors to sell more of their output in each successive year, to the extent that U.S. demand for sugar grows.

Implementation

On September 9, 2008, USDA announced that the FY2009 OAQ will be 8.925 million STRV. This complies with the new statutory requirement that USDA establish the OAQ at not less than 85% of estimated U.S. human sugar consumption (projected at 10.5 million STRV for FY2009).

The FY2009 OAQ level is considerably higher than USDA's latest estimate of FY2009 sugar production. Though cane sugar output is projected to increase by 4% over FY2008, beet sugar production is expected to be almost 11% lower because of spring 2008 weather problems in North Dakota and Minnesota, a major beet-producing region, and reduced planted beet acreage. As of mid-January 2009, USDA projected 2008/2009 U.S. sugar production at 7.8 million ST, 1.1 million ST below the OAQ that USDA announced last September. With the OAQ split between the beet and cane producing sectors using the percentage shares laid out in law, each sector is expected to fully market all of the sugar that USDA projects will be produced during FY2009 (**Table 2**).

Table 2. Comparison of National Sugar Allotment to USDA-Projected Sugar Production, FY2009

		Overall Allotment Quantity	Estimated Production[a]	Shortfall (Allotment Deficit)
	share	*short tons, raw value*		
National	100.00 %	8,925,000	7,800,000	-1,125,000
Beet Sugar	54.35 %	4,850,738	4,225,000	-625,738
Cane Sugar	45.65 %	4,074,262	3,575,000	-499,262

Source: USDA, "USDA Announces Fiscal Year 2009 Sugar Program," Release No. 0226.08, September 8, 2008; USDA, World Agricultural Outlook Board, *World Agricultural Supply and Demand Estimates*, January 12, 2009.

a. As of January 12, 2009.

USDA is currently projecting historically low ending stocks at the end of the 2008/2009 marketing year (i.e., September 30, 2009). This normally would imply higher than average wholesale sugar prices in the late summer period, but that is not the case across the board this year. The raw cane sugar futures price has been below the loan forfeiture levels for Florida and Hawaii since late October 2008, and below the loan forfeiture levels for Louisiana and Texas from mid-November to early December 2008. By contrast, current spot U.S. refined sugar (cane and beet) prices are well above the historical average. Various explanations for this price divergence are offered. Some point out that continued imports of raw-sugar-equivalent product from Mexico keeps downward pressure on the U.S. raw cane sugar futures price. However, USDA projects that imports from Mexico this year will be somewhat lower than last year. Others note that the loss of the production capacity of the cane sugar refinery in Savannah, Georgia, caused by an explosion a year ago, has reduced demand for raw cane sugar, keeping the raw cane sugar price lower than might be the case otherwise. Higher-than-average refined sugar prices are also attributed to the loss of refined product that this refinery would normally supply.[14]

In the interim, any call made by sugar users for USDA to increase the end-of-year U.S. sugar supply (to meet their desire for lower sugar prices) is limited by what USDA can do under the enacted 2008 farm bill to increase imports (see "Implementation" under "Import Quotas", above). If USDA were to determine before the midpoint of the marketing year (April 1) that a sugar shortage exists, it would first have to reassign existing allotment deficits to imports of raw cane sugar (i.e., increase the raw cane sugar TRQ to accommodate the deficit amount). If a sugar shortage still existed after taking such action, USDA would only then be able to increase the refined sugar TRQ if two conditions are met and this second increase in supply will not result in sugar loan forfeitures. Before taking this second step, USDA must take into account that both the sales of domestic sugar and the refining capacity of domestic raw cane sugar have "been maximized." On or after April 1, USDA can only allow for imports of raw cane sugar to address an allotment deficit.

Taking this year's current price outlook into account, any USDA decision to allow for additional imports of raw cane sugar would likely depress raw cane prices. As a result, USDA could see

[14] Imperial Sugar's Savannah operations supply almost 10% of the refined sugar consumed in the United States. The firm plans to resume refining raw cane sugar to produce bulk granulated sugar in early 2009, and to completely restore its packaging facilities by the fall of 2009.

some sugarcane processors forfeit on their loans in late summer 2009—a development that would run counter to the enacted farm bill's intent that USDA operate the sugar program at no cost. For this reason, USDA may be reluctant to allow such an increase, even though refined sugar prices are expected to remain high by historical comparison.

Sugar for Ethanol

Background

Sugar producers and processors have had an ongoing interest in exploring the potential for using sugar crops and processed sugar as a feedstock to produce ethanol (a gasoline additive). In the 2002-2003 period, they encouraged USDA to explore selling forfeited sugar stocks to corn-based ethanol processors. A few ethanol producers experimented by adding sugar to speed up the ethanol fermentation process, but the results were disappointing.

In 2005, Congress approved the Dominican Republic-Central American Free Trade Agreement (DR-CAFTA) that gives six countries increased access for their sugar to the U.S. market. During the debate, producers and processors sought a deal with the Bush Administration on a sugar-for-ethanol package. Their objective was to have this option available to divert additional sugar imports under DR-CAFTA whenever domestic prices fall below support levels.[15] With Congress mandating in 2005 that the use of renewable fuels be doubled by 2012,[16] some advocated that sugar be considered as a feedstock along with other agricultural crops and waste. Separately, Hawaii mandated (effective April 2006) that 85% of the gasoline sold must contain 10% ethanol. This requirement assumes that over time, the sugarcane produced on the islands will be available to use as the prime feedstock for ethanol.

If the cost of feedstock is excluded, producing ethanol from sugar cane can be less costly than producing it from corn. This is because the starch in corn must first be broken down into sugar before it can be fermented. This extra step adds to the cost of processing corn into ethanol, when contrasted to using sugarcane or processed sugar. Further, sugar cane waste (bagasse) also can be burned to provide energy for an ethanol plant, reduce associated energy costs, and improve sugar ethanol's energy balance relative to corn ethanol.

Brazil's success at integrating sugar ethanol into its passenger vehicle fuel supply has stimulated interest in exploring prospects for sugar-based ethanol in the United States. However, wide differences in sugar production costs and market prices in the two countries cause the economics of sugar-based ethanol to differ significantly. In investigating the economics of ethanol from sugar, USDA concluded that producing sugar cane ethanol in the United States would be more than twice as costly as U.S. corn ethanol and nearly three times as costly as Brazilian sugar ethanol.[17] Feedstock costs accounted for most of this price differential.[18] The USDA study

---

[15] Though the Administration did not agree to such a package, the Secretary of Agriculture pledged to divert surplus sugar imports—through purchases—for ethanol and other non-food uses, to ensure that the sugar program operates as authorized only through FY2008.

[16] For more information, see CRS Report R40168, *Alternative Fuels and Advanced Technology Vehicles: Issues in Congress*, by Brent D. Yacobucci.

[17] Office of Economics, *The Economic Feasibility of Ethanol Production from Sugar in the United States*, July 2006.

[18] In Brazil, the cost of producing raw cane sugar reportedly ranges from 6¢ to 9¢ per pound (i.e., 9¢ to 12¢/lb. when converted to refined basis). In the United States, raw cane sugar production costs range from 12¢ to 20¢/lb. U.S. (continued...)

showed that while sugar ethanol may be a positive energy strategy in such countries as Brazil, it may not be economical in the United States.[19]

Legislation

The enacted 2008 farm bill incorporates a proposal presented to the Agriculture Committees by the U.S. sugar production sector. The "Feedstock Flexibility Program for Bioenergy Producers" requires USDA to administer a sugar-for-ethanol program using sugar intended for food use but deemed to be in surplus. USDA will sell both surplus sugar that it purchases if determined necessary to maintain prices above support levels, and the sugar acquired as a result of loan forfeitures, to bioenergy producers for processing into fuel grade ethanol and other biofuel. Competitive bids would be used by USDA to purchase sugar from processors, at a price not less than sugar program support levels, which it would then sell to ethanol firms. USDA would implement this program only in those years where purchases are required to operate the sugar program at no cost. USDA's CCC would provide open-ended funding.

Because it would cost much more to produce ethanol from U.S.-priced sugar than from corn, this new program would require a considerable subsidy to operate as intended. The prime market for such sugar likely would be existing corn-based ethanol facilities close to sugar beet producing areas (e.g., the Upper Midwest) and new plants constructed in sugarcane-producing states (Hawaii and Louisiana). Producers of ethanol from corn in the continental United States, though, would likely need to adjust their fermentation process and/or invest in new equipment to handle sugar. As a result, they may not be as interested in purchasing sugar as a feedstock unless the price is significantly discounted further (e.g., requiring even more of a subsidy) to reflect the additional costs of processing sugar instead of corn. However, the availability of this subsidy could facilitate the development of the ethanol sector in Hawaii and partially reduce the islands' dependence on importing gasoline for its vehicle transportation needs.

As designed, this program will rely on U.S.-produced (rather than foreign) sugar. The amount that USDA decides to purchase would approximate its estimate of the extent that imports under trade agreements reduce the U.S. sugar price below support levels. Producers supported this provision, viewing it as an insurance policy for receiving the benefits of a guaranteed minimum price for sugar marketed for food use. Sugar users opposed this program "to ostensibly manage surplus supplies." In their July 13, 2007, letter to Members of Congress, they argued that this authority "will likely be used to short domestic markets, further restricting the availability of sugar for food use in the U.S. market." They characterized this approach as "wasteful of taxpayer resources" because sugar is not price competitive with corn as a feedstock, and will require large subsidies to ethanol producers "to induce them to accept the sugar." The Bush Administration opposed this sugar-for-ethanol component, commenting that it would not allow USDA to dispose of surplus sugar to end uses other than ethanol production, even if "those uses would yield a much higher return for taxpayers."[20]

(...continued)

production costs for refined beet sugar range from 17¢ to 33¢/lb. For additional perspective, see "Costs of Production and Sugar Processing" in USDA, Economic Research Service, *Sugar Backgrounder*, July 2007, pp. 17-21.

[19] This discussion is adapted from "Sugar Ethanol" in CRS Report RL33928, *Ethanol and Biofuels: Agriculture, Infrastructure, and Market Constraints Related to Expanded Production*, by Brent D. Yacobucci and Randy Schnepf.

[20] Office of Management and Budget, "Statement of Administration Policy" on the Senate bill (Food and Energy Security Act of 2007), November 6, 2007, p. 3.

Outlook

The current U.S. sugar market outlook (i.e., demand considerably above current supply, implying that market prices will be above loan forfeiture levels) suggests that, at present, USDA will likely be able to meet the program's no-cost directive without having to activate the new sugar-for-ethanol program in FY2009. USDA confirmed this with its September 2008 determination that no sugar will be available for this program, taking into account its forecast that sugar loan forfeitures are unlikely in FY2009. The status of this program in subsequent years will depend on whether U.S. sugar production returns to more normal levels, and on how sugar users (particularly in the beverage sector) in both the United States and Mexico respond to higher HFCS prices caused by (until recently) high corn prices. If HFCS prices are higher than Mexican sugar prices, the likely result will be a smaller displacement of Mexican-produced sugar by HFCS imports from the United States, and thus a smaller surplus available to be exported without restriction to the U.S. market. This reportedly did occur during FY2008.[21]

Sugar Program Costs

For the six years covered by the 2002 farm bill (FY2003-FY2008), USDA succeeded in operating the sugar program at no cost, as directed by law. Budget forecasts in early 2007 had projected that the sugar program, if continued without change, would cost from almost $700 million (Congressional Budget Office (CBO)) to about $800 million (USDA) over the FY2008-FY2012 five-year period. Projected outlays reflected estimates of the budgetary impact of additional sugar imports from Mexico and from other countries with additional access to the U.S. market for their sugar under bilateral FTAs. Each cost projection assumed that the additional supplies depress the domestic sugar price below price support levels, and lead processors to forfeit on a portion of their loans.[22]

Outlook

The policy changes enacted in the 2008 farm bill are intended to head off these potential costs and ensure that USDA can operate the program at no cost. However, estimating future budgetary impacts is difficult, considering that market conditions can change quickly and dramatically, and can differ significantly from historical experience. In its latest baseline budget projection (January 2009), CBO assumes that USDA essentially succeeds in operating the sugar price support program at no cost. It estimates outlays of $2 million in each of FY2009-FY2012, or a total of $6 million over the farm bill's five-year period. For the sugar-for-ethanol program, which is an integral part of the sugar program but scored separately as a non-commodity activity, CBO estimates outlays of $325 million over the farm bill period. The latter estimate assumes that USDA sells sugar to bioenergy producers at a much lower price than the sugar program's minimum guaranteed price which USDA is required to pay for the surplus sugar purchased or acquired from domestic sugar processors.

---

[21] Inside U.S. Trade, "USDA Projects Lower Mexican Sugar Exports; Corn Syrup Price Link Seen," June 6, 2008, pp. 11-12.

[22] The forfeiture of a price support loan results in a budget outlay, because the credit that had been extended is not paid back by the processor (resulting in a loss to the U.S. government). To the extent USDA succeeds in selling forfeited sugar, proceeds flow back to USDA and reduce the loss.

Appendix. Comparison of 2008 Farm Bill Sugar Program Provisions with Previous Law and House and Senate Bills

Prior Law/Policy	House-Passed Bill (H.R. 2419)	Senate-Passed Substitute Amendment (H.R. 2419)	2008 Farm Bill (P.L. 110-246)
No Net Cost Directive			
Requires USDA to the maximum extent practicable to operate the sugar non-recourse loan program at no net cost by avoiding sugar forfeitures to the CCC. *[7 U.S.C. 7272 (g), 7 U.S.C. 1359bb (b), 7 U.S.C. 1359cc (b)(2)]*	Retains current no net-cost requirement. *[Secs. 1301 and 1303(b)]*	Same as the House bill. *[Secs. 1501 and 1504(b)]*	Continues no-cost requirement found in prior law. *[Secs. 1401, 1403]* Requires USDA to operate sugar-for-ethanol program (in Energy title) to ensure this no-cost directive is met. *[Sec. 9001]*
Price Support Levels, Loans and Payments			
Sets raw cane and refined beet sugar loan rates at 18.0¢/lb. and 22.9¢/lb., respectively, through FY2008. Expands loan eligibility to in-process sugars and syrups at 80% of the applicable cane or beet loan rates. Makes nonrecourse loans available to processors that meet specified conditions. Sets 9-month repayment term for such loans. *[7 U.S.C. 7272 (a, b, d, e, f)]*	Increases raw cane sugar and refined beet sugar loan rates to 18.5¢/lb. and 23.5¢/lb, respectively, for FY2009 through FY2013. *[Sec. 1301]*	Increases raw cane sugar loan rate to 19.0¢/lb. by FY2013, in 1/4¢ increments beginning in FY2010, as follows: ¢ / lb. FY2009—18.00 FY2010—18.25 FY2011—18.50 FY2012—18.75 FY2013—19.00 Increases beet sugar loan rate, beginning in FY2010, to be set at 128.5% of the raw cane loan rate in effect each year (e.g., reaching 24.42¢/lb. in FY2013), or as follows: ¢ / lb. FY2009—22.90 FY2010—23.45 FY2011—23.77 FY2012—24.09 FY2013—24.42 *[Sec. 1501]*	Increases raw cane sugar loan rate to 18.75¢/lb. by FY2012, in 1/4¢ increments beginning in FY2010, as follows: ¢ / lb. FY2009—18.00 FY2010—18.25 FY2011—18.50 FY2012—18.75 FY2013—18.75 Sets refined beet sugar loan at 22.9¢/lb. in FY2009. Starting in FY2010, sets beet sugar rate equal to 128.5% of the raw cane loan rate in effect (e.g., rising to 24.09¢/lb. by FY2012, or as follows: ¢ / lb. FY2009—22.90 FY2010—23.45 FY2011—23.77 FY2012—24.09 FY2013—24.09 Continues other provisions found in prior law. *[Sec. 1401]*

CRS-11

Prior Law/Policy	House-Passed Bill (H.R. 2419)	Senate-Passed Substitute Amendment (H.R. 2419)	2008 Farm Bill (P.L. 110-246)
Authorizes CCC to accept bids from sugar processors to purchase USDA-owned sugar in conjunction with reduced production of new sugar crops. *[7 U.S.C. 7272 (g)]*	Continues in-kind authority. Stipulates that planted beets or cane diverted from production can only be used as bioenergy feedstock. *[Sec. 1301]*	Similar to the House bill. *[Sec. 1501]*	Continues in-kind authority and adds House/Senate provision. *[Sec. 1401]*
USDA now pays storage rates of 8¢ per 100 lbs. for raw cane sugar and 10¢ per 100 lbs. for refined beet sugar that has been forfeited under the nonrecourse loan program. *[15 U.S.C. 714b & 714c; 7 CFR Part 1423]*	No comparable provision.	Requires (only through crop year 2011) USDA minimum storage payment rates of 10¢/cwt. and 15¢/cwt. on forfeited raw cane and refined beet sugar. *[Sec. 1503]*	Adopts Senate provision. *[Sec. 1405]*
Authorizes CCC to provide financing to processors of domestic sugar to construct or upgrade storage and handling facilities. *[Sec. 1402]*	No comparable provision.	Retains authority, but stipulates that loans shall not require any prepayment penalty. *[Sec. 1502]*	Continues prior law and adds Senate provision. *[Sec. 1404]*
Marketing Allotments and Allocations			
To avert loan forfeitures, requires USDA to limit the amount of sugar processors can sell each year. This is done through a national "overall allotment quantity" (OAQ) that is split between beet and cane sectors (54.35% and 45.65%, respectively), and then allocated to individual processors. The OAQ must accommodate WTO and NAFTA import commitments (1.532 million short tons). If imports are greater, USDA's authority to implement allotments is suspended. *[7 U.S.C. 1359aa, 1359bb, 1359cc, and 1359dd]*	Continues purpose and structure of marketing allotments and allocations, but changes some key provisions. Changes formula to require USDA to set OAQ at not less than 85% of estimated human food and beverage sugar use. Eliminates allotment suspension provision. *[Sec. 1303(a)-(d)]*	Similar to the House bill. *[Sec. 1504(a)-(d)]*	Continues marketing allotment authority and adopts House/Senate provisions that: —require USDA to set OAQ at not less than 85% of estimated U.S. human consumption, and —eliminate allotment suspension trigger. *[Sec. 1403(a)-(d)]*

Prior Law/Policy	House-Passed Bill (H.R. 2419)	Senate-Passed Substitute Amendment (H.R. 2419)	2008 Farm Bill (P.L. 110-246)
Directs USDA to reassign unused raw cane and beet sugar marketing allocations *first* to other cane states and beet processors, respectively; *second* to cane processors within each state; *third* to sales of sugar in CCC's inventory; and *fourth* to imports. *[7 U.S.C. 1359ee]*	Requires that any reassignment of unused cane and beet allocations to imports in the fourth step must be met by imports "of raw cane sugar." *[Sec. 1303(e)]*	Similar to the House bill. *[Sec. 1504(e)]*	Adopts House/Senate change to prior law. *[Sec. 1403(e)]*
Trade-Related Provisions			
In accord with a 1994 trade commitment, USDA sets an annual global sugar import quota of not less than 1.256 million short tons. USTR allocates the quota among eligible countries, and also administers preferential sugar import quotas for free trade agreement partner countries. Effective January 1, 2008, Mexico can ship duty free an unlimited amount of sugar to the U.S. market.	Makes no changes to import quota commitments found in various trade agreements and laws.	Makes no changes to import quota commitments.	Makes no change to current U.S. trade commitments.
Requires USTR in 2002-07 to reallocate unused country quota allocations to other quota-holding countries with sugar to sell. *[7 U.S.C. 1359kk]*	Repeals requirement for reallocating sugar import quota shortfalls. *[Sec. 1303(i)]*	Similar to the House bill. *[Sec. 1504(i)]*	Adopts House/Senate repeal provision. *[Sec. 1403(i)]*
USDA has discretion to increase the size of global raw cane and refined sugar import quotas when domestic sugar supplies are inadequate to meet U.S. demand at reasonable prices. *[Chapter 17, additional note 5, of the U.S. Harmonized Tariff Schedule; 19 CFR Part 2001, Subpart A]*	Requires USDA to set quotas for raw cane and refined sugar at the minimum level necessary to comply with U.S. trade agreement obligations. In cases of emergency sugar shortages, before April 1 of each marketing year, requires USDA to increase supplies *first* by reassigning allotment deficits to imports of raw cane sugar (i.e., increase the raw sugar quota), and *second* the refined sugar quota, if certain conditions are met. On or after April 1, allows USDA only to increase the raw cane sugar quota, if specified conditions are met. *[Sec. 1303(l)]*	Similar to the House bill. *[Sec. 1504(j)]*	Adopts House/Senate provision on setting initial import quotas at minimum levels and laying out steps to be followed to increase imports in the event of a sugar shortage. *[Sec. 1403(j)]*

CRS-13

Prior Law/Policy	House-Passed Bill (H.R. 2419)	Senate-Passed Substitute Amendment (H.R. 2419)	2008 Farm Bill (P.L. 110-246)
To protect domestic sugar prices, USDA regulated the flow of sugar imports from large quota holders (through 2005).	Requires USDA to establish "orderly shipping patterns" for major suppliers of sugar to the U.S. market. *[Sec. 1303(i)]*	No comparable provision.	Deletes House "shipping patterns" provision.
The U.S.-Mexican agreement on bilateral market access for sugar and high-fructose corn syrup (HFCS) created an industry and government task force to address problems that might arise after the elimination of tariffs on sweeteners on January 1, 2008. *[Exchange of Letters between USTR and Mexico's Secretariat of Economy, July 27, 2006]*	No comparable provision.	Expresses sense of Senate that U.S. & Mexican governments should coordinate their sugar policies to be consistent with U.S. international commitments, to avoid disruptions of each country's sweetener markets (sugar and HFCS). *[Sec. 1505]*	Deletes Senate provision.
The U.S. withdrew from the International Sugar Organization (ISO) in 1992 because of opposition to the allocation of country contributions to ISO's budget.	Requires the Secretary of Agriculture to work with the Secretary of State to restore U.S. membership in the ISO within one year. *[Sec. 1302]*	Similar to the House bill. *[Sec. 1504]*	Adopts House provision. *[Sec. 1402]*
Sugar-for-Ethanol Program (Feedstock Flexibility Program)			
No comparable provision.	Requires USDA (for FY2008-FY2012) to purchase sugar from those firms that sell sugar (equal to the quantity of imports that USDA estimates exceeds U.S. food demand), and to resell such sugar as a biomass feedstock to produce bioenergy, in a way to ensure that sugar price support program provisions (see above) operate at no cost and avoid loan forfeitures. Requires USDA to use CCC resources, including "such sums as are necessary," to implement this new authority. *[Sec. 9013]*	Similar to the House bill. *[Sec. 1501]*	Adopts House/Senate provisions, and extends program by one year (FY2013). Prescribes how CCC-inventory sugar is to be disposed for this Program and other purposes, and allows for the sale of CCC-inventory sugar in the case of emergency shortages of sugar for food use. *[Sec. 9001]*

Author Contact Information

Remy Jurenas
Specialist in Agricultural Policy
rjurenas@crs.loc.gov, 7-7281

Biofuels Provisions in the 2007 Energy Bill and the 2008 Farm Bill: A Side-by-Side Comparison

Randy Schnepf
Specialist in Agricultural Policy

Brent D. Yacobucci
Specialist in Energy and Environmental Policy

January 26, 2009

Congressional Research Service
7-5700
www.crs.gov
RL34239

CRS Report for Congress
Prepared for Members and Committees of Congress

Summary

The Energy Independence and Security Act of 2007 (EISA, P.L. 110-140), also known as the 2007 energy bill, significantly expands existing programs to promote biofuels. The Food, Conservation, and Energy Act of 2008 (P.L. 110-246), also known as the 2008 farm bill, contains a distinct energy title (Title IX) that covers a wide range of energy and agricultural topics with extensive attention to biofuels, including corn-starch based ethanol, cellulosic ethanol, and biodiesel. Research provisions relating to renewable energy are found in Title VII and tax provisions are found in Title XV of the farm bill.

Key biofuels-related provisions of EISA and the 2008 farm bill include:

- a major expansion of the renewable fuel standard (RFS) established in the Energy Policy Act of 2005 (P.L. 109-58) [EISA];
- expansion and/or modification of tax credits for ethanol [farm bill];
- grants and loan guarantees for biofuels (especially cellulosic) research, development, deployment, and production [EISA, farm bill];
- studies of the potential for ethanol pipeline transportation, expanded biofuel use, market and environmental impacts of increased biofuel use, and the effects of biodiesel on engines [EISA, farm bill];
- expansion of biofuel feedstock availability [farm bill];
- reauthorization of biofuels research and development at the U.S. Department of Energy [EISA] and the U.S. Department of Agriculture and Environmental Protection Agency [farm bill]; and
- reduction of the blender tax credit for corn-based ethanol, a new production tax credit for cellulosic ethanol, and continuation of the import duty on ethanol [farm bill].

This report includes information from CRS Report RL34130, *Renewable Energy Policy in the 2008 Farm Bill*, by Tom Capehart, and CRS Report RL34136, *Biofuels Provisions in the Energy Independence and Security Act of 2007 (P.L. 110-140), H.R. 3221, and H.R. 6: A Side-by-Side Comparison*, by Brent D. Yacobucci.

Contents

Tables

Contacts

Introduction

Recent high energy prices, concerns over energy security, and the desire to promote rural business and to reduce air pollutant and greenhouse gas emissions have sparked congressional interest in promoting greater use of alternatives to petroleum fuels. Biofuels—transportation fuels produced from plant and animal materials—have attracted particular interest. Ethanol and biodiesel, the two most widely used biofuels, receive significant federal support in the form of tax incentives, loan and grant programs, and regulatory programs.[1]

The Energy Policy Act of 2005 (EPAct, P.L. 109-58) established a renewable fuel standard (RFS). This initial RFS required the increasing use of renewable fuel in gasoline, starting at 4.0 billion gallons in 2006 and increasing to 7.5 billion gallons in 2012. However, the RFS was significantly expanded on December 19, 2007, when President Bush signed the Energy Independence and Security Act of 2007 (EISA). Instead of requiring 5.4 billion gallons of renewable fuel in 2008, the new law requires 9.0 billion gallons. Further, the 2007 law requires that the RFS be expanded to 36 billion gallons of renewable fuel by 2022, as compared to an estimated 8.6 billion gallons under EPAct. Although this is not an explicit ethanol mandate, it is expected that much of this requirement will be met using corn-based ethanol.[2] The U.S. ethanol industry expanded rapidly in response to EPAct, outpacing the required growth in the earlier RFS and leading some proponents of corn-based ethanol to support an increase in the mandated levels of the RFS.

The Food, Conservation, and Energy Act of 2008 (2008 farm bill, P.L. 110-246) promotes the development of cellulosic ethanol production through new tax credits, reduces slightly the blender's tax credit for corn-derived ethanol beginning in 2009, and continues the tariff on imported ethanol. It also expands research on agricultural renewable energy and encourages infrastructure development needed for cellulosic ethanol production.

During the final months of the farm bill debate, both food and fuel prices increased dramatically, and the role of corn-based ethanol in food price inflation became the subject of intense debate. Because of the rapid expansion of U.S. corn ethanol capacity—as of December 2008, existing capacity was an estimated at 10.4 billion gallons per year, while an additional 1.8 billion gallons was under construction[3]—some are concerned that the United States will soon reach the limit of ethanol that can be produced from corn. Critics of corn-based ethanol argue that the industry does not need continued government support, and that current corn demand for ethanol is putting a strain on corn and other grain markets, leading to increases in other commodity prices, such as livestock feed, which then leads to higher dairy and meat prices.[4] Critics also argue that the environmental costs of corn-based ethanol may outweigh the benefits. Proponents of corn-based ethanol production assert that increased acreage and upward-trending yields will enable corn producers to satisfy the demand for corn for feed, fuel, and exports.

---

[1] For more information on federal biofuels incentives, see CRS Report RL33572, *Biofuels Incentives: A Summary of Federal Programs*, by Brent D. Yacobucci.

[2] For more information on ethanol, see CRS Report RL33290, *Fuel Ethanol: Background and Public Policy Issues*, by Brent D. Yacobucci.

[3] Renewable Fuels Association, http://www.ethanolrfa.org/industry/locations.

[4] For more information on the issues surrounding rapid ethanol expansion, see CRS Report RL33928, *Ethanol and Biofuels: Agriculture, Infrastructure, and Market Constraints Related to Expanded Production*, by Brent D. Yacobucci and Randy Schnepf.

Advanced biofuels based on non-food feedstocks are generating much interest. Feedstocks that could be grown on marginal land with reduced inputs compared with corn would solve the food versus fuel issue, it has been argued. However, biofuels that rely on other sources of biomass, including agricultural wastes, municipal solid waste, and dedicated non-food energy crops such as perennial grasses, fast-growing trees, and algae are still years from commercial production. Nonetheless, this interest has led to proposals to support and/or mandate biofuels produced from feedstocks other than corn starch through explicit requirements, research, development and extension funding, and/or tax incentives.[5] Non-corn biofuels could include fuels produced from cellulosic material (such as perennial grasses), ethanol produced from sugarcane or beets, and biodiesel or renewable diesel produced from vegetable or animal oils.[6] Under EISA, eligible corn-based ethanol production is capped at 15 billion gallons beginning in 2015. Starting in 2009, the RFS will require that an increasing amount of the mandate be met through the use of "advanced biofuels"—biofuels produced from feedstocks other than corn starch.

Currently, cellulosic ethanol is not produced on a commercial scale. As of July 2008, there was one commercial-scale refinery under construction, with two demonstration-scale plants and three pilot-scale plants completed. Industry sources expect commercial production of cellulosic ethanol to begin in 2010 or 2011. In January 2009, USDA announced funding for a cellulosic biofuels plant under the Biorefinery Program with projected output of 20 million gallons annually, beginning in 2010. The RFS mandates cellulosic ethanol production of 100 million gallons in 2010. For more information on cellulosic ethanol, see CRS Report RL34738, *Cellulosic Biofuels: Analysis of Policy Issues for Congress*, by Tom Capehart.

Key Elements of EISA and the 2008 Farm Bill

The following table provides a side-by-side comparison of biofuels-related provisions in EISA with the enacted farm bill—the Food, Conservation, and Energy Act of 2008. President Bush signed EISA on December 19, 2007, and the Food, Conservation, and Energy Act of 2008 became law on June 18, 2008, after President Bush's veto was overridden by both the Senate and the House.[7] Both bills cover a wide range of energy and agricultural topics in addition to biofuels.

The table is organized in the same order as EISA, followed by provisions that are exclusively in the enacted farm bill.

Key biofuels-related provisions of EISA and the 2008 farm bill include:

- a major expansion of the renewable fuel standard (RFS) established in the Energy Policy Act of 2005 (P.L. 109-58) [EISA];

[5] Non-corn-starch feedstocks include other parts of the corn plant, such as the husks and the stalks, which are high in cellulose.

[6] For more information on biodiesel, see CRS Report RL32712, *Agriculture-Based Renewable Energy Production*, by Randy Schnepf.

[7] The conference agreement on the 2008 farm bill was originally approved by the House and the Senate as H.R. 2419and vetoed by the President in May 2008. Both chambers overrode the veto, making the bill law (P.L. 110-234). However, the trade title was inadvertently excluded from the enrolled bill. To remedy the situation, both chambers repassed the farm bill conference agreement (including the trade title) H.R. 6124. The President vetoed the measure in June 2008 and both chambers again overrode the veto, which made H.R. 6124 law (P.L. 110-246), and repealed P.L. 110-234.

- expansion and/or modification of tax credits for ethanol [farm bill];
- grants and loan guarantees for biofuels (especially cellulosic) research, development, deployment, and production [EISA, farm bill];
- studies of the potential for ethanol pipeline transportation, expanded biofuel use, market and environmental impacts of increased biofuel use, and the effects of biodiesel on engines [EISA, farm bill];
- expansion of biofuel feedstock availability [farm bill];
- reauthorization of biofuels research and development at the U.S. Department of Energy [EISA] and the U.S. Department of Agriculture and Environmental Protection Agency [farm bill]; and
- reduction of the blender tax credit for corn-based ethanol, a new production tax credit for cellulosic ethanol, and continuation of the import duty on ethanol [farm bill].

Table 1. Comparison of Current or Prior Law with Biofuels Provisions in EISA and the Enacted Farm Bill

Topic	Current or Prior Law	Energy Independence and Security Act (P.L. 110-140)	2008 Farm Bill (P.L. 110-246)
P.L. 110-140, Title I—Energy Security Through Improved Fuel Economy			
Consumer Information	No prior provision.	The Dept. of Transportation is required to carry out an educational program to inform consumers about the fuel savings and emissions benefits of new vehicles, including the benefits from the use of alternative fuels. [Sec. 105]	No comparable provision.
Fuel Tank Labeling Requirement	No prior provision.	Requires the Dept. of Transportation to issue a final rule by June 2011 requiring automakers to clearly label the fuel compartment of alternative fuel vehicles with the form of alternative fuel stated on the label. [Sec. 105]	No comparable provision.
Extension of Flexible Fuel Vehicle Credit Program / Biodiesel as Alternative fuel for CAFE Purposes	Under the Corporate Average Fuel Economy (CAFE) program, automakers may generate credits toward their compliance for the production and sale of alternative fuel vehicles, as defined in law. Currently, B20 (a blend of 20% biodiesel and 80% petroleum diesel) vehicles are not considered alternative fuel vehicles. [49 U.S.C. 32901 et seq.]	Amends the CAFE program to extend alternative fuel vehicle credits through model year 2019, at a declining rate. Also allows vehicles capable of operating on B20 to be treated as vehicles eligible for CAFE credits. Expanding the definition of alternative fuel vehicle to include B20 could make all diesel passenger cars and light trucks eligible for credits under CAFE. Currently, some diesel passenger vehicles are warrantied to run on B5, but few technical barriers exist to make new diesel vehicles B20-capable. [Sec. 109]	No comparable provision.
P.L. 110-140, Title II—Energy Security Through Increased Production of Biofuels			
Renewable Fuel Standard	The Energy Policy Act of 2005 established a Renewable Fuel Standard (RFS) which requires the use of an increasing amount of renewable fuels in gasoline. The mandate increases from 4.0 billion gallons in 2006 to 7.5 billion gallons in 2012. Starting in 2013, the proportion of renewable fuel to gasoline	Amends RFS to include all transportation fuels (except for fuels used in ocean-going vessels). Expands the existing requirement to 9.0 billion gallons in 2008, increasing to 36 billion gallons in 2022. Requires renewable fuels produced at new facilities to have at least 20% lower lifecycle greenhouse gas (GHG) emissions than	No comparable provision.

CRS-4

Topic	Current or Prior Law	Energy Independence and Security Act (P.L. 110-140)	2008 Farm Bill (P.L. 110-246)
	must equal or exceed the proportion in 2012. Starting in 2013, of the amount mandated above, at least 250 million gallons must be fuel derived from cellulosic material. [P.L. 109-58, Sec. 1501], [42 U.S.C. 7545]	petroleum fuels. Starting in 2009, requires that an increasing amount of the above mandate be met using "advanced biofuels," defined as biofuels derived from feedstocks other than corn starch with 50% lower lifecycle GHG emissions. By 2022, requires 21 billion gallons of advanced biofuel. Of the advanced biofuel mandate, specific carve-outs are made for cellulosic fuels and biomass-derived diesel substitutes. [Sec. 202]	
Study of Impact of Increased Renewable Fuel Use	By August 2009, EPA must publish a draft analysis of the effects of the fuels provisions in P.L. 109-58 on air pollutant emissions and air quality. [P.L. 109-58, Sec. 1505], [42 U.S.C. 7545(b)] EPA is required to conduct a survey to determine the market share of gasoline containing ethanol and other renewable fuels. [P.L. 109-58, Sec. 1501(c)], [42 U.S.C. 7545] DOE is required to collect and publish monthly survey data on the production, blending, importing, demand, and price of renewable fuels, both on a national and regional basis. [P.L. 109-58, Sec. 1508], [42 U.S.C. 7135]	DOE, in consultation with USDA and EPA, is required to enter into an agreement with the National Academy of Sciences (NAS) to study the impacts of the RFS on industries related to feed grains, livestock, food, forest products, and energy. The NAS study must assess the likely effects on domestic animal agriculture and policy options to alleviate negative effects; identify agricultural conditions that would warrant a waiver of the RFS requirements; and make recommendations to limit adverse economic impacts from the RFS. [Sec. 203]	Comprehensive Study of Biofuels—The Dept. of Treasury, with DOE, USDA, and EPA shall have the National Academy of Sciences analyze scientific findings on current and future biofuels production, impacts, trends, and policies. [Sec. 15322]
Environmental and Resource Conservation Impacts	No prior provision.	EPA, in consultation with the USDA and DOE, must study the impacts of the RFS on environmental issues, resource conservation issues, and invasive or noxious species. [Sec. 204]	No comparable provision.
Biomass Based Diesel and Biodiesel Labeling	No prior provision.	The Federal Trade Commission is required to promulgate rules requiring diesel retailers to label their pumps with the percentage of biomass-based diesel or biodiesel that is offered for sale. [Sec. 205]	No comparable provision.

CRS-5

Topic	Current or Prior Law	Energy Independence and Security Act (P.L. 110-140)	2008 Farm Bill (P.L. 110-246)
Study of Credits for Use of Renewable Electricity in Electric Vehicles / Production of Renewable Fuel Using Renewable Energy	Under the original RFS, cellulosic biofuels are eligible for additional credits under the mandate. A gallon of cellulosic biofuel is considered equal to 2.5 gallons of ethanol. For this section, "cellulosic biofuels" includes both biofuels produced from cellulose and biofuels produced from sugars or starches (e.g., corn ethanol) if biomass is used to displace fossil energy in the refining of the fuel. [P.L. 109-58, Sec. 1501], [42 U.S.C. 7545]	EPA is required to study the feasibility of issuing credits under the RFS for electric vehicles powered by electricity from renewable resources. Within 180 days of enactment, EPA must report to Congress on the findings of the study. [Sec. 206]	No comparable provision.
Grants for Production of Advanced Biofuels	The DOE may provide grants for the construction of facilities to produce renewable fuels (including ethanol) from cellulosic biomass, agricultural byproducts, agricultural waste, and municipal solid waste. Discretionary appropriations of $100 million for FY2006, $250 million for FY2007, and $400 million for FY2008 are authorized. [P.L. 109-58, Sec. 1512] An existing grant program finances the development and construction of biorefineries and biofuel production plants and implements projects to demonstrate the commercial viability of converting biomass to fuels or chemicals. No funds have been appropriated for the program. [P.L. 107-171, Sec. 9003], [7 USC 8103]	Requires DOE to establish a grant program for the production of advanced biofuels that have at least an 80% reduction in lifecycle greenhouse gas emissions relative to current fuels. Authorizes discretionary appropriation of a total of $500 million for FY2008-FY2015. [Sec. 207]	Biorefinery Assistance. New Sec. 9003. Provides competitive grants and loan guarantees for construction and retrofitting of biorefineries for the production of advanced biofuels. Biorefinery grants up to 30% of total cost. Loan guarantees limited to $250 million or 80% of project cost. Mandatory funding of $75 million in FY2009 and $245 million in FY2010, available until expended for loan guarantees. Discretionary funding of $150 million annually is authorized for FY2009-FY2012. [Sec. 9001] Repowering Assistance. New section 9004. Provides USDA funds for repowering assistance to reduce or eliminate their use of fossil fuels for biorefineries in existence at enactment. Mandatory Commodity Credit Corporation (CCC) funding of $35 million for FY2009, available until expended. Authorizes discretionary funding of $15 million annually for FY2009-FY2012. [Sec. 9001]

CRS-6

Topic	Current or Prior Law	Energy Independence and Security Act (P.L. 110-140)	2008 Farm Bill (P.L. 110-246)
Integrated Consideration of Water Quality in Determinations on Fuels and Fuel Additives	Section 211(c) of the Clean Air Act allows EPA to control or prohibit the production and/or sale of any engine, vehicle, fuel, or fuel additive that causes or contributes to air pollution "that may be reasonably anticipated to endanger the public health or welfare." [42 U.S.C. 7545(c)]	Expands EPA's authority to control engines, vehicles, fuels, and fuel additives under Sec. 211(c) of the Clean Air Act to include effects on water pollution. [Sec. 208]	No comparable provision.
Anti-Backsliding	No provision.	Requires EPA to study the potential adverse effects to air quality from the expanded RFS, and to promulgate regulations to mitigate those effects. [Sec. 209]	No comparable provision.
Effective Date, Savings Provision, and Transition Rules	No provision.	For 2008 and 2009, any ethanol plant powered by natural gas, biomass, or a combination of the two is treated as having a 20% reduction in lifecycle greenhouse gas emissions (See Sec. 202). For 2008, all current EPA regulations on the RFS are unchanged, except for the increase in the volume mandated by Sec. 202. [Sec. 210]	No comparable provision.
Biodiesel Report	No provision.	Requires the DOE to report to Congress on the R&D challenges to expanding biodiesel use (to an unspecified level) [Sec. 221]	No comparable provision.
Biogas Report	No provision.	Requires the DOE to report to Congress on the R&D challenges to expanding biogas and biogas/natural gas blends (to an unspecified level). [Sec. 222]	No comparable provision.
Grants for Biofuel Production and R&D in Certain States	DOE is authorized to receive $25 million annually for FY2006-FY2010 for R&D and implementation of renewable fuel production technologies in states with low rates of ethanol production that are under the federal reformulated gasoline (RFG) program. [P.L. 109-58, Sec. 1511(d)], [42 U.S.C. 7411]	Discretionary appropriations of $25 million authorized annually for FY2008-FY2010 for R&D and commercial application of biofuel production in states with low rates of ethanol and cellulosic ethanol production (this could in effect apply to all states). [Sec. 223]	No comparable provision.

Topic	Current or Prior Law	Energy Independence and Security Act (P.L. 110-140)	2008 Farm Bill (P.L. 110-246)
Biorefinery Energy Efficiency	The DOE is directed to conduct research on commercial applications of biomass and bioenergy. [P.L. 109-58, Sec. 932], [42 U.S.C. 16232]	Amends Sec. 932 of P.L. 109-58 to include research on energy efficiency at biorefineries and on technology to convert existing corn-based ethanol plants to process cellulosic materials. [Sec. 224]	No comparable provision.
Study of Optimization of Flexible Fueled Vehicles to use E-85 Fuel	No provision.	DOE is directed to study whether optimizing flexible fuel vehicles (FFVs) to run on E85 would increase their fuel efficiency. Current FFVs are optimized to run on gasoline, since that tends to be their primary fuel. [Sec. 225]	No comparable provision.
Study of Engine Durability and Performance Associated with the Use of Biodiesel	No current provision.	DOE, in consultation with EPA, is directed to study the effects of various biodiesel/diesel blends on engine performance and durability. [Sec. 226]	No comparable provision.
Study of Optimization of Biogas Used in Natural Gas Vehicles	No prior provision.	DOE is directed to study the potential for optimizing natural gas vehicles to run on biogas (methane produced from biological feedstocks). [Sec. 227]	No comparable provision.
Algal Biomass	Various statutes promote biofuels R&D, including the development of biofuels from algae, at the Department of Energy. [42 U.S.C. 16232]	DOE is required to report to Congress on progress toward developing algae as a feedstock for biofuel production. [Sec. 228]	No comparable provision.
Biofuels and Biorefinery Information Center	No prior provision.	Directs DOE to establish a technology transfer center to provide information on biofuels and biorefineries. [Sec. 229]	No comparable provision.
Cellulosic Ethanol and Biofuels Research	No prior provision.	Authorizes the DOE to provide biofuels R&D grants to 10 institutions from land-grant colleges, Historically Black Colleges or Universities, tribal serving institutions, or Hispanic serving institutions. $50 million for FY2008 is authorized to be appropriated, to be available until expended. [Sec. 230]	No comparable provision.
Bioenergy R&D Authorization of Appropriation	DOE is directed to conduct R&D on biomass, bioenergy, and bioproducts. Discretionary appropriations of $213	Amends Sec. 931 of P.L. 109-58 to authorize a total of $1.2 billion in discretionary appropriations for FY2008-	No comparable provision.

CRS-8

Topic	Current or Prior Law	Energy Independence and Security Act (P.L. 110-140)	2008 Farm Bill (P.L. 110-246)
	million are authorized for FY2007, $251 million for FY2008, and $274 million for FY2009. [P.L. 109-58, Sec. 931(c)], [42 U.S.C. 16232]	FY2010 for R&D on biomass, bioenergy, and bioproducts. [Sec. 231]	
Environmental Research and Development	DOE is required to establish a program of research, development, and demonstration in microbial and plant systems biology, protein science, and computational biology. Biomedical research and research related to humans are not permitted as part of the program. [P.L. 109-58, Sec. 977], [42 U.S.C. 16232]	DOE is required to expand the biological R&D program established in Sec. 977 of P.L. 109-58 to include environmental effects, potential for greenhouse gas reductions, and the potential for more sustainable agriculture. See also Sec. 233 of EISA (below). [Sec. 232(a)]	No comparable provision.
Lifecycle Analysis Tools for Evaluating the Energy Consumption and Greenhouse Gas Emissions from Biofuels	No prior provision.	DOE is required to study and develop tools for evaluating the lifecycle energy consumption and the potential for greenhouse gas emissions from biofuels. [Sec. 232(b)]	Requires USDA to support research on making a farm or ranch energy-neutral. [Sec. 7207]
Small-Scale Production and Use of Biofuels	No prior provision.	Amends the Biofuels Research and Development Act of 2000 to require the Secretary of Agriculture to establish a R&D program to facilitate small-scale production and local and on-farm use of biofuels. [Sec. 232(c)]	Requires USDA support of on-farm energy conservation and renewable energy production. [Sec 7207]
Bioenergy Research Centers	DOE is required to establish a program of R&D and demonstration of microbial and plant systems biology, protein science, and computational biology. Biomedical research and research related to humans are not permitted as part of the program. [Energy Policy Act of 2005, P.L. 109-58, Sec. 977], [42 U.S.C. 16232]	Requires the establishment of at least seven research centers that focus on bioenergy to be included in the R&D program established in Sec. 977 of P.L. 109-58. [Sec. 233]	No comparable provision.
University Based Research and Development Grant Program	No prior provision.	Requires DOE to establish a program of competitive grants to institutions of higher education for research on renewable energy technologies. Each grant may not exceed $2 million. A total of $25 million in discretionary	No comparable provision.

CRS-9

Topic	Current or Prior Law	Energy Independence and Security Act (P.L. 110-140)	2008 Farm Bill (P.L. 110-246)
		appropriations is authorized for the program. [Sec. 234]	
Prohibition on Franchise Agreement Restrictions Related to Renewable Fuel Infrastructure	No prior provision.	Amends the Petroleum Marketing Practices Act (15 U.S.C. 2801 et seq.) to make it unlawful for a franchiser to prohibit a franchisee from installing E85 or B20 tanks and pumps within the franchise agreement. [Sec. 241]	No comparable provision.
Renewable Fuel Dispenser Requirements—Report to Congress	No prior provision.	DOE is required to report to Congress on the market penetration of flexible fuel vehicles and on the feasibility of requiring fuel retailers to install E85 infrastructure. [Sec. 242]	No comparable provision.
Ethanol Pipeline Feasibility Study	No prior provision.	DOE, in consultation with DOT, is required to report on the feasibility of constructing dedicated ethanol pipelines. $1 million in discretionary funds is authorized annually for FY2008 and FY2009, to remain available until expended. [Sec. 243]	No comparable provision..
Renewable Fuel Infrastructure Development	No prior provision.	Directs DOE to provide grants for conversion assistance, technical and marketing assistance, and pilot programs to expand infrastructure for ethanol/gasoline blends of between 11% and 84% ethanol, and renewable fuel/diesel fuel blends of at least 10% renewable diesel. Discretionary funds of $200 million is authorized annually for FY2008-FY2014. [Sec. 244]	No comparable provision.
Study of the Adequacy of Transportation of Domestically Produced Renewable Fuel by Railroads and Other Modes of Transportation	No prior provision.	DOE, jointly with DOT, are required to report on the adequacy of railroads and modes for transportation of domestically produced renewable fuel. [Sec. 245]	No comparable provision.

CRS-10

Topic	Current or Prior Law	Energy Independence and Security Act (P.L. 110-140)	2008 Farm Bill (P.L. 110-246)
Federal Fleet Refueling Centers	No prior provision.	Requires the head of each federal agency to install at least one renewable fuel pump at each federal fleet refueling center by January 1, 2010. Further, the Administration is required to report each October 31 on progress toward meeting this requirement. The requirement does not apply to Department of Defense fueling centers with less than 100,000 gallons in annual fuel turnover. [Sec. 246]	No comparable provision.
Standard Specifications for Biodiesel	No prior provision.	If ASTM International (originally the American Society for Testing and Materials) has not adopted standards for B5 and B20 within one year of enactment, the EPA Administrator is required to do so. No new funding is authorized. [Sec. 247]	No comparable provision.
Biofuels Infrastructure	No prior provision.	Directs DOE to conduct an R&D program on the effects of biofuels on existing transportation fuel distribution systems. [Sec. 248]	Requires joint USDA, DOE, EPA study on the infrastructure needs associated with significant expansion in biofuels production and use. [Sec. 9002]
Waiver for Fuel or Fuel Additives	Under Sec. 211(f) of the Clean Air Act, no new fuels or fuel additives may be introduced into commerce unless granted a waiver by EPA. If EPA has not acted within 180 days of receipt of a waiver request, the waiver is treated as granted. [42 U.S.C. 7545(f)]	Prohibits the introduction of new renewable fuels or renewable fuel additives unless EPA explicitly grants a waiver under Sec. 211(f) of the Clean Air Act. EPA is required to take final action within 270 days of receipt of the waiver request. Before the passage of EISA, inaction or failure to complete review of an additive by EPA allowed a fuel to receive the waiver. Under EISA, no waiver would be granted without *explicit* approval by EPA. [Sec. 251]	No comparable provision.

Topic	Current or Prior Law	Energy Independence and Security Act (P.L. 110-140)	2008 Farm Bill (P.L. 110-246)
P.L. 110-140, Title V—Energy Savings in Government and Public Institutions			
Capitol Complex E-85 Refueling Station	No prior provision.	The Architect of the Capitol is authorized to install an E85 tank and pumping system on or near the Capitol Grounds Fuel Station. $640,000 in discretionary funds is authorized for FY2008. [Sec. 502]	No comparable provision.
Procurement and Acquisition of Alternative Fuels	No prior provision.	Federal agencies are prohibited from procuring alternative or synthetic transportation fuels if the lifecycle emissions exceed those of petroleum-based fuels. [Sec. 526]	No comparable provision.
P.L. 110-140, Title VIII—Improved Management of Energy Policy			
Sense of Congress Relating to the Use of Renewable Resources to Generate Energy	No prior provision.	Expresses the Sense of the Congress that renewable resources from agriculture and forestry should provide at least 25% of all U.S. energy needs by 2025. [Sec. 806]	No comparable provision.
P.L. 110-246, Title VII 2008 Farm Bill—Energy Provisions (excluding those cited in above sections)			
Bioenergy Research	Provides for research and development appropriations for bioenergy. [7 U.S.C. 3154]		Section 1419 of the National Agricultural Research, Extension, and Teaching Policy Act of 1977 (7 U.S.C. 3154) is repealed. Funding for related biomass research through USDA is contained in Title VII Sec. 7207 of the farm bill. [Sec. 7110]
Biochar Research, Development and Demonstration	No current provision.	No comparable provision.	Research on biochar (biomass charcoal) production and sequestration is included as a high-priority research and extension area the Research Title. [Sec. 7204]

CRS-12

Topic	Current or Prior Law	Energy Independence and Security Act (P.L. 110-140)	2008 Farm Bill (P.L. 110-246)
Agricultural Bioenergy Feedstock and Energy Efficiency Research and Extension Initiative	The Biomass Research and Development Act of 2000 (reauthorized by the 2002 farm bill) provides competitive funding for R&D and demonstration projects on biofuels and bio-based chemicals and products, administered jointly by USDA and DOE. Specified mandatory CCC funding of $5 million in FY2002 and $14 million annually for FY2003- FY2007 to remain available until expended. Also authorized appropriations of $200 million for each of FY2006-FY2015. [P.L. 107-171, Sec. 9008], [7 U.S.C. 5925]	No comparable provision.	Establishes the Agricultural Bioenergy Feedstock and Energy Efficiency Research and Extension Initiative in Title VII (Research) to improve biomass, production, biomass conversion in biorefineries, and biomass use. Provides grants of up to 50% of cost for energy efficient research and extension projects. Establishes a best practices database of biomass crops. Authorized appropriations of $50 million annually for FY2008-12. [Sec. 7207]
Research, Extension, and Educational Programs on Biobased Energy Technologies and Products	The "Sun Grant" program established 5 national sun grant research centers based at land-grant universities and each covering a different region. The purpose is to enhance coordination and collaboration between USDA, DOE, and land-grant universities in the development, distribution, and implementation of biobased energy technologies. Authorized appropriations of $25 million in FY2005, $50 million in FY2006, and $75 million annually for FY2007-FY2010. [7 U.S.C. 8109]	No comparable provision.	Continues sun grant program. Provides matching grants to land grant institutions to develop, distribute, and implement biobased energy technologies and to promote diversification and sustainability of agricultural production, and economic diversification in rural areas through biobased energy and product technologies. Establishes a Sun Grant Information Analysis Center. Requires annual reports. Discretionary funds of $75 million for FY2008-12 are authorized. [Sec. 7526]
P.L. 110-246, Title IX 2008 Farm Bill—Energy Provisions (excluding those cited in above sections)			
Federal Procurement of Biobased Products	Under the 2002 farm bill, federal agencies are required to purchase biobased products under certain conditions. Current law authorizes a voluntary biobased labeling program. USDA regulations define biobased products, identify biobased product categories, and	No comparable provision.	New section 9002. Renames as the Biobased Markets Program. Extends the program through FY2012 and refines federal procurement rules for biobased products. Requires federal agencies to maximize

CRS-13

Topic	Current or Prior Law	Energy Independence and Security Act (P.L. 110-140)	2008 Farm Bill (P.L. 110-246)
	specify the criteria (including testing) for qualifying those products for preferred procurement. Mandatory Commodity Credit Corporation (CCC) funding of $1 million was authorized for each of FY2002-FY2007 for testing biobased products. [P.L. 107-171, Sec. 9002], [7 U.S.C. 8102]		procurement of biobased products and submit reports to Congress. Continues voluntary labeling. Establishes testing centers and education grants. Authorizes mandatory funding of $1 million for FY2008 and $2 million annually for FY2009-12. Discretionary funding of $2 million in annual appropriations is authorized for FY2009-12. [Sec. 9001]
Adjustments to the Bioenergy Program	Originally a Clinton Administration initiative, the Bioenergy Program was made statutory by the 2002 farm bill. Provides CCC incentive payments to biofuels producers based on year-to-year increases in the quantity of biofuel produced. Mandatory CCC funding of $150 annually for FY2002-FY2006. No funding was available for FY2007. [P.L. 107-171, Sec. 9010], [7 U.S.C. 8108]	No comparable provision.	New section 9005. Establishes the Bioenergy Program for Advanced Biofuels. Provides payments to producers to support and expand production of advanced biofuels. Mandatory funding of $55 million for FY2009, $55 million for FY2010, $85 million for FY2011, and $105 million for FY2012. Authorizes additional appropriation of $25 million annually for FY2009-12. [Sec. 9001]
Biodiesel Fuel Education Program	Awards competitive grants to nonprofit organizations that educate governmental and private entities operating vehicle fleets, and educate the public about the benefits of biodiesel fuel use. Mandatory CCC funding of $1 million annually was authorized for FY2003-FY2007. [P.L. 107-171, Sec. 9004], [7 U.S.C. 8104]	No comparable provision.	New section 9006. Extends the Biodiesel Fuel Education Program through FY2012. Provides mandatory CCC funding of $1 million annually for FY2008-12. [Sec. 9001]
Energy Audit and Renewable Energy Development Program	The 2002 farm bill authorized a competitive grant program for eligible entities to carry out a program to assist farmers, ranchers, and rural small businesses in becoming more energy efficient and in using renewable energy technology and resources. Authorized	No comparable provision.	New section 9007. Folds the Energy Audit and Renewable Energy Development Program into the Rural Energy for America Program. (See below.) [Sec. 9001]

CRS-14

Topic	Current or Prior Law	Energy Independence and Security Act (P.L. 110-140)	2008 Farm Bill (P.L. 110-246)
	appropriations of such sums as are necessary to carry out the program for FY2002-FY2007. [P.L. 107-171, Sec. 9005], [7 U.S.C. 8105]		
Renewable Energy Systems and Energy Efficiency Improvements	Authorizes loans, loan guarantees, and grants to farmers, ranchers, and rural small businesses to purchase and install renewable energy systems and to make energy efficiency improvements. Mandatory CCC funding of $23 million annually for FY2003-FY2007. [P.L. 107-171, Sec. 9006], [7 U.S.C. 8106]	No comparable provision.	New section 9007. Renamed as the "Rural Energy for America Program," Funds energy audits for state agencies, cooperatives, educational institutions and utilities. Provides grants, loan guarantees and incentive payments for energy efficiency and renewable energy, and manure-to-energy projects. Reserves 20% for small projects. Mandatory funds of $55 million for FY2009, $60 million for FY2010, $70 million for FY2011, and $70 million for FY2012. Discretionary appropriations of $25 million annually for FY2009-12 are authorized. [Sec. 9001]
Biomass Research and Development	Section 9008. The program—created originally under the Biomass Research and Development Act (BRDA) of 2000—provides competitive funding for research, development, and demonstration projects on biofuels and bio-based chemicals and products, administered jointly by USDA and DOE. Specified mandatory CCC funding of $5 million in FY2002 and $14 million for each of FY2003 through FY2007 (available until expended). Additional appropriation authority of $200 million for each of FY2006 through FY2015. [7 U.S.C. 8101]	No comparable provision.	New section 9008. Defines biobased product. Provides for coordination of biomass research and development between USDA and DOE. Establishes the Biomas Research and Development Board and the Biomass Research and Development technical Advisory Committee to assist the Board in coordinating biomass research in the Federal government. USDA and DOE are to establish a Biomass Research and Development Initiative to competitively award grants, contracts, and financial assistance for research on biofuels and biobased products production,

CRS-15

Topic	Current or Prior Law	Energy Independence and Security Act (P.L. 110-140)	2008 Farm Bill (P.L. 110-246)
			and biobased feedstocks and development. Grants are to be awarded to universities, national laboratories, state and federal research agencies, private businesses, and nonprofits. Mandatory funding is authorized of $20 million for FY2009, $28 million for FY2010, $30 million for FY2011, and $40 million for FY2012. Discretionary funding of $35 million annually is authorized to be appropriated for FY2009-12. [Sec. 9001]
Rural Energy Self-Sufficiency Initiative Grant Program	No current provision.	No comparable provision.	New section 9009. Establishes the Rural Energy Self-Sufficiency Initiative, providing cost- share (up to 50%) grants to assist rural communities with community-wide energy systems that reduce conventional energy use and increase the use of energy from renewable sources. Grants are made available to assess energy use in a rural community, evaluate ideas for reducing energy use, and develop and install integrated renewable energy systems. Authorizes $5 million in discretionary funds annually for FY2009-FY2012. [Sec. 9001]
Feedstock Flexibility Program for Bioenergy Producers	No current provision.	No comparable provision.	New section 9010. Establishes the Feedstock Flexibility Program, authorizing the use of such sums as necessary of CCC funds to purchase surplus sugar, to ensure the sugar program operates at no-net-cost, to be resold as a biomass feedstock to produce bioenergy. [Sec. 9001]

CRS-16

Topic	Current or Prior Law	Energy Independence and Security Act (P.L. 110-140)	2008 Farm Bill (P.L. 110-246)
Financial Assistance for the Production of Biomass Energy Crops and Infrastructure for Harvesting, Storage, and Transportation of Biomass to Local Biorefineries	No current provision.	No comparable provision.	New section 9011. Establishes the Biomass Crop Assistance Program (BCAP) to encourage biomass production or biomass conversion facility construction with contracts which will enable producers to receive financial assistance for crop establishment costs and annual payments for biomass production. Producers must be within economically practicable distance from a biomass facility. Also provides payments to eligible entities to assist with costs for collection, harvest, storage and transportation to a biomass conversion facility. A report is required no later than 4 years after enactment. CCC funds of such sums as necessary are to be made available for each of FY2008-12. [Sec. 9001]
Forest Biomass for Energy Program	No current provision.	No comparable provision.	New section 9012. Requires the Forest Service to conduct a competitive research and development program to encourage use of forest biomass for energy. Priority given to projects that utilize low-value forest by-products, integrate the production of energy from forest biomass with existing manufacturing streams, develop new transportation fuels from forest biomass, or improve the production of forest biomass feedstocks. Appropriations of $15 million per year are authorized for FY2009-FY2012. [Sec. 9001]

CRS-17

Topic	Current or Prior Law	Energy Independence and Security Act (P.L. 110-140)	2008 Farm Bill (P.L. 110-246)
Community Wood Energy R&D Program	No current provision.	No comparable provision.	New section 9013. Establishes the Community Wood Energy R & D Program, providing grants of up to $50,000 for up to 50% of the cost for communities to develop wood energy plans and purchase systems for public buildings. Authorizes $5 million in discretionary funds annually (FY2009-FY2012). [Sec. 9001]
Rural Nitrogen Fertilizer Study	No current provision.	No comparable provision.	Requires a report with 1 year of appropriations on the production of fertilizer from renewable energy sources in rural areas. Must identify challenges to commercialization of rural fertilizer production, processes and technologies and potential impacts of renewable fertilizer on fossil fuel use and the environment. Appropriations of $1 million are authorized for FY2009. [Sec. 9003]]
P.L. 110-246, Title XI 2008 Farm Bill—Energy Provisions (excluding those cited in above sections)			
Study on Bioenergy Operations	No comparable provision.	No comparable provision.	Directs USDA to produce a report on the potential economic issues (including costs) associated with animal manure used in normal agricultural operations and as a bioenergy feedstock. [Sec. 11014]

CRS-18

Topic	Current or Prior Law	Energy Independence and Security Act (P.L. 110-140)	2008 Farm Bill (P.L. 110-246)
P.L. 110-246, Title XV 2008 Farm Bill—Energy Provisions (excluding those cited in above sections)			
Tax Credit for Production of Cellulosic Biofuels	All fuel ethanol is allowed a tax credit of $0.54 per gallon, regardless of feedstock. Small producers may claim an additional credit of $0.10 per gallon. [26 U.S.C. 40], [26 U.S.C. 40(d)(4)]	No comparable provision.	Establishes a credit of $1.01 for producers of cellulosic biofuels alcohol through December 31, 2012. For cellulosic ethanol, value of the credit, plus the existing small ethanol producer credit and alcohol fuels credits cannot exceed $1.01 per gallon. [Sec. 15321]
Modification of Alcohol Tax Credit	The American Jobs Creation Act of 2004 established a blender's tax credit for the use of ethanol used as motor fuel. The credit is valued at $0.51 per gallon of ethanol blended into gasoline. [P.L. 108-357, Sec. 301], [26 U.S.C. 40(d)(4)]	No comparable provision.	In the first calendar year after EPA certifies that 7.5 billion gallons of renewable fuel have been blended into gasoline, the credit is reduced from $0.51 to $0.45 per gallon. In 2008, the volume blended exceeded 9 billion gallons and the reduction became effective January 1, 2009. [Sec. 15331]
Calculation of Volume of Alcohol for Fuel Tax Credits	Currently, any denaturant added to alcohol (up to 5%) is considered as part of the volume of alcohol for tax purposes. [26 U.S.C. 40(d)(4)]	No comparable provision.	For the purposes of calculating the per-gallon credit for the volume of alcohol used as a fuel or in a qualified mixture, the volume of alcohol includes any denaturant, including gasoline. This provision reduces the amount of allowable denaturant (added to make it unfit for human consumption) to 2% of the volume of the alcohol. [Sec. 15332]
Ethanol Tariff Extension	In general, fuel ethanol imports are subject to a $0.54 per gallon duty and a 2.5% ad valorem tariff. The duty expires January 1, 2009. [P.L. 99-499] [19 U.S.C. 3001 et seq.]	No comparable provision.	Extends the $0.54 cent per gallon duty through December 31, 2010. [Sec. 15333]

CRS-19

Topic	Current or Prior Law	Energy Independence and Security Act (P.L. 110-140)	2008 Farm Bill (P.L. 110-246)
Elimination and Reductions of Duty Drawback on Certain Imported Ethanol	Currently, if a manufacturer imports an intermediate product then exports the finished product or a similar product, that manufacturer may be eligible for a refund (drawback) of up to 99% of the duties paid. The duty drawback provisions include special provisions for the production of petroleum derivatives. [19 U.S.C. 1313(p)]	No comparable provision.	Duty drawback will not be available on exports which do not contain ethanol. Ends the practice in which imported ethanol is blended with gasoline and jet fuel (containing no ethanol, but considered a "like commodity" to the finished gasoline) is then exported to qualify for the drawback in lieu of finished gasoline containing the originally imported ethanol. [Sec. 15334]

Source: CRS.

CRS-20

Author Contact Information

Randy Schnepf
Specialist in Agricultural Policy
rschnepf@crs.loc.gov, 7-4277

Brent D. Yacobucci
Specialist in Energy and Environmental Policy
byacobucci@crs.loc.gov, 7-9662

Economic Research Service November 2009 U.S. Department of Agriculture

This is a summary of an ERS report.

Find the full report at *www.ers.usda.gov/publications/err86*

Ethanol and a Changing Agricultural Landscape

Scott A. Malcolm, Marcel Aillery, and Marca Weinberg

U.S. policy to expand the production of biofuel for domestic energy use has significant implications for agriculture and resource use. While ongoing research and development investment may radically alter the way biofuel is produced in the future, for now, corn-based ethanol continues to account for most biofuel production. As corn ethanol production increases, so does the production of corn. The effect on agricultural commodity markets has been national, but commodity production adjustments, and resulting environmental consequences, vary across regions. Changes in the crop sector have also affected the cost of feed for livestock producers. As the Nation demands more biofuel production, and markets for new biofuel feedstocks, such as crop residues, emerge, the agricultural landscape will be further transformed.

What Is the Issue?

The Energy Independence and Security Act of 2007 (EISA) specifies a minimum total amount of U.S. biofuel production through 2022, and also sets target levels for fuels produced from specific feedstock categories. Together with volatile energy prices, this and earlier Federal legislation supporting biofuel processing have increased demand for biofuels and the agricultural feedstocks used to produce them. Greater demand for biofuel increases pressure on the agricultural land base as more land is put into production and/or more inputs, such as fertilizer, water, and pesticides, are applied to cropland. Rising demand for corn, the principal biofuel feedstock in the United States, changes the profitability of growing corn and other "energy crops". Farmers respond by changing their planting decisions, which alter crop mix, land use, and use of inputs, such as fertilizer, which then influence water quality, soil erosion, and other environmental indicators. The environmental consequences of shifts in agricultural production vary by region.

This report also looks at the economic and environmental implications should crop residues, such as corn stover and wheat straw, become commercially viable as biofuel feedstocks. Widespread harvesting of crop residues as an alternative biofuel feedstock has implications for input use, nutrient runoff, erosion control, and soil productivity.

What Did the Study Find?

Land for new biofuel feedstock production comes from two main sources: acreage not currently in production and acreage shifted from other crops. The amount of additional land and displaced crops associated with increased biofuel production differs by region. If the RFS targets are met, total cropland is projected to increase by 1.6 percent over baseline conditions by 2015, with corn acreage expanding by 3.5 percent and accounting for most of the cropland increase. While corn acreage expands in every region, traditional corn-growing areas would likely see the largest

ERS is a primary source of economic research and analysis from the U.S. Department of Agriculture, providing timely information on economic and policy issues related to agriculture, food, the environment, and rural America.

www.ers.usda.gov

increases—up 8.6 percent in the Northern Plains, 1.7 percent in the Corn Belt, and 2.8 percent in Lake States. Prices are expected to increase slightly for most crops compared with the baseline, although the price increase could be reduced if corn yields increase at a faster rate than expected.

Change in continuous corn acres, by scenario, 2015

Million acres

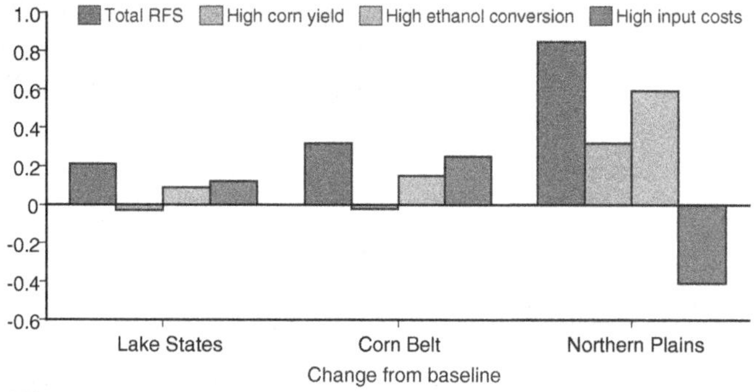

RFS = Renewable fuel standard.

Source: USDA, Economic Research Service calculations based on Regional Environment and Agriculture Programming (REAP) model data.

Corn is a heavy user of nitrogen fertilizer. Given the RFS targets, the resulting increase in fertilizer use and shift from corn-soybean rotations to continuous corn production leads to deterioration of key environmental performance measures. Nitrogen losses to surface water and groundwater increase by 1.7 and 2.8 percent, respectively, while soil runoff increases by 1.6 percent from the baseline. Differences in geography, soil type, and prevailing agricultural production activities lead to considerable variation in environmental effects among regions. The increases in leaching to groundwater are greatest in the Lake States and Southeast, while increases in runoff to surface water are greatest in the Corn Belt and Northern Plains.

As energy feedstocks that are also used as animal feed move more toward biofuel use, higher costs of animal feed reduce returns to animal production. Production of livestock declines slightly by 2015 relative to the baseline—0.6 percent for farm-fed cattle and 0.5 percent for poultry—which may result in reduced manure nutrient runoff and leaching in some areas.

Technical advances in biofuel production may soon allow other plant material to be used as energy feedstock. One of the most readily available sources of "cellulosic" feedstock is crop residues. Increased use of residue could reduce demand for corn, reducing requirements for most agricultural inputs. But replacing corn-based ethanol with biofuel created from crop residues could have mixed results on environmental quality. Removal of large amounts of crop residues requires replacement of nutrients through increased application of fertilizer and increases runoff and soil erosion. Replacing 3 billion gallons of corn ethanol with crop residue ethanol could increase nitrogen runoff and leaching in the Corn Belt, although reduced corn plantings in other regions cause these measures to decline in much of the United States.

How Was the Study Conducted?

A regionalized agricultural sector mathematical programming model with linked environmental process models was used to simultaneously estimate profit-maximizing decisions on land use, livestock production, crop mix, crop rotations, tillage practices, and fertilizer application rates. In essence, we compare the market equilibrium prior to EISA's passage with the market equilibrium expected if the new RFS production targets are met in 2015, the year that the corn-ethanol target peaks. The environmental impacts of land use and agronomic practices were estimated by applying coefficients derived from a crop biophysical simulation model that incorporates soil, weather, and management information to estimate crop yields, erosion, and chemical (pesticide and fertilizer) discharges to the environment under various crop rotation and soil management regimes. Changes to U.S. agriculture and environmental outputs from meeting EISA's biofuel production targets for 2015 were evaluated against a baseline case that reflects 2007 U.S. Department of Agriculture (USDA) projections for biofuel demand in 2015 (developed just prior to EISA's passage).

www.ers.usda.gov

Cellulosic Biofuels: Analysis of Policy Issues for Congress

Kelsi Bracmort
Analyst in Agricultural Conservation and Natural Resources Policy

Randy Schnepf
Specialist in Agricultural Policy

Megan Stubbs
Analyst in Agricultural Conservation and Natural Resources Policy

Brent D. Yacobucci
Specialist in Energy and Environmental Policy

February 1, 2010

Congressional Research Service
7-5700
www.crs.gov
RL34738

CRS Report for Congress
Prepared for Members and Committees of Congress

Summary

Cellulosic biofuels are produced from cellulose (fibrous material) derived from renewable biomass. They are thought by many to hold the key to increased benefits from renewable biofuels because they are made from potentially low-cost, diverse, non-food feedstocks. Cellulosic biofuels could also potentially decrease the fossil energy required to produce ethanol, resulting in lower greenhouse gas emissions.

Cellulosic biofuels are produced on a very small scale at this time—significant hurdles must be overcome before commercial-scale production can occur. The renewable fuels standard (RFS), a major federal incentive, mandates the use of 100 million gallons per year (mgpy) of cellulosic biofuels in 2010. After 2015, most of the increase in the RFS is intended to come from cellulosic biofuels, and by 2022, the mandate for cellulosic biofuels will be 16 billion gallons. Whether these targets can be met is uncertain. Research is ongoing, and the cellulosic biofuels industry may be on the verge of rapid expansion and technical breakthroughs. However, at this time, only a few small refineries are scheduled to begin production in 2010, with an additional nine expected to commence production by 2013 for a total output of 389 mgpy, compared with an RFS requirement of 500 mgpy in 2012 (a year earlier).

The federal government, recognizing the risk inherent in commercializing this new technology, has provided loan guarantees, grants, and tax credits in an effort to make the industry competitive by 2012. In particular, the Food, Conservation, and Energy Act of 2008 (the 2008 farm bill, P.L. 110-246) supports the nascent cellulosic industry through authorized research programs, grants, and loans exceeding $1 billion. The enacted farm bill also contains a production tax credit of $1.01 per gallon for ethanol produced from cellulosic feedstocks. Private investment, in many cases by oil companies, also plays a major role in cellulosic biofuels research and development.

Three challenges must be overcome if the RFS is to be met. First, cellulosic feedstocks must be available in large volumes when needed by refineries. Second, the cost of converting cellulose to ethanol or other biofuels must be reduced to a level to make it competitive with gasoline and corn-starch ethanol. Third, the marketing, distribution, and vehicle infrastructure must absorb the increasing volumes of renewable fuel, including cellulosic fuel mandated by the RFS.

Congress will continue to face questions about the appropriate level of intervention in the cellulosic industry as it debates both the risks in trying to pick the winning technology and the benefits of providing start-up incentives. The current tax credit for cellulosic biofuels is set to expire in 2012, but its extension may be considered during the 111th Congress. Congress may continue to debate the role of biofuels in food price inflation and whether cellulosic biofuels can alleviate its impacts. Recent congressional action on cellulosic biofuels has focused on the definition of renewable biomass eligible for the RFS, which is considered by some to be overly restrictive. To this end, legislation has been introduced to expand the definition of renewable biomass eligible under the RFS.

Contents

Figures

Tables

Contacts

Introduction

Cellulosic biofuels are produced from cellulose[1] derived from renewable biomass feedstocks such as corn stover (plant matter left in the field after harvest), switchgrass, wood chips, and other plant or waste matter. Current production consists of a few small scale pilot projects—and significant hurdles must be overcome before industrial-scale production can occur.

Ethanol produced from corn starch and biodiesel produced from vegetable oil (primarily soybean oil) are currently the primary U.S. biofuels.[2] High oil and gasoline prices, environmental concerns, rural development, and national energy security have driven interest in domestic biofuels for many years. However, the volume of fuel that can be produced using traditional row crops such as corn and soybeans without causing major market disruptions is limited; to fulfill stated goals, biofuels must also come from other sources that do not compete for the same land used by major food crops. Proponents see cellulosic biofuels as a potential solution to these challenges and support government incentives and private investment to hasten efforts toward commercial production. Some federal incentives—grants, loans, tax credits, and direct government research—attempt to push cellulosic biofuels technology to the marketplace. Demand-pull mechanisms such as the renewable fuel standard (RFS) mandate the use of biofuel blends—creating an incentive for the development of a new technology to enter the marketplace.

In contrast, petroleum industry critics of biofuel incentives argue that technological advances such as seismography, drilling, and extraction continue to expand the fossil-fuel resource base, which has traditionally been cheaper and more accessible than biofuel supplies. Other critics argue that current biofuel production strategies can only be economically competitive with existing fossil fuels in the absence of subsidies if significant improvements to existing technologies are made or new technologies are developed. Until such technological breakthroughs are achieved, critics contend that the subsidies distort energy markets and divert research funds from the development of other renewable energy sources not dependent on internal combustion technology, such as wind, solar, or geothermal, which offer potentially cleaner, more bountiful alternatives. Still others debate the rationale behind policies that promote biofuels for energy security, questioning whether the United States could ever produce and manage sufficient feedstocks of starches, sugars, vegetable oils, or even cellulose to permit biofuel production to meaningfully offset petroleum imports. Finally, there are those who argue that the focus on development of alternative energy sources undermines efforts to score energy savings through lower consumption.

The Renewable Fuel Standard: A Mandatory Usage Mandate

Principal among the cellulosic biofuels goals to be met is a biofuels usage mandate—the renewable fuel standard (RFS) as expanded by the Energy Independence and Security Act of 2007 (EISA, P.L. 110-140, Section 202)—that includes a specific carve-out for cellulosic biofuels.[3] The RFS is a demand-pull mechanism that requires a minimum usage of biofuels in the nation's fuel supply. This mandate can be met using a wide array of technologies and fuels.

[1] Cellulose is the structural component of the primary cell wall of green plants.

[2] For more information on ethanol, see CRS Report R40488, *Ethanol: Economic and Policy Issues*, by Randy Schnepf.

[3] For more information on the RFS, see CRS Report R40155, *Selected Issues Related to an Expansion of the Renewable Fuel Standard (RFS)*, by Randy Schnepf and Brent D. Yacobucci.

Although most of the RFS is expected to be met using corn ethanol initially, over time the share of advanced (non-corn-starch derived) biofuels in meeting the mandate increases. The RFS specifies three non-corn-starch carve outs: cellulosic biofuels, biomass-based diesel fuel, and other (or unspecified), which could potentially be met by imports of sugar-cane-based ethanol. The RFS mandate for cellulosic biofuels begins at 100 million gallons per year in 2010 and rises to 16 billion gallons per year in 2022 (**Figure 1**). This mandate represents a prodigious challenge to the biofuels industry in light of the fact that no commercial production of cellulosic biofuels yet exists in the United States.

Figure 1. Renewable Fuel Standard Under EISA as of November 2009

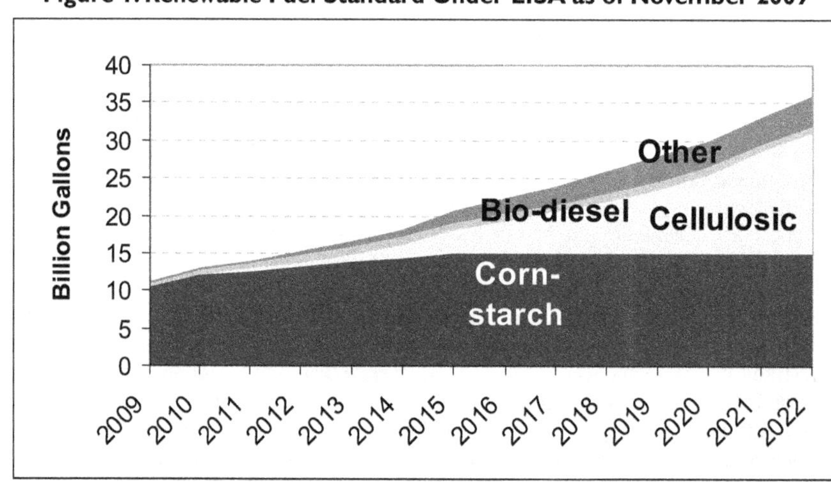

Source: EISA, (P.L. 110-140, Section 202).

Notes: Corn-starch ethanol volume is a cap, whereas other categories are floors. Biodiesel includes any type of biomass-based diesel substitute.

The RFS also mandates maximum lifecycle greenhouse gas emissions for each type of biofuel. Lifecycle greenhouse gas emissions encompass emissions[4] at all levels of production, from the field to retail sale, including emissions resulting from land use changes, that is, the clearing of forests for cropland due to increased energy crop production elsewhere. Under the law, GHG emissions for cellulosic biofuels qualifying for the RFS are limited to 60% of the GHG emissions from extracting, refining, distributing, and consuming gasoline.[5]

Challenges Facing the Industry

Cellulosic biofuels have potential, but there are significant hurdles to overcome before competitiveness is reached. In his 2007 State of the Union Address, President Bush announced the "Twenty in Ten" initiative, calling for the rapid expansion of renewable biofuels production as

[4] Greenhouse gases include carbon dioxide, methane, and nitrous oxide (CO_2, CH_4, and N_2O respectively).

[5] For more information on the lifecycle analysis of greenhouse gas emissions under the RFS, see CRS Report R40460, *Calculation of Lifecycle Greenhouse Gas Emissions for the Renewable Fuel Standard (RFS)*, by Brent D. Yacobucci and Kelsi Bracmort.

a major part of an effort to reduce U.S. gasoline use by 20% through biofuels and conservation. This goal was given substance in December 2007, when Congress passed EISA, mandating the RFS for the use of specific volumes of renewable biofuels through 2022 and setting a goal of commercial-scale cellulosic biofuels production by 2012.

This report provides background on the current effort to develop industrial-scale, competitive technology to produce biofuels from cellulosic feedstocks. It outlines the three major challenges faced in the context of the RFS: (1) feedstock supply, (2) extraction of fuel from cellulose, and (3) biofuel distribution and marketing issues. It then examines the current role of government (in cooperation with private industry) in developing that technology. Finally, the report reviews the role of Congress with respect to the emerging cellulosic biofuels industry, reviews recent congressional actions affecting the industry, and discusses key questions facing Congress.

Cellulosic Feedstock Supplies

Feedstocks used for cellulosic biofuels are potentially abundant and diverse. Initially it was thought that a major advantage of cellulosic biofuels over corn-starch ethanol was that they could be derived from potentially inexpensive feedstocks that could be produced on marginal land.[6] Corn, on the other hand, is a resource-intensive crop that requires significant use of chemicals, fertilizers, and water, and is generally grown on prime farmland. However, field research now suggests that establishment costs, as well as collection, storage, and transportation costs, associated with the production of bulky biomass crops are likely to be more challenging than originally thought.[7]

Cellulose, combined with hemicellulose and lignin, provides structural rigidity to plants and is also present in plant-derived products such as paper and cardboard. Feedstocks high in cellulose come from agricultural, forest, and even urban sources (see **Table 1**). Agricultural sources include crop residues and biomass crops such as switchgrass; forest sources include tree plantations, natural forests, and cuttings from forest management operations. Municipal solid waste, usually from landfills, is the primary urban source of renewable biomass.

Cellulosic feedstocks may have some environmental drawbacks. Some crops suggested for biomass are invasive species when planted in non-native environments. Municipal solid wastes may likely require extensive sorting to segregate usable material and may also contain hazardous material that is expensive to remove. In general, calculation of the estimated cost of biofuels production does not reflect environmental or related impacts, but such impacts are relevant to overall consideration of biofuels issues.

Biomass feedstocks are bulky and difficult to handle, presenting the industry with a major challenge. Whether feedstocks are obtained from agriculture or forests, specialized machinery would need to be developed to harvest and handle large volumes of bulky biomass. For instance, harvesting corn for both grain and stover would be more efficient with a one-pass machine capable of simultaneously segregating and processing both—a combination forage and grain

[6] *Breaking the Link between Food and Biofuels*, Bruce A. Babcock, Briefing Paper 08-BP 53, July 2008.

[7] Preliminary draft of the feedstock and technology chapters from a Purdue University study on "Cellulosic Biofuels: Technology, Market, and Policy Assessment," September 7, 2009. This study is being conducted by Purdue University under contract with CRS and is supported by Joyce Foundation funding.

harvester. Currently, machines such as these are being developed to handle biomass crops, but few are commercially available.[8] Storage facilities capable of keeping immense volumes of cellulosic material in optimal conditions must also be developed, if an industry is to grow.

Table 1. Potential Cellulosic Feedstock Sources

Source	Feedstock
Agricultural crop residues	Crop residues—stover, straw, etc.
Agricultural commercial crops	Perennial prairie grasses
Forest woody biomass	Logging residues from conventional harvest operations and forest management and land clearing operations
	Removal of excess biomass from timberlands and other forest lands
	Fuelwood from forest lands
	Perennial woody crops
Agricultural or forest processing by-products	Food/feed processing residues
	Pulping (black) liquor from paper mills
	Primary and secondary wood processing mill residues
Urban	Municipal solid waste
	Packaging wastes and construction debris

Source: CRS.

Crop Residues

Crop residues are by-products of production processes (such as producing grain), and so their production costs are minimal. Corn stover[9] and rice and wheat straw are abundant agricultural residues with biomass potential.[10] Among different residues, corn stover has attracted the most attention for biofuels production. However, an important indirect cost associated with using crop residue as a biomass feedstock is a potential loss of soil fertility. When harvesting stover, sufficient crop residue must be left in place to prevent erosion and maintain soil fertility. Research suggests that under the right soil conditions up to 60% of some residuals can be removed without detrimental soil nutrition or erosion effects. Results from early trials suggest the potential ethanol yield from corn stover (not including the grain harvested, which could be used for feed or fuel) is approximately 180 gallons of ethanol per acre. This compares with roughly 425 gallons of corn-starch ethanol[11] (from grain) and 662 gallons per acre of sugar cane (in Brazil), when grown as dedicated energy crops.[12]

[8] *Growing and Harvesting Switchgrass for Ethanol Production in Tennessee*, Clark D. Garland, Tennessee Biofuels Initiative, University of Tennessee Institute of Agriculture, UT Extension SP-701a.

[9] Corn stover consists of the cob, stalk, leaf, and husk left in the field after harvest.

[10] Bioenergy Feedstock Information Network, Biomass Resources, http://bioenergy.ornl.gov/main.aspx.

[11] USDA Economic Research Service, *Ethanol Reshapes the Corn Market*, Amber Waves, April 2006, http://www.ers.usda.gov/AmberWaves/April06/Features/Ethanol.htm.

[12] Cellulosic Ethanol: A Greener Alternative, by Charles Stillman, June 2006, http://www.cleanhouston.org/energy/features/ethanol2.htm.

Prairie Grasses

Perennial prairie grasses include native species, which were common before the spread of agriculture, and non-indigenous species, some of which are now quite common. Switchgrass is a native perennial grass that once covered American prairies and is a potential source of biomass. Its high density and native immunity to diseases and pests have caused many to focus on its use as a cellulosic feedstock. According to research at the University of Tennessee, the 10-foot tall grass, if harvested after frost, will produce for 10 to 20 years. However, like other perennials, switchgrass takes some time to establish—according to field trials, in the first year of production, yields are estimated at 30% (two tons per acre) of the full yield potential. In the second year, yield is about 70% (five tons per acre), and in the third year yields reach full potential at seven tons per acre,[13] the equivalent of 500[14] to 1,000 gallons of ethanol.[15]

Miscanthus is another fast-growing perennial grass. Originally from Asia, it is now common in the United States. Miscanthus produces green leaves early in the planting season and retains them through early fall, maximizing the production of biomass.[16] Like switchgrass, it grows on marginal lands with minimal inputs. Research in Illinois shows miscanthus can produce 2½ times the volume of ethanol (about 1,100 gallons) per acre as corn—under proper conditions.[17]

At South Dakota State University, field trials with mixtures of native grasses produced biomass yields slightly lower than switchgrass monocultures, but suggest that such mixtures result in better soil health, improved water quality, and better wildlife habitat.[18] Similar research at the University of Minnesota with mixtures of 18 native prairie species resulted in biomass yields three times greater than switchgrass.[19]

Forest Sources of Biomass

Forest resources for biomass include naturally occurring trees, residues from logging and other removals, and residue from fire prevention treatments. Extracting and processing forest biomass can be expensive because of poor accessibility, transportation, and labor availability. More efficient and specialized equipment than currently exists is needed for forest residual recovery to become cost effective.[20]

---

[13] *Growing and Harvesting Switchgrass for Ethanol Production in Tennessee*, Clark D. Garland, Tennessee Biofuels Initiative, University of Tennessee Institute of Agriculture, UT Extension SP-701a.

[14] Ibid.

[15] "DuPont Danisco and University of Tennessee Partner to Build Innovative Cellulosic Ethanol Pilot Facility," press release, Nashville, TN, July 23, 2008.

[16] Science News, "Giant Grass Miscanthus Can Meet US Biofuels Goal Using Less Land Than Corn Or Switchgrass," *Science Daily*, August 4, 2008, http://www.sciencedaily.com/releases/2008/07/080730155344.htm.

[17] Ibid.

[18] *Perennial Bioenergy Feedstocks Report to Chairman Collin Peterson*—House Agriculture Committee, April 5, 2007, North Central Bio-economy Consortium.

[19] *Carbon-Negative Biofuels from Low-Input High-Diversity Grassland Biomass*, by David Tilman, Jason Hill, and Clarence Lehman, Science, December 8, 2006: vol. 314, p. 1598.

[20] *Biomass as a Feedstock for a Bioenergy and Bioproducts Industry: The Technical Feasibility of a Billion-Ton Annual Supply*, Environmental Sciences Division, Oak Ridge National Laboratory, 2005, http://feedstockreview.ornl.gov/pdf/billion_ton_vision.pdf, p36.

Commercial tree plantations (perennial woody crops) are another source of woody biomass. Compared to prairie grasses, perennial woody crops such as hybrid poplar, willow, and eucalyptus trees, are relatively slow to mature and require harvesting at long intervals (2-4 year intervals for willow or 8-15 years for poplar). Using specialized equipment, harvesting usually occurs in the winter, when trees are converted to chips on site and then transported to the refinery for processing. Some trees, such as willow, re-sprout after cutting and can be harvested again after a few years.[21] An acre of woody biomass (i.e., hybrid poplar) yields an estimated 700 gallons of biofuel on an annual basis.[22]

Secondary and Tertiary Feedstocks

Secondary and tertiary feedstocks are derived from manufacturing (secondary) or consumer (tertiary) sources. In many cases their use as feedstocks recovers value from low- or negative-value materials. Food and feed processing residues such as citrus skins are major agricultural residues often suitable as renewable biomass. Residues from wood processing industries such as paper mills or from feed processing are major secondary sources. Tertiary sources include urban wood residues such as construction debris, urban tree trimmings, packaging waste, and municipal solid waste. One ton of dry woody biomass produces approximately 70 gallons of biofuels.[23]

Feedstock Issues

Volumes Required

Ethanol plants are intended to operate 24/7, that is, year-round with only a brief temporary stoppage for maintenance. As a result, accumulating and storing enough feedstock to supply a commercial-scale refinery producing 10-20 million gallons per year (mgpy) would require as much as 700 tons of feedstock a day—nearly the volume of 900 large round bales of grass or hay—or about 240,000 tons annually.[24] In contrast, a (much larger) 100 mgpy corn ethanol plant requires about 2,500 tons of corn per day, but corn is much denser and easier to handle than most renewable biomass sources.[25] The U.S. Department of Energy (DOE) is currently focusing research efforts on harvest and collection, preprocessing, storage and queuing, handling, and transportation of feedstocks.[26] These are major challenges facing an emerging biofuels industry due to the sheer bulk of the biomass and divergent growth cycles of different biomass crops. Pelletizing and other methods for compressing feedstocks reduce transportation costs but increase processing costs. According to a Purdue University study, the total per ton costs for transporting biomass 30 miles range from \$39 to \$46 for corn stover and \$57 to \$63 for switchgrass—

---

[21] IAE Bioenergy, *Sustainable Production of Woody Biomass for Energy*, March 2002, http://www.energycommission.org/files/finalReport/IV.4.c%20-%20Cellulosic%20Ethanol%20Fact%20Sheet.pdf.

[22] "Fast-Growing Trees Could Take Root as Future Energy Source," press release of Purdue University Study funded by DOE, http://www.purdue.edu/UNS/html4ever/2006/060823.Chapple.poplar.html.

[23] *Producing Ethanol from Wood*, presentation by Alan Rudie, USDA Forest Service Forest Products Laboratory, Madison, WI, http://www.csrees.usda.gov/nea/plants/pdfs/rudie.pdf.

[24] DOE refinery feedstock estimates and CRS calculations into large round bales.

[25] *Analysis of the Efficiency of the U.S. Ethanol Industry 2007*, by May Wu, Center for Transportation Research Argonne National Laboratory Delivered to Renewable Fuels Association on March 27, 2008.

[26] *From Biomass to Biofuels*, National Renewable Energy Laboratory; NREL/BR-510-39436, August 2006, http://www.nrel.gov/biomass/.

compared with roughly $10 for corn.[27] The USDA-DOE goal is to reduce the total feedstock cost at the plant (production, harvest, transport, and storage) from $60 per ton (the 2007 level) to $46 per ton in 2012.[28]

A 2005 USDA-DOE study undertaken by the Oak Ridge National Laboratory estimates that just over 1.3 billion tons of biomass (**Figure 2**) could be available annually in the United States for all forms of bioenergy production (including electricity and power from biomass, and fuels from cellulose).[29] If processed into biofuel, this 1.3 billion tons of biomass could replace 30% of U.S. transportation fuel consumption at 2004 levels, according to USDA. However, this estimate has been heavily criticized for several reasons, including the claim that it ignores the costs and difficulties likely to be associated with harvesting or collecting woody biomass and urban waste, as well as that it uses optimistic yield growth assumptions to achieve its biomass tonnages. The USDA estimate also predates the definition of renewable biomass eligible for the RFS. Current provisions restrict the use of woody biomass to trees grown in plantations or pre-commercial thinnings from non-federal lands, while USDA's study included woody biomass from federal and private forests as well as commercial forests. As a result, the potential volume of biomass available for conversion is substantially less than the USDA-DOE estimate of 1.3 billion tons.

Figure 2. Annual Biomass Resource Potential According to USDA

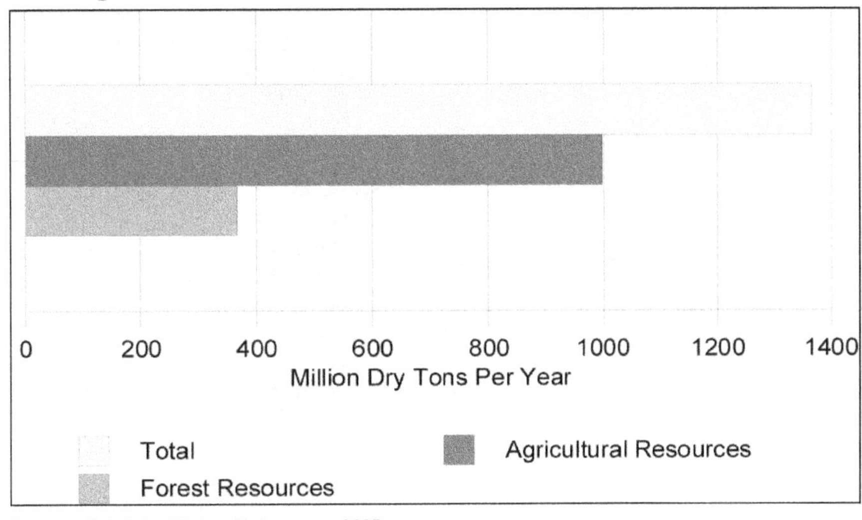

Source: Oak Ridge National Laboratory, 2005.

Note: Total is roughly equivalent to 42 billion gallons of gasoline.

[27] *The Economics of Biomass Collection, Transportation, and Supply to Indiana Cellulosic and Electric Utility Facilities*, Working Paper #08-03, by Sarah Brechbill and Wallace Tyner Purdue University, April 25, 2008.

[28] *Biomass Multi-Year Plan*, DOE Office of the Biomass Program, March 2006. http://www1.eere.energy.gov/biomass/pdfs/biomass_program_mypp.pdf.

[29] *Biomass as a Feedstock for a Bioenergy and Bioproducts Industry: The Technical Feasibility of a Billion-Ton Annual Supply*, Environmental Sciences Division, Oak Ridge National Laboratory, 2005, http://feedstockreview.ornl.gov/pdf/billion_ton_vision.pdf.

Impacts on Food Supplies

Compared with corn, cellulosic feedstocks are thought to have smaller impacts on food supplies.[30] By refining corn into ethanol, food markets are indirectly affected via cattle and dairy feed markets. In contrast, cellulosic feedstocks are non-food commodities and thus do not reduce food output unless they displace food crops on cropland.[31] However, many cellulosic feedstocks do not need prime farmland. Waste streams such as municipal solid waste, most crop residues, wood pulp residues, and forest residues are potential sources of biomass that have no impact on food crop acreage.[32] Corn stover, removed in appropriate quantities, could also be refined into ethanol without affecting food supplies. Feedstocks such as switchgrass and fast-growing trees appear to do well in marginal conditions and would likely have a minimal impact on food supplies, particularly in the case of woody biomass feedstocks from forested areas not suitable for crops.[33]

Establishment Costs and Contracting Arrangements

In the United States, crops are traditionally grown on an annual basis. Thus, contracts, loans, and other arrangements are generally established for a single growing year. Arrangements for producing perennial crops would necessarily reflect their multi-year cycles. Producers, whether they own or rent land, can expect reduced returns while the crop becomes established. Producers renting land would need long-term agreements suitable for multi-year crops. Some suggest a legal framework would have to be developed for multi-year harvests. For example, the University of Tennessee has entered into three-year contracts with producers to ensure switchgrass availability for a pilot refinery scheduled to have started ethanol production in 2009.[34]

Extracting Fuel from Cellulose: Conversion

Breaking down cellulose and converting it into fuel requires complex chemical processing—technology that is now rudimentary and expensive (see **Table 2**). Starches (such as corn) and sugars (such as cane sugars) are easily fermented into alcohol, but cellulose must first be separated from hemicellulose and lignin and then broken down into sugars or starches through enzymatic processes.[35] Alternatively, biomass can be thermochemically converted into synthesis gas (syngas),[36] which can then be used to produce a variety of fuels. Regardless of the pathway, as discussed below, at the present time processing cellulose into fuels is expensive relative to other conventional and alternative fuel options.

---

[30] *Breaking the Link between Food and Biofuels*, Bruce A. Babcock, Briefing Paper 08-BP 53, July 2008, Center for Agricultural and Rural Development, Iowa State University, http://www.card.iastate.edu.

[31] *Ibid.*

[32] *Ibid.*

[33] For more information on biofuels and food supplies see CRS Report RL34474, *High Agricultural Commodity Prices: What Are the Issues?*, by Randy Schnepf.

[34] "DuPont Danisco and University of Tennessee Partner to Build Innovative Cellulosic Ethanol Pilot Facility," TN, July 23, 2008, http://www.utbioenergy.org/NR/rdonlyres/7D4527E3-00F7-4574-92B1-E6749851AA41/1262/072308FINALUTDDCEPressReleaseDDCELetterallogos.pdf.

[35] *Biofuels Energy Program 2007*, DOE http://www1.eere.energy.gov/biomass/publications.html#vision.

[36] A mixture of hydrogen and carbon monoxide.

Table 2. Basic Steps Required to Produce Ethanol

Product	Feedstock	Refining Required after Milling			Cost (per gal.)
Sugar ethanol	Sugar cane	Fermentation into ethanol			$0.30
Corn-starch ethanol	Corn starch	Hydrolysis makes fermentable sugars	Fermentation into ethanol		$0.53
Cellulosic ethanol (biochemical process)	Switchgrass, corn stover, woody bio-mass, municipal solid waste	Pre-treatment makes cellulose accessible	Hydrolysis makes fermentable sugars	Fermentation into ethanol	$1.59
Cellulosic ethanol (therm-chemical process)	Switchgrass, corn stover, woody bio-mass, municipal solid waste	Pre-treatment makes cellulose accessible	Syngas production through therm-chemical processes	Conversion of syngas to products including ethanol	$1.21

Source: National Renewable Energy Laboratory (NREL), Research Advances in Cellulosic Ethanol, costs from the *Economic Feasibility of Ethanol Production From Sugar in the United States*, Hosein Shapouri, USDA, and Michael Salassi, Nelson Fairbanks, LSU Agricultural Center, March 2005.

Production Processes

Three basic methods can be used to convert cellulose into ethanol: (1) acid hydrolysis (dilute or concentrated), (2) enzymatic hydrolysis, and (3) thermochemical gasification and pyrolysis. There are many different variations on these, depending on the enzymes and processes used. Currently all these methods are limited to pilot or demonstration plants, and all comprise the "pre-treatment" phase of ethanol production.

Acid Hydrolysis

Dilute and concentrated acid hydrolysis pre-treatments use sulphuric acid to separate cellulose from lignin and hemicellulose. Dilute acid hydrolysis breaks down cellulose using acid at high temperature and pressure. Only about 50% of the sugar is recovered because harsh conditions and further reactions degrade a portion of the sugar. In addition, the combination of acid, high temperature, and pressure increase the need for more expensive equipment.

On the other hand, concentrated acid hydrolysis occurs at low temperature and pressure and requires less expensive equipment. Although sugar recovery of over 90% is possible, the process is not economical, due to extended processing times and the need to recover large volumes of acid.[37] [38]

[37] J. D. Wright and C. G. d'Agincourt, "Evaluation of Sulfuric Acid Hydrolysis Processes for Alcohol Fuel Production," *Biotechnology and Bioengineering Symposium*, no. 14, (New York: John Wiley and Sons, 1984), pp 105-123, http://qibioenergy.wordpress.com/2008/03/08/acid-hydrolysis/.

[38] P. C. Badger, "Ethanol From Cellulose: A General Review," in J. Janick and A. Whipkey, eds., *Trends in New Crops and New Uses* (ASHS Press, 2002).

Enzymatic Hydrolysis

DOE suggests that enzymatic hydrolysis, a biochemical process that converts cellulose into sugar using cellulase enzymes, offers both processing advantages as well as the greatest potential for cost reductions.[39] However, the cost of cellulase enzymes remains a significant barrier to the conversion of lignocellulosic biomass to fuels and chemicals. Enzyme cost primarily depends on the direct cost of enzyme preparation ($/kg enzyme protein) and the enzyme loading required to achieve the target level of cellulose hydrolysis (gram enzyme protein/gram cellulose). According to DOE, the near-term goal is to reduce the cost of cellulase enzymes from $0.50 to $0.60 per gallon of ethanol to approximately $0.10 per gallon.[40] The National Renewable Energy Laboratory (NREL) of DOE is conducting research to lower enzyme costs by allowing cellulase yeasts and fermenting yeasts to work simultaneously—with significant savings.

The total conversion cost (excluding feedstock cost) for biochemical conversion of corn stover to ethanol is estimated to be about $1.59 per gallon[41]—compared with the USDA-DOE goal of $0.82 per gallon in 2012.[42]

Thermochemical Gasification and Pyrolysis

Thermochemical processes such as gasification and pyrolysis convert lignocellulosic biomass into a gas or liquid intermediate (syngas) suitable for further refining to a wide range of products including ethanol, diesel, methane, or butanol.[43] Recovery rates of up to 50% of the potentially available ethanol have been obtained using synthesis gas-to-ethanol processes. Two-stage processes producing methanol as an intermediate product have reached efficiencies of 80%. However, developing a cost-effective thermochemical process has been difficult.[44] The Fischer-Tropsch (FT) process uses gasification to produce syngas that is then converted into biofuels such as diesel, methane, or butanol. It is possible to produce diesel and other fuels using syngas from coal or natural gas, but biomass-derived syngas is technically challenging because of impurities that must be removed during processing.

The cost of gasification conversion (excluding the cost of feedstock) in 2005 was estimated at $1.21 per gallon (2007 dollars).[45] The USDA-DOE goal for 2012 is $0.82 cents per gallon.

---

[39] DOE, EERE, Biomass Program, "Cellulase Enzyme Research," at http://www1.eere.energy.gov/biomass/biomass_feedstocks.html.

[40] *Development of New Sugar Hydrolysis Enzymes: DOE, Novozymes Biotech, Inc.* http://search.nrel.gov/cs.html?url=http%3A//www1.eere.energy.gov/biomass/fy04/new_sugar_hydrolysis_enzymes.pdf&charset=utf-8&qt=cellulase+enzymes&col=eren&n=2&la=en.

[41] *Biomass Multi-Year Plan*, DOE Office of the Biomass Program, March 2006. http://www1.eere.energy.gov/biomass/pdfs/biomass_program_mypp.pdf.

[42] *Biomass Multi-Year Plan*, DOE Office of the Biomass Program, March 2006. http://www1.eere.energy.gov/biomass/pdfs/biomass_program_mypp.pdf.

[43] J. D. Wright, "Evaluation of Sulfuric Acid Hydrolysis Processes for Alcohol Fuel Production," in *Biotechnology and Bioengineering Symposium*, no. 14, (New York: John Wiley and Sons, 1984), pp 103-123.

[44] P. C. Badger, "Ethanol From Cellulose: A General Review," in J. Janick and A. Whipkey, eds., *Trends in New Crops and New Uses* (ASHS Press, 2002)..

[45] *Biomass Multi-Year Plan*, DOE Office of the Biomass Program, March 2006. http://www1.eere.energy.gov/biomass/pdfs/biomass_program_mypp.pdf.

Distribution and Absorption Constraints

Distribution and absorption constraints may hinder the use of cellulosic biofuels even if they are ultimately produced on an industrial scale. In the coming years, greater volumes of advanced biofuels (i.e., cellulosic or non-corn-starch ethanol, biodiesel, or imported sugar ethanol) would need to be blended into motor fuel to fulfill the rising advanced biofuel mandate.

Distribution Bottlenecks

Distribution issues may hinder the efficient delivery of ethanol to retail outlets. Ethanol, mostly produced in the Midwest, would need to be transported to more populated areas for sale. It cannot currently be shipped in pipelines designed for gasoline because it tends to attract water in gasoline pipelines.

In addition, ethanol must be stored in unique storage tanks and blended prior to delivery to the retail outlet, because it tends to separate if allowed to sit for an extended period after blending. This would require further infrastructure investments.

The current ethanol distribution system is dependent on rail cars, tanker trucks, and barges. Because of competition, options (especially rail cars) are often limited. As non-corn biofuels play a larger role, some infrastructure concerns may be alleviated as production is more widely dispersed across the nation. If biomass-based diesel substitutes are produced in much larger quantities, some of these infrastructure issues may be mitigated.

The Blend Wall[46]

The blend wall refers to the volume of ethanol required if all gasoline used in the United States contained 10% ethanol (E-10)[47]—or roughly 14 billion gallons. However, because of infrastructure issues associated with transporting and storing midwestern ethanol in coastal markets, the effective blend wall is probably about 12 billion gallons per year. U.S. ethanol production is rapidly approaching this level. Once the blend wall is reached, the market will have difficulty absorbing further production increases, even if they are mandated by the RFS. Although greater use of E-85 could absorb additional volume, it is limited by the lack of E-85 infrastructure (including the considerable expense of installing or upgrading tanks and pumps) and the size of the flex-fuel fleet. These concerns could be sidestepped if additional non-ethanol biofuels are introduced into the market, especially "drop-in" fuels that are chemically similar to petroleum fuels and could be blended directly with those fuels.[48]

To maximize ethanol use, proposals to raise the ethanol blend level for conventional vehicles from E-10 to E-15 or E-20 are being considered by EPA. DOE is conducting tests to determine

---

[46] For more information on the blend wall, see CRS Report R40445, *Intermediate-Level Blends of Ethanol in Gasoline, and the Ethanol "Blend Wall"*, by Brent D. Yacobucci.

[47] E-10 refers to a fuel blend of 10% ethanol and 90% gasoline. Likewise, E-15 is a blend of 15% ethanol, 85% gasoline; E-20 is 20% ethanol, 80% gasoline; and E-85 is 85% ethanol, 15% gasoline.

[48] Potential drop-in fuels include synthetic gasoline or diesel fuel produced from biomass, as well as butanol or other chemicals that may not have some of the blending limitations faced by ethanol.

different blends' compatibility with conventional automobiles. These blends could potentially be supplied by existing conventional infrastructure (storage tanks and fuel pumps)[49] but would require U.S. Environmental Protection Agency (EPA) approval, as well as approval by automakers (using higher blend ratios could currently void a manufacturer's warranty), fuel pump manufacturers, and other equipment suppliers.

Economic and Environmental Issues

Economic Efficiency

Cellulosic biofuels are generally thought to have favorable economic efficiency potential over corn-starch ethanol primarily because of the low costs of production for feedstocks. However, current NREL estimates of the total cost of producing cellulosic ethanol, including feedstock production, marketing, and conversion, are $2.40 per gallon, more than twice the cost of producing corn ethanol.

A major impediment to the development of a cellulose-based ethanol industry is the state of cellulosic conversion technology (i.e., the process of gasifying cellulose-based feedstocks or converting them into fermentable sugars).[50] DOE's goal of competitiveness in 2012 assumes $1.30 (2007 dollars) per gallon costs for corn stover ethanol based on a feedstock price of $13 per ton. This compares with USDA's estimated cost of producing corn-based ethanol in 2002 of $0.958 per gallon (about $1.07 per gallon in 2007 dollars).[51] In addition, the cost of harvesting, transporting, and storing bulky cellulosic biomass is not well understood and consequently is often undervalued. As a result, even though cellulosic ethanol benefits from the $1.01 production tax credit (discussed below), which is $0.56 per gallon higher than the blender's tax credit of $0.45 per gallon for corn ethanol, it remains at a substantial cost disadvantage compared with corn-starch ethanol.

Energy Balance

The net energy balance (NEB) is a comparison of the ratio of the per-unit energy produced versus the fossil energy used in a biofuel's production process. The use of cellulosic biomass in the production of biofuels yields an improvement in NEB compared with corn ethanol. Corn ethanol's NEB was estimated at 67% by USDA in 2004—67% more energy was available in the ethanol than was contained in the fossil fuel used to produce it. This is at the upper range of estimates for corn ethanol's energy balance. By contrast, estimates of the NEB for cellulosic biomass range from 300%[52] to 900%.[53] As with corn-based ethanol, the NEB varies based on the production process used to grow, harvest, and process feedstocks.

---

[49] National Biofuels Action Plan, October 8, 2008.

[50] *Research Advances in Cellulosic Ethanol*, DOE-NREL, at http://www.nrel.gov/biomass/pdfs/40742.pdf, NOEL/BR-510-40742, March 2007.

[51] *The Energy Balance of Corn Ethanol: An Update*, Shapouri, Hosein; James A. Duffield, and Michael Wang. USDA, Office of the Chief Economist, Office of Energy Policy and New Uses, Agricultural Economic Report (AER) No. 813, July 2002; available at http://www.usda.gov/oce/reports/energy/index.htm.

[52] *Cellulosic Ethanol Fact Sheet*, by Lee R. Lynd, presented at the National Commission on Energy Policy Forum: The Future of Biomass and Transportation Fuels, June 13, 2003.

Another factor that favors cellulosic ethanol's energy balance over corn-based ethanol relates to byproducts. Corn-based ethanol's co-products are valued as animal feeds, whereas cellulosic ethanol's co-products, especially lignin, are expected to serve directly as a processing fuel at the plant, substantially increasing energy efficiencies.

Additionally, switchgrass uses less fertilizer than corn, by a factor of two or three,[54] and its perennial growth cycle reduces field passes for planting. Some suggest that ethanol from switchgrass has at least 700% more energy output per gallon than fossil energy input.[55] The same is largely true of woody biomass that, even in plantations, requires minimal fertilizer and infrequent planting operations.

Environment

Ethanol and biodiesel produced from cellulosic feedstocks, such as prairie grasses and fast-growing trees, have the potential to improve the energy and environmental effects of U.S. biofuels. As previously stated, a key potential benefit of cellulosic feedstocks is that they can be grown without the need for chemicals. Reducing or eliminating the need for chemical fertilizers could address one of the largest energy inputs for corn-based ethanol production. Fast-growing trees and woody crops could offer additional environmental benefits of improved soil and water quality, reduced CO_2 emissions, and enhanced biodiversity.[56]

Despite potential environmental benefits, additional concerns about cellulosic feedstocks exist, including concerns that required increases in per-acre yields to obtain economic feasibility could require the use of fertilizers or water resources, and that availability of sufficient feedstock supply is limited and expansion could generate additional land use pressures for expanded production (see "Greenhouse Gas Emissions" discussion, below). In addition to these concerns, some groups say that other potential environmental drawbacks associated with cellulosic fuels should be addressed, such as the potential for soil erosion, increased runoff, the spread of invasive species, and disruption of wildlife habitat.

Greenhouse Gas Emissions

Greenhouse gas emissions differ among types of ethanol because of a number of factors, including the feedstock crop converted into ethanol, the fuel used to power the refinery (fossil or renewable), and the original state of the land on which the feedstock was produced. For instance, if virgin forest land were cleared and planted with switchgrass, higher greenhouse gas emissions would result than if switchgrass were grown on previously cleared cropland, mainly because GHG emissions associated with clearing and plowing the virgin soil would have to be included as part of the production process. Likewise, a cellulosic refinery powered by coal or natural gas

(...continued)

[53] Worldwatch Institute, *Biofuels for Transportation, Global Potential and Implications for Sustainable Agriculture and Energy in the 21st Century*. Table 10-1, p. 127, June 2006.

[54] *Ethanol From Biomass: Can It Substitute for Gasoline?* Michael B. McElroy, book chapter draft.

[55] *Net Energy of Cellulosic Ethanol from Switchgrass*, M.R. Schimer, K.P. Vogel, R.B. Mitchell, and R.B. Perrin, PNAS January 15, 2008, vol. 105, no. 2, available at http://www.pnas.org/cgi/doi/10.1073/pnas.0704767105.

[56] Timothy A. Volk, Theo Verwijst, and Pradeep J. Tharakan et al., "Growing Fuel: A Sustainability Assessment of Willow Biomass Crops," *Frontiers in Ecology & Environment Journal*, vol. 2, no. 8 (2004), pp. 411-418.

would have higher greenhouse gas emissions than one powered by recovered feedstock co-products.

Multi-year harvesting of perennial crops decreases greenhouse gas emissions by minimizing field passes. Prairie grasses and woody crops require reduced inputs compared with corn—and have lower greenhouse gas emissions. Also, because cellulosic feedstocks require less fertilizer for their production, the energy balance benefit of cellulosic ethanol could be significant. A study by the Argonne National Laboratory concluded that with advances in technology, the use of herbaceous[57]-feedstock cellulose-based E-10 could reduce fossil energy consumption per mile by 8%, while herbaceous-feedstock cellulose-based E-85 could reduce fossil energy consumption by roughly 70%.[58]

According to the EPA's Office of Transportation and Air Quality, for every unit of energy measured by British Thermal Units (BTU) of gasoline replaced by cellulosic ethanol, the total lifecycle greenhouse gas emissions (including carbon dioxide, methane, and nitrous oxide) would be reduced by an average of about 90%. In comparison, the reduction from corn ethanol averages 22%.[59]

Private Investment

Private investment is viewed by many to be critical to the development of the cellulosic biofuels industry. However, the aggregate value of required private investment is difficult to determine. Anecdotal evidence suggests the main sources of capital are venture capitalists and petroleum companies—commercial banks have a minor role. Venture capitalists generally have an extended (10-year) perspective, which fits well with nascent technologies and is insulated from shorter-term financial volatility. Petroleum companies, faced with mandatory blending of biofuels with gasoline, have been eager to invest in the cellulosic industry. Numerous partnerships have been formed: British Petroleum (BP) and Verenium announced a partnership in August 2008 to accelerate the commercialization of cellulosic ethanol, with BP investing $90 million in the deal.[60] In another collaboration, Royal Dutch Shell has teamed up with Imogen Corporation to develop cellulosic ethanol processes.[61] Mascoma, a major ethanol producer, raised $30 million to support its investment in cellulosic feedstock conversion with technical support from General Motors and Marathon Oil.[62] A collaboration between Monsanto and Mendel Biotechnology Inc. will focus on the breeding and development of crops for production of cellulosic biofuels.[63]

---

[57] A herbaceous plant is a plant that has leaves and stems that die down at the end of the growing season to the soil level.

[58] Wang, et al., table 7.

[59] *Greenhouse Gas Impacts of Expanded Renewable and Alternative Fuels Use*, EPA Office of Transportation and Air Quality, EPA-420-F-07-035, April 2007.

[60] *BP and Verenium Partner To Commercialize Cellulosic Ethanol*, RenewableEnergy-World.com, August 11, 2008, http://www.renewableenergyworld.com/rea/news/story?id=53280.

[61] *Shell Boosts Second Generation Biofuels*, by Ed Crooks, Financial Times, July 16, 2008, http://www.ft.com/cms/s/7f9f1ee8-52d0-11dd-9ba7-000077b07658.html.

[62] *U.P. Biofuel Plant Lands $50m in State, Fed Aid*, by Gary Heinlein and David Shepardson, The Detroit News, October 8, 2008, http://www.detnews.com/apps/pbcs.dll/article?AID=/20081008/BIZ/810080393.

[63] *Monsanto Company and Mendel Biotechnology, Inc. Announce Cellulosic Biofuels Collaboration*, BioSpace.com, April 28, 2008, http://www.biospace.com/news_story.aspx?NewsEntityId=94113.

Federal Cellulosic Biofuels Policies

USDA and DOE are currently engaged in a variety of activities to encourage development and demonstration of cellulosic biofuels technologies. The Energy Independence and Security Act of 2007 (EISA, P.L. 110-140), the Food, Conservation, and Energy Act of 2008 (the 2008 farm bill, P.L. 110-246), and other legislation support research and development of a broad range of cellulosic technologies through USDA and DOE programs. Many of these programs extend the goals of the Energy Policy Act of 2005 (EPAct, P.L. 109-58) and President Bush's *20 in 10* initiative.[64] The Biomass Research and Development Initiative (BRDI) coordinates federal interagency *technology-push* efforts, such as R&D, loans, and grants, under the guidance of the Biomass Research and Development Board. The Board was authorized in the Biomass Research and Development Act of 2000 and is co-chaired by USDA and DOE. BRDI plays a major role in R&D for the cellulosic biofuels industry.[65]

In October 2008, then-USDA Secretary Ed Schafer and DOE Secretary Samuel W. Bodman released the National Biofuels Action Plan (NBAP), which provided an outline of the major challenges facing the cellulosic biofuels industry: feedstock production and logistics; conversion science and technology; distribution infrastructure and blending. The plan reflects current federal and industry efforts to develop the cellulosic biofuels industry.[66]

Direct Federal Spending on R&D

Recognizing that cellulosic biofuels can contribute to improving national energy security, reducing greenhouse gas emissions, and boosting rural economic development, discretionary DOE spending on bioenergy R&D (including a major cellulosic component) under the Biomass and Biorefinery Systems Program (Energy Efficiency and Renewable Energy Programs) was $196 million in FY2007.[67] DOE appropriations for this purpose totaled $198 million in FY2008, of which 33% was spent on conversion R&D, 7% on feedstock infrastructure, and 52% on biorefinery development.[68] DOE appropriations for 2009 amounted to $217 million, of which $215 million was directed toward cellulosic biofuels, plus an FY2009 stimulus of $786 million. DOE appropriations for FY2010 are $220 million.[69] Approximately $3 million in congressionally directed spending for FY2010 will be directed toward cellulosic biofuel initiatives.[70]

USDA discretionary outlays for Bioenergy and Renewable Energy Programs, which funded cellulosic biofuels in part, were nearly $100 million in FY2008, and the FY2009 budget request is

---

[64] "2007 State of the Union Address" 20 in 10: *Strengthening America's Energy Security*, http://www.whitehouse.gov/stateoftheunion/2007/initiatives/energy.html.

[65] For more information on federal biofuels incentives see CRS Report R40110, *Biofuels Incentives: A Summary of Federal Programs*, by Brent D. Yacobucci.

[66] The National Biofuels Action Plan is available at http://www.eere.energy.gov/biomass/pdfs/nbap.pdf.

[67] FY2009 DOE Budget Request to Congress, available at http://www.cfo.doe.gov/budget/09budget/start.htm#Detailed%20Budget%20Justifications.

[68] Presentation at Platts Cellulosic Ethanol Conference, by Valri Lightner, DOE Biomass Program, October 2008, http://www.autobloggreen.com/photos/platts-conference-doe-presentation/1099010/.

[69] CRS Report R40669, *Energy and Water Development: FY2010 Appropriations*, coordinated by Carl E. Behrens.

[70] U.S. Congress, House, *Energy and Water Development and Related Agencies Appropriations Act, 2010*, conference report to accompany H.R. 3183, 111th Cong., 1st sess., September 30, 2009, H.Rept. 111-278, pp. 282, 283, 290.

$82 million.[71] USDA R&D expenditures were estimated at $39 million in FY2008, and budgeted at $59 million in FY2009. Commercialization outlays (primarily the Bioenergy Program) totaled an estimated $59 million in FY2008, and are budgeted at $18 million in FY2009. These totals are modest in comparison to the $5 to $8 billion in annual federal support for corn ethanol. Over time, as the corn ethanol industry matures, the focus of USDA efforts is likely to increasingly shift to cellulosic biofuels.

Federal-Private Partnerships

Private sector investment received a substantial federal policy boost on February 28, 2007, when the DOE announced the awarding of up to $385 million in mandatory cost-share funding for the construction of six cellulosic ethanol plant projects over a four-year period under Section 932 of the EPAct of 2005 as expanded by EISA of 2007. When fully operational, the six plants combined were expected to produce up to 100 mgpy of cellulosic ethanol. These demonstration-scale biorefinery projects focus on near-term commercial processes. The combined cost-share plus federal funding for the projects represents total planned investment of more than $1.2 billion.

The uncertainties of moving an industry from laboratory to commercial reality were highlighted when two recipients with total grant funding of $113 million dropped out of the program, one because of a substantially higher offer from the Canadian government,[72] and the other after determining that the risks involved outweighed any anticipated benefits.[73]

Renewable Energy Provisions in the 2008 Farm Bill (P.L. 110-246)

Renewable energy policy in the 2008 farm bill (Food, Conservation, and Energy Act of 2008, P.L. 110-246) builds on earlier programs originally authorized in the 2002 farm bill (P.L. 107-171) or the EPAct of 2005 (P.L. 109-58) but provides greater emphasis on cellulosic biofuels.[74] Title IX, the energy title, authorizes or reauthorizes grants, loans, and loan guarantees to foster research on agriculture-based renewable energy, to share development risk, and to promote the adoption of renewable energy systems.[75] Implementation of the farm bill provisions is underway, and regulations for new programs have not been finalized. Funding for the cellulosic component of renewable energy programs is difficult to determine because most provide support to a wide range of biofuels. Title VII, the research title, contains provisions supporting R&D in cellulosic biofuels, and Title XV, the tax and trade title, contains tax incentives and tariffs. The following programs provide support to cellulosic biofuels research, demonstration, and production.

For information on additional related provisions in the 2008 farm bill, see CRS Report RL34130, *Renewable Energy Programs in the 2008 Farm Bill*, by Megan Stubbs.

---

[71] The values in this paragraph are from personal correspondence with USDA's Office of Budget and Policy Analysis.

[72] "Iogen Suspends U.S. Cellulosic Ethanol Plant Plans," Earth2tech, http://earth2tech.com-/2008/06/04/iogen-suspends-us-cellulosic-ethanol-plant-plans/.

[73] "Alico to Discontinue Ethanol Efforts," June 2, 2008, press release, http://www.aliconinc.-com/june208.asp.

[74] CRS Report RL34130, *Renewable Energy Programs in the 2008 Farm Bill*, by Megan Stubbs

[75] Title IX cites are amendments to the 2002 farm bill (P.L. 107-171).

Tax Credit for Cellulosic Biofuels (Section 15321)

Tax and trade provisions in the 2008 farm bill benefit cellulosic biofuels. One significant incentive is a production tax credit of $1.01 per gallon that applies to cellulosic biofuels production, more than twice the blenders' tax credit of 45 cents per gallon that applies to corn ethanol. In the case of cellulosic ethanol, the $1.01 credit amount is reduced by (1) the ethanol blender's credit and (2) the small ethanol producer credit, making the total value of tax incentives for cellulosic ethanol equal $1.01.[76]

Ethanol Tariff Extension (Section 15333)

In addition to tax credits, an ethanol tariff benefits the U.S. industry by reducing the competitiveness of imported ethanol sold in this country. Domestic ethanol benefits from a 54-cent-per-gallon duty (and a smaller *ad valorem* tariff) on imported ethanol (except for limited imports under the Caribbean Basin Initiative).[77] The original intent of the tariff was to prevent imported ethanol from benefitting from the U.S. tax credit.

Agricultural Bioenergy Feedstock and Energy Efficiency Research and Extension Initiative (Section 7207)

This new program awards competitive matching (up to 50%) grants for projects supporting on-farm biomass crop research and the dissemination of results to enhance the production of biomass energy crops and their integration with the production of bioenergy. It consists of elements of earlier initiatives that were moved to the research title (Title VII) in the 2008 farm bill. Discretionary funding of $50 million annually is authorized for FY2008 through FY2012, subject to appropriations. Funding has not been appropriated for this program in FY2008, FY2009, or FY2010.

Biorefinery Assistance (Section 9003)

This initiative provides loan guarantees for the development, construction, and retrofitting of commercial-scale biorefineries and provides grants to help pay for the development and construction costs of demonstration-scale biorefineries. Loan guarantees are limited to $250 million per project, subject to the availability of funds. The program received mandatory funding of $320 million ($75 million for FY2009 and $245 million for FY2010) for commercial-scale biorefinery loan guarantees, and discretionary funding, subject to appropriations, of $150 million annually for FY2009 through FY2012 for both demonstration and commercial-scale biorefineries. This program was originally authorized by the 2002 farm bill, but received no funding from FY2002 through FY2007.

Of the five applications (worth over $1 billion) received in FY2009, USDA funded two. In January 2009, USDA funded the first cellulosic biofuels plant under the Biorefinery Program with a loan guarantee of $80 million. The recipient, Range Fuels Inc., in Soperton, GA, also received a

---

[76] For more information, see CRS Report R40110, *Biofuels Incentives: A Summary of Federal Programs*, by Brent D. Yacobucci.

[77] For more information about ethanol imports under the CBI, see CRS Report RS21930, *Ethanol Imports and the Caribbean Basin Initiative (CBI)*, by Brent D. Yacobucci.

$76 million grant from DOE for this plant. The biorefinery project will use wood chips to produce a projected output of 20 million gallons of ethanol annually, beginning in 2010. The second application under this program was approved on June 24, 2009, for a $25 million biodiesel plant. USDA anticipates receiving and reviewing applications in FY2010, but not awarding funds until FY2011, pending regulations.

Repowering Assistance (Section 9004)

The Repowering Assistance program encourages existing biorefineries to replace fossil fuels used to produce heat or power at their facilities by making payments for installing new systems that use renewable biomass, or to produce new energy from renewable biomass. The 2008 farm bill provided mandatory funding of $35 million in FY2009 to remain available until expended. The program is also authorized to receive $15 million in discretionary funding for each of FY2009 through FY2012, pending appropriations. On June 12, 2009, USDA issued a Notice of Funding Availability in the *Federal Register* to make $20 million in payments to biorefineries converting to renewable biomass.[78] The remaining $15 million of mandatory funding is expected to be available in FY2010 pending regulations. There has been no appropriation of discretionary funding.

Bioenergy Program for Advanced Biofuels (Section 9005)

The Bioenergy Program is the lead program under Title IX providing support for the development of conversion technologies for cellulosic biofuels. It was originally established by Executive Order in 1999 and provided Commodity Credit Corporation (CCC) incentive payments to ethanol and biodiesel producers on the basis of yearly increases in production. Eligibility is now limited to producers of advanced biofuels. Eligible producers entering into a contract with USDA are paid based on quantity and duration of advanced biofuel production and on net renewable energy content of the advanced biofuel. Under the 2002 farm bill (P.L. 107-171), the Bioenergy Program received total funding of $426 million during FY2003 to FY2006 but no appropriations for FY2007 or FY2008. The 2008 farm bill provides a total of $300 million in mandatory funding for FY2009 to FY2012 ($55 million annually in FY2009 and FY2010, $85 million in FY2011, and $105 million in 2012), and also authorizes $25 million annually, subject to appropriations, from FY2009 to FY2012.

The first Notice of Contract Proposals was published in the *Federal Register* on June 12, 2009.[79] Over 180 applications were received and 161 were deemed eligible. USDA anticipates making payments totaling $30 million to eligible biorefineries in early 2010. The remaining $25 million unexpended from FY2009 and the $55 million of mandatory funding for FY2010 will be available for payments in FY2011 following publication of proposed regulations.[80]

---

[78] USDA, Rural Business-Cooperative Service, "Notice of Funding Availability for Repowering Assistance Payments to Eligible Biorefineries," 74 *Federal Register* 112, June 12, 2009.

[79] Rural Business-Cooperative Service, USDA, "Notice of Contract Proposal for Payments to Eligible Advanced Biofuel Producers," 74 *Federal Register* 112, June 12, 2009.

[80] U.S. Congress, House Committee on Agriculture, Subcommittee on Conservation, Credit, Energy, and Research, *Statement of Dallas Tonsager, Under Secretary for Rural Development, USDA*, hearing to review the next generation biofuels, 111th Cong., 1st sess., October 29, 2009, p. 6.

Biomass Research and Development Initiative (Section 9008)

This program was originally authorized in the 2002 farm bill (P.L. 107-171) and is administered jointly by USDA and DOE. It supports research on and development and demonstration of biofuels and biobased products, and the methods, practices, and technologies for their production. The 2008 farm bill provides mandatory funding of $118 million for FY2009 to FY2012 ($20 million for FY2009, $28 million for FY2010, $30 million for FY2011, and $40 million for FY2012). The farm bill also authorizes the appropriation of $35 million for each of FY2009 through FY2012. The program received $5 million in FY2002, $14 million for each of FY2003, FY2004, and FY2005, and $12 million for FY2006, and was not funded in FY2007 and FY2008. On January 30, 2009, USDA and DOE announced the FY2009 funding opportunity of $25 million ($20 million of mandatory funding from the 2008 farm bill and $5 million from DOE). Award recipients have been selected, but no announcement has been made.

Biomass Crop Assistance Program (Section 9011)

This is a new program intended to help farmers establish and produce crops for conversion to bioenergy and assist with the collection, harvest, storage and transportation of eligible material for use in a biomass conversion facility that produces heat, power, biobased products, or advanced biofuels. The program is implemented by the Farm Service Agency with support from other federal and local agencies and has mandatory CCC funding of such sums as necessary.

The program is implemented in two phases. On June 11, 2009, a Notice of Funds Availability was published in the *Federal Register* announcing the implementation of the 2009 collection, harvest, storage, and transportation (CHST) portion of the program.[81] As of October 23, 2009, 114 biomass conversion facilities have been qualified and listed as approved. Over 125 facility agreements and applications are pending at the FSA national office with another 72 in progress at FSA state offices. A total of $23.6 million was obligated in FY2009 and $514 million is requested for FY2010. The FSA has also issued a draft environmental impact statement[82] and is expected to implement the final phase of the program following the publication of a proposed rule in 2010.[83]

Forest Biomass for Energy (Section 9012)

Under this new program, USDA's Forest Service is authorized to conduct a comprehensive research and development program to use forest biomass for energy. Other federal agencies, state and local governments, Indian tribes, land-grant colleges and universities, and private entities are eligible to compete for program funds. No mandatory funding is available, but discretionary appropriations of $15 million annually for FY2009 to FY2012 are authorized. This program has not yet been implemented or appropriated funding. Priority research projects include the following:

---

[81] Commodity Credit Corporation, USDA, "Notice of Funds Availability for the Collection, Harvest, Storage, and Transportation of Eligible Material," 74 *Federal Register* 11, June 11, 2009.

[82] Commodity Credit Corporation, USDA, "Notice of Availability of the Draft Programmatic Environmental Impact Statement for the Biomass Crop Assistance Program," 74 *Federal Register* 152, August 10, 2009.

[83] Office of Budget and Program Analysis, *Highlights: Farm Bill Working Group*, USDA, Title IX - Energy, October 26, 2009.

- the use of low-value forest biomass for energy from forest health and hazardous fuels reduction treatment;
- the integrated production of energy from forest biomass into biorefineries or other existing manufacturing;
- the development of new transportation fuels from forest biomass; and
- the improved growth and yield of trees for renewable energy production.

Energy Improvement and Extension Act of 2008 (P.L. 110-343)

The Emergency Economic Stabilization Act of 2008 (P.L. 110-343), which incorporates the Energy Improvement and Extension Act of 2008, contains tax and trade incentives for renewable energy production. Enacted on October 3, 2008, it expands federal benefits for renewable energy to fuels and processes that previously did not qualify and limits trade practices that benefit foreign producers but do not enhance U.S. energy independence.

Expansion of the Allowance for Cellulosic Ethanol Property (Division B, Section 201)

Previous federal tax law limited the eligibility for first-year bonus depreciation of cellulosic biofuels to facilities producing ethanol; those producing non-ethanol fuels from cellulosic feedstocks did not qualify for the allowance. P.L. 110-343 does not limit the allowance to any particular type of cellulosic fuel or production process. Taxpayers can immediately write off 50% of the cost of facilities that produce cellulosic biofuels if such facilities are placed in service before January 1, 2013.

Legislative Proposals

Congress has shown a strong interest in the development of biofuels in general and cellulosic biofuels in particular. Debate may continue on the appropriate level of incentives needed to jump start the industry. Perhaps the most critical emerging issue is the federal mandate for cellulosic biofuels under the RFS—and the industry's potential to meet that mandate. In the long term, Congress might also consider the ongoing level of government support that is appropriate for the cellulosic biofuels industry—considered by some to be essential, especially if the RFS is to be met. Others contend such support could distort market signals. The general level of support in the form of grants and loans has been determined in the 2008 farm bill but will be revisited as appropriations are considered. The cellulosic biofuels tax credit applies to fuel produced from 2009 through 2012 and extension of this credit could be the subject of debate during the 111th Congress. In addition, Congress has considered the definition of biofuels and biofuel feedstocks that qualify for federal incentives.

Legislative Changes in the RFS Volume

Citing the RFS and corn ethanol production as contributing to rising food prices and high input costs for livestock and poultry producers, some are calling for a reduction of the RFS. During the 110th Congress, S. 3031 was introduced in May 2008, to limit the corn-starch component of the

RFS to 9 billion gallons compared with 15 billion under the current law. Opponents of the reduction claim it would set back efforts to reduce the nation's dependence on foreign oil and achieve environmental goals. Reducing the corn-starch component of the RFS would increase the importance of advanced fuels, primarily cellulosic biofuels, in meeting the mandate. Legislation to alter RFS volume requirements has not been introduced in the 111th Congress.

Expanding Biomass Eligible under the RFS

The definition of forest-based renewable biomass under the RFS is considered by some to be too restrictive because it limits eligible woody biomass to privately planted trees and tree residue from actively managed tree plantations, and slash[84] and pre-commercial thinnings from non-federal forests.

The definition of renewable biomass specifically excludes biomass from federal forests. Some suggest that this exclusion eliminates much potential biomass and creates regional disparities. One-third of the 755 million acres of forest in the United States is owned by the federal government—and this acreage is concentrated in the western states. Likewise, the exclusion of private, naturally regenerated forests affects the northern and southeastern parts of the country where other feedstocks eligible under the RFS may not be as readily available. According to some, biomass extraction could become a powerful tool for improving federal land management.[85] Markets for small-diameter trees would enable a wider range of options for management of wildlife habitat, forest hydrology, hazardous fuels reduction, and pest infestations. These markets are not likely to appear if federal forests remain excluded from the RFS.

The House Committee on Agriculture Subcommittee on Conservation, Credit, Energy, and Research held hearings during July 2008 on producer eligibility under the RFS. The subcommittee heard from government officials, researchers, and producers who provided an update on the implementation of the RFS and shared concerns on barriers to eligibility for many agricultural producers. Subsequently, a Senate Energy and Natural Resources Committee field hearing on forest waste for biofuels was held in South Dakota on August 18, 2008.

During the 111th Congress, H.R. 2454 would allow for modification of the non-federal lands portion of the definition of "renewable biomass." Low-value materials are frequently removed during fire or disease reduction efforts or ecosystem health supporting activities. Waste materials such as wood waste and wood residues from private forests are also included.

Also during the 110th Congress, the House passed the Comprehensive American Energy Security and Consumer Protection Act (H.R. 6899). It contained a sense of Congress provision recommending a broad definition of renewable biomass to "encourage cellulosic biofuels ... produced from a highly diverse array of feedstocks, allowing every region of the country to be a potential producer of this fuel." No Senate action was taken. No similar action has been taken in the 111th Congress thus far.

---

[84] The accumulation of limbs, tops, and miscellaneous residue left by forest management activities, such as thinning, pruning, and timber harvesting.

[85] *Federal Forests and the Renewable Fuel Standard Factsheet*, Environmental and Energy Study Institute, July 17, 2008.

Time Frame for Cellulosic Biofuels Production

Estimates for commercial production of cellulosic biofuels vary widely. At least 16 plants are under construction as planned for the United States, bringing projected cumulative capacity to 389 mgpy by 2013. However, the pace of plant construction falls short of DOE's stated goal to make cellulosic ethanol competitive as a mature technology by 2012.[86] Some analysts have predicted a growth trend for the cellulosic ethanol industry similar to that for corn-starch ethanol. However, there is a major difference between the two: the basic process for making corn-starch ethanol (fermentation) is thousands of years old, whereas that for cellulosic is very new.

The USDA Office of Energy and New Uses projects that cellulosic biofuels are not expected to be commercially viable on a large scale until at least 2015.[87] However, the cellulosic biofuel portion of the RFS mandate is set at 3 billion gallons by 2015, a substantial amount. In its January 2009 baseline, the Food and Agricultural Policy Research Institute (FAPRI) of the University of Missouri assumes cellulosic biofuel production will fall behind the RFS and, as a consequence, the mandate will be waived by EPA.[88] In an August 2009 baseline update, FAPRI projects cellulosic ethanol production in 2013 at 245 mgpy, about a third of the 1 billion gallons in the RFS for that year.[89]

Author Contact Information

Kelsi Bracmort
Analyst in Agricultural Conservation and Natural Resources Policy
kbracmort@crs.loc.gov, 7-7283

Megan Stubbs
Analyst in Agricultural Conservation and Natural Resources Policy
mstubbs@crs.loc.gov, 7-8707

Randy Schnepf
Specialist in Agricultural Policy
rschnepf@crs.loc.gov, 7-4277

Brent D. Yacobucci
Specialist in Energy and Environmental Policy
byacobucci@crs.loc.gov, 7-9662

[86] *Biomass Multi-year Program Plan*, DOE, March 2008, http://www1.eere.energy.gov/biomass/publications.html#vision.

[87] *U.S. Biobased Products: Market Potential and Projections Through 2025*, Office of the Chief Economist, Office of Energy Policy and New Uses, USDA, the Center for Industrial Research and Service of Iowa State University, Informa Economics, Michigan Biotechnology Institute, and The Windmill Group. OCE-2008-1, http://www.oce.usda.gov.

[88] FAPRI, Iowa State University, and University of Missouri-Columbia, *FAPRI 2009: U.S. and World Agricultural Outlook*, Report 09-FSR 1, January 2009.

[89] *Baseline Update for Agricultural Markets*, FAPRI, FAPRI MU-Report #06-09, August 2009. http://www.fapri.missouri.edu.

Order Code RS22859
April 10, 2008

Food Price Inflation: Causes and Impacts

Tom Capehart
Specialist in Agricultural Policy
Resources, Science and Industry Division

Joe Richardson
Specialist in Domestic Social Policy
Domestic Social Policy Division

Summary

U.S. food prices rose 4% in 2007 and are expected to gain 3.5% to 4.5% in 2008. Higher farm commodity prices and energy costs are the leading factors behind higher food prices. Farm commodity prices have surged because (1) demand for corn for ethanol is competing with food and feed for acreage; (2) global food grain and oilseed supplies are low due to poor harvests; (3) the weak dollar has increased U.S. exports; (4) rising incomes in large, rapidly emerging economies have changed eating habits; and (5) input costs have increased. Higher energy costs increase transportation, processing, and retail costs.

Although the cost of commodities such as corn or wheat are a small part of the final retail price of most food products, they have risen enough to have an impact on retail prices. Generally, price changes at the farm level have a diminished impact on retail prices, especially for highly processed products.

The impact of higher food prices on U.S. households varies according to income. Lower-income households spend a greater portion of their income on food and feel price hikes more acutely than high-income families. Higher food costs impact domestic food assistance efforts in numerous ways depending on whether benefits are indexed, enrollments are limited, or additional funds are made available. Higher food and transportation costs also reduce the impact of U.S. contributions of food aid under current budget constraints.

Introduction

U.S. food prices are increasing. According to USDA, the Consumer Price Index (CPI) for "all food" increased 4% in 2007, the largest annual jump since 1990. In 2008,

Congressional Research Service ❧ *The Library of Congress*
Prepared for Members and Committees of Congress

CRS-2

this trend is expected to continue: the "all food" CPI is forecast to increase 3.5% to 4.5%.[1] This rapid inflation follows an extended period of stable food prices. From 1987 through 2007, food prices increased an average of 2.7% per year, excluding the drought years of 1989 and 1990.[2] During 2005 and 2006, food prices rose 2.4%. This report examines the cause of food price increases and evaluates their impacts on U.S. consumers.

Key Factors Behind Higher Commodity Prices

Robust Domestic Demand. Corn, soybean, and wheat prices all reached 10-year highs during the 2006-2007 crop year. High prices for corn reflected increased use for ethanol (22% of the 2007 crop) and strong exports. High corn prices in turn encouraged growers to move acres from wheat and soybeans into corn, contributing to tight supplies and higher prices for those crops. U.S. farm prices in 2007 for corn are estimated at $3.75 to $4.00 per bushel, compared with $2.00 in 2005; soybean prices are estimated at $10.00 to $10.80 per bushel, up from $5.66 in 2005; and wheat prices are estimated at $6.50 to $6.80 per bushel, up from $3.42 in 2005.[3]

Global Stocks Are at Low Levels. Globally, stocks of corn, wheat, and soybeans are at historically low levels. Drought in Australia and Eastern Europe and poor weather in Canada, Western Europe, and the Ukraine have reduced available quantities. With world stocks for wheat at a 30-year low,[4] buyers are turning to the U.S. for supplies.

Global Consumption Patterns Are Changing. Higher incomes are boosting demand for processed foods and meat in countries such as India and China. These shifts require more feed grains and edible oil. Even in low-income countries of sub-Saharan Africa, Asia, and Latin America, the vegetable oil share of diets has risen as processed food consumption rises. In China, consumption of meats, other livestock products, and fruits has increased while consumption of grain-based foods (such as bread) has slipped.[5] Improving food distribution systems are altering Chinese food preferences by introducing non-local foods. In India, per capita consumption of grains has fallen, while that of animal products, edible oils, vegetables and fruits has increased.[6] Better food distribution systems are altering Chinese food preferences by introducing non-local foods.

---

[1] "Food CPI, Prices, and Expenditures Briefing Room," U.S. Department of Agriculture (USDA) Economic Research Service (ERS), at [http://www.ers.usda.gov/Briefing/CPIFoodAndExpenditures/consumerpriceindex.htm].

[2] USDA/ERS, *Amber Waves*, "Corn Prices Near Record High," by Ephraim Leibtag, February 2008.

[3] U.S. Department of Agriculture, "World Agricultural Supply and Demand Estimates," March 11, 2008.

[4] For more information, see CRS Report RS22824, *High Wheat Prices: What are the Issues?*

[5] Center for Agriculture and Rural Development, "Changing Diets in China's Cities: Empirical Fact or Urban Legend?" by Fexgzia Dong and Frank H. Fuller, at [http://www.card.iastate.edu/publications/synopsis.aspx?id=1031].

[6] USDA/ERS, *Amber Waves*, "Rising Food Prices Intensify Food Insecurity in Developing Countries," by Stacy Rosen and Shahla Shapouri, February 2008.

Weak Dollar Boosts Demand for U.S. Exports. As the dollar depreciates against foreign currencies, U.S. exports become more competitive, boosting demand and prices. The dollar, adjusted for relative inflation rates, is expected to depreciate 7% against the euro, 6% against the Chinese yuan, and 8% against the Brazilian real in 2008.[7] The exchange rate is an important determinant of agricultural trade. The depreciation of the U.S. dollar since 2002 has helped improve U.S. agricultural export performance. According to the U.S. Department of Agriculture's (USDA's) Economic Research Service (ERS), the dollar is forecast to be up 2% versus the yen, unchanged against the Canadian dollar, down 2% against the Mexican peso, and down 6% against the Argentine peso in 2008.

How Do Higher Commodity Prices Impact Consumers?

As commodity prices rise, food prices follow, but to a lesser extent. On average, about 20 cents of each dollar spent on food is the farm share — the retail cost less the value-added after the product leaves the farm and moves along the marketing chain to the retail outlet.[8] In less processed foods, the farm component of the final product is larger and changes in the farm price have a greater impact at the retail level. For instance, eggs, and fresh fruits and vegetables undergo minimal processing after they leave the farm — they are consumed in essentially their original form. The retail value of such products tends to have a large farm component and changes at the farm level have a greater impact on the consumer. On the other hand, in highly processed products, such as breakfast cereal, the corn, wheat, or rice used is completely transformed and the final product bears little resemblance to the original commodity. An 18-ounce box of corn flakes contained about 3.3 cents worth of corn in 2006.[9] Higher corn prices in 2007 increased the corn share to 4.9 cents. This is a small part of the retail value of a box of corn flakes. Most of the retail price represents packaging, processing, advertising, transportation, profit, and other costs.

Energy Costs

Energy costs affect all levels of the food production sector. Recent record crude oil prices in excess of $110 per barrel affect costs throughout the marketing chain.[10] Producers spend more for fertilizer (for which natural gas is a major input), crop drying, and transportation — raising production costs. At the processing, wholesale, and retail levels, the cost of transportation and operating packing houses, manufacturing plants, and retail stores has increased. Some of these costs are passed on to consumers in the form of higher prices. In addition, high petroleum prices increase the competitiveness of ethanol, further boosting demand for corn.

---

[7] USDA/ERS, "Outlook for U.S. Agricultural Trade," February 21, 2008.

[8] USDA/ERS, "Price Spreads from Farm to Consumer," by Howard Elitzak, at [http://www.ers.usda.gov/Data/FarmToConsumer/Data/marketingbiltable1.htm].

[9] USDA/ERS, *Amber Waves*, "Corn Prices Near Record High," by Ephriam Leibtag, February 2008.

[10] West Texas Intermediate (WTI), a crude oil price traded at Cushing, OK, reached $110 per barrel for the first time on March 13, 2008.

Food Price Changes Vary by Food Type

Meat, Poultry, Dairy, and Eggs. The CPI for all meats advanced 3.3% during 2007.[11] Beef increased 4.4%, pork 2%, broilers 5.2%, and eggs 29.2%, and dairy products advanced 7.4% in 2007. The farm share of these products is large compared with other foods, so changes at the farm level are passed, to a greater extent, to the consumer. In many cases, higher feed and energy costs were behind these increases. Strong export demand — spurred by the weak dollar — and reduced flocks played a role in the price hikes for poultry and eggs. The CPI for meats is forecast to increase by 1.5% to 2.5% in 2008. Compared with other food categories, these high-value items also account for a large share (11.1%) of the average consumer's food budget.

Fruits and Vegetables. Prices for fruits and vegetables gained 3.8% in 2007 and are forecast to increase 3% to 4% in 2008. Production shortfalls affected some varieties, especially bananas, the largest by volume. Supplies of oranges were strong, offsetting other declines. Energy costs were a large factor in higher fruit and vegetable price increases. Fruits and vegetables account for 8.4 cents of the consumer food dollar.

Cereals and Bakery Products. The CPI for these items advanced 4.4% in 2007 and is projected to rise 6.5% to 7.5% next year. Tight global wheat supplies and acreage reductions to promote ethanol production have caused a spike in wheat prices. However shifts in wheat prices have a relatively small impact on grocery store prices because the farm share of these products is small. Prices for these product are affected more by marketing factors such as transportation, labor, and energy costs than the cost of basic inputs.

Oilseeds and Related Products. Low stocks and strong export demand for soybeans are reflected in the CPI for these products, which gained 2.9% in 2007. While much of this category is supplied by soybeans, substitutes exist and will help moderate increases. In 2008, the CPI is set to rise 7% to 8% due to continued strong export demand from countries where changing diets require more vegetable oil.

Impact on Low-Income Households

Although U.S. consumers generally spend a smaller share of their income on food compared with many other countries, that share varies widely across income levels. Overall, U.S. households spend 12.6% of their income on food,[12] so changes in the price of food have to be large to affect their total budget. However, the picture is vastly different for low-income households. In 2006, households with incomes in the lowest reported income category spent 17.1% of their income on food. Households with incomes greater than $70,000 spent 11.3% of their income on food. When food prices rise, families with lower incomes feel the pinch more acutely since food expenditures make up a larger share of their total expenditures. Also, higher-income families can shift food

---

[11] Food CPI's for 2007 and 2008 are from the USDA/ERS *Food CPI, Prices, and Expenditures Briefing Room*, at [http://www.ers.usda.gov/briefing/cpifoodandexpenditures/].

[12] U.S. Department of Labor, Bureau of Labor Statistics, *Consumer Expenditure Survey*, "Table 46, Income Before Taxes," at [http://stats.bls.gov/cex/].

consumption to the home from restaurants, saving money without reducing consumption. A 4% to 5% increase in food expenditures has a significant impact on purchasing power for low-income families.

Federal Spending for Domestic Food Assistance Programs

Food price inflation increases spending on domestic assistance efforts. Increasing prices encourage those who are eligible but not participating to enroll. Increasing prices translate directly into benefit payments and per-meal subsidies for entitlement programs in which benefits are indexed to food-price inflation (e.g., food stamps, school meal programs). Increasing prices place pressure on appropriators to provide more funding to support caseloads for discretionary programs like the Special Supplemental Nutrition Program for Women, Infants, and Children (the WIC program).

Food Stamps. The Food Stamp program is the largest of the federally supported domestic food assistance programs. Its benefits are indexed annually for changes in the cost of USDA's least costly food plan, the "Thrifty Food Plan" (TFP). For a number of years and well into 2006, annual increases in the TFP typically ranged between 1.5% and 2.5%, with a few exceptions. However, starting in late 2006, food prices (as reflected in the cost of items in the TFP) began to increase at a faster rate. The last benefit increase, effective October 2007, was 4.6%. As a result, the average monthly benefit will be $6 per person higher in FY2008.

The impact of benefit increases on food stamp costs also depends on participation. For FY2008, the benefit increase noted above (combined with estimated growth in enrollment) yields a likely $2 billion cost attributable to adjustments for food price increases (out of total spending of $36.7 billion), about double the $1 billion that would have occurred based on pre-2007 price increases. Costs are expected to increase even more in FY2009.

Child Nutrition. Federal payments for lunches and breakfasts served to children in participating school meal programs are the second largest federal commitment to domestic food assistance, about $11 billion per year. These per-meal subsidies — now ranging as high as $2.83 a meal, including the value of USDA commodity donations — are indexed every July to food-price changes reflected in the "Food Away From Home" component of the CPI over the 12-month period ending each May.

Indexed maximum subsidy rates (those paid for the majority of school meals that are served free or at a reduced price to children from lower-income families) have increased by some 25 cents a meal between the 2005-2006 school year and the current 2007-2008 school year. The annual increase in subsidies has gone from 2.9% for the 2005-2006 school year to 3.3% for the 2007-2008 school year, increasing federal support by about $300 million above spending if earlier food price increases had prevailed. According to ERS, this trend is expected to continue into FY2009.

The WIC Program. Unlike food stamps and child nutrition programs, the WIC program is discretionary. Spending depends on annual appropriations, based largely on estimates of participation and the cost of the food packages that are purchased with WIC vouchers. The value of benefits is not indexed, per se. Rather, WIC vouchers are redeemable at whatever the participating retailer charges for the items covered by the vouchers, which differ according to the type of recipient (e.g., pregnant mother, infant, child). As a result, the cost of WIC vouchers reflect food price changes without the time lag built into other nutrition programs like food stamps. Just as important, WIC vouchers are highly specific as to the food items they cover and have a relatively heavy emphasis on certain types of food — dairy items and infant formula being a major component.

In recent years, the cost of WIC food vouchers has varied a great deal, largely because of changes in dairy-related food prices. The average per-participant monthly cost of vouchers has ranged from $34.80 in FY2002 to $39.15 in FY2007. However the annual percentage increase has been very small for some years (1% or less for FY2003, FY2005, and FY2006) and more substantial for other years (6.6% for FY2004 and 5.6% for FY2007). Most recently, monthly per-participant WIC food costs averaged $42.50 for the first three months of FY2008. Given this significant volatility, it is difficult to produce specific estimates of the effect of food price inflation on WIC program costs. However, the ERS forecasts of increases in egg and dairy product prices in the 2% to 4% range in 2008 indicate that relatively high WIC food costs are likely in the near term.

Foreign Food Aid

Higher commodity and food prices reduce our ability to provide food aid to other countries without additional appropriations. Food aid usually takes the form of basic food grains such as wheat, sorghum, and corn, and vegetable oil — commodities critical to developing-country diets. Since there is very little value added for these commodities, shifts in prices translate directly into higher prices for food-insecure countries or reduced food aid contributions per dollar spent. Also, higher energy costs have increased shipping costs for both food purchases and food aid. Unlike some domestic nutrition programs, foreign food aid is not adjusted to account for changing costs. After a long period of declining food costs, developing countries are facing increased food import bills — for some countries as high as 25% in 2007.[13]

The U.S. Agency for International Development (USAID) has indicated that rising food and fuel prices would result in a significant reduction in emergency food aid. According to press reports in March 2008, USAID expects a $200 million shortfall in funding to meet emergency food aid needs. For FY2008, Congress appropriated $1.2 billion for P.L. 480 food aid, the same as FY2007. For FY2009, the President's budget again requested $1.2 billion. In six out of ten years since 1999, supplemental funding for P.L. 480 Title II food aid has been appropriated.

Last year, the U.N. World Food Program (WFP) estimated it would need $2.9 billion to cover 2008 food aid needs. Recent commodity, energy, and food cost increases have boosted this estimate to $3.4 billion. According to the WFP, the current price increases force the world's poorest people to spend a larger proportion of their income on food.

[13] USDA/ERS, Rising Food Prices Intensify Food Insecurity in Developing Countries," *Amber Waves*, February 2008.

Order Code RS22908
Updated September 17, 2008

CRS Report for Congress

Livestock Feed Costs: Concerns and Options

Geoffrey S. Becker
Specialist in Agricultural Policy
Resources, Science, and Industry Division

Summary

Livestock producers in 2008 have seen sharply higher feed costs, fueled by competing use demands for corn and soybeans and by higher energy prices. Some analysts argue that current public policies, including financial incentives that divert corn from feed uses into ethanol production, exacerbated if not caused these higher costs. Other factors, which some believe to be at least as significant, include crop production declines due to weather, and higher global demand for commodities. Proposed options aimed at easing the impacts of higher feed costs include changes in ethanol incentives, use of conservation land for forage use, and direct aid to producers.

Economic Situation[1]

Production costs through the first half of 2008 climbed in every segment of animal agriculture. The main driver was feed, which may account for 60%-70% of total livestock production costs in any given year. Overall, total U.S. feed expenses were forecast to reach a record-high $48 billion in 2008, a jump of nearly $10 billion or 26% over 2007 — a year that was $6.7 billion higher than 2006. Corn accounts for about 90% of feed grains used for feed, and soybean meal is the principal oil crop product used as feed, according to the U.S. Department of Agriculture's (USDA's) Economic Research Service (ERS). Farm-level corn prices have jumped, from an average of $3.04 per bushel (bu.) in the 2006/2007 marketing year (September 1-August 31) to a record $4.20 per bu. in the 2007/2008 marketing year and a projected $5.00-$6.00 in 2008/2009 — a boon for farmers who grow the crops, but bane for animal producers who must buy them.

Soybean farm prices averaged $6.43 per bu. in 2006/2007 and climbed to an estimated record of $10.15 per bu. in the 2007/2008 marketing year. USDA projects that average farm prices for soybeans could reach $11.60 to $13.10 per bu. in 2008/2009.

---

[1] Unless noted, sources for this section include USDA, ERS, "Farm Income and Costs: 2008 Farm Sector Income Forecast," August 2008; USDA's *World Agricultural Supply and Demand Estimates*, September 12, 2008; testimony by USDA Chief Economist Joe Glauber before the Senate Committee on Energy and Natural Resources, June 12, 2008; and "Feed Grains and Livestock: Impacts on Meat Supplies and Prices," *Choices Magazine Online!*, 2nd quarter 2008.

Congressional Research Service ❧ The Library of Congress
Prepared for Members and Committees of Congress

CRS-2

That has pushed up average soybean meal prices from $205 per short ton in 2006/2007 to $335 for 2007/2008 and a projected $330 to $390 per ton in 2008/2009.

For many animal producers, returns from their own sales were not covering these higher production costs. Total feed costs for finishing a feeder pig in Iowa were about $42, or 30% of the total cost of production, in May 2006; they reached $70 and 46% of total production costs in May 2008. From November 2007 through April 2008, the average loss for each Iowa feeder pig finished for slaughter were nearly $22 per head.[2]

Cattle feeders (who feed young cattle up to slaughter weight) in the Southern Plains, where the major U.S. cattle feeding states include Texas, Nebraska and Kansas, have been in the red since June 2007, according to the Livestock Marketing Information Center (LMIC). Southern Plains feeders have been losing, on average, over $100 per head for much of 2008.[3] CattleFax, which provides market analysis for the cattle industry, estimates that the total cost to "produce cattle and beef from pasture to plate" increased from $726 per head in 2005 to $1,131 in 2008, or by 56%.[4]

Dairy producers have experienced cost pressures, although farm milk prices have remained relatively firm in 2008 due to strong, export-fueled demand. Milk industry analysts also noted that feed costs have moderated somewhat from earlier levels, and that cost impacts may have been muted because "[a] large share of dairy feed in 2008 seems to have either been grown on the dairy or contracted in advance."[5]

A short-term effect when grain prices spike can be more production of red meat and poultry, as herd sizes are reduced and/or as more animals and milk are sold to maintain cash flow to cover higher prices. This can depress farm (and wholesale) prices at least temporarily, further exacerbating the cost-price squeeze. If current market conditions persist, meat supplies will decline — and prices rise — through producer attrition and reduced capacity. Retail prices, too, will eventually adjust accordingly.[6]

"To date, rising costs have largely been absorbed by livestock and poultry producers, often with significant financial loss. However, higher costs of production will ultimately have to be reflected in higher prices for meat, milk, and eggs at retail counters in the United States and elsewhere. This adjustment process is complex, lengthy, painful, and not without unintended consequences," Lawrence and others wrote in *Choices* (see footnote 1). They added that "speed of adjustment will vary significantly as industries with shorter production cycles, such as poultry, are able to respond in a matter of months

---

[2] Cost of production data by John D. Lawrence and Shane Ellis, Iowa State University, accessed June 2008, at [http://www.econ.iastate.edu/faculty/lawrence/Lawrence_Website/estreturns.htm]. Lawrence has noted that these losses follow 10 years that were generally profitable for producers.

[3] LMIC website accessed on September 17, 2008, at [http://www.lmic.info/].

[4] "Something's Got to Give," *CattleFax Update*, June 20, 2008. Also see David P. Anderson et al., *The Effects of Ethanol on Texas Food and Feed*, Agricultural and Food Policy Center, Texas A&M University, April 10, 2008.

[5] National Milk Producers Federation (NMPF), July/August 2008 Dairy Market Report.

[6] Testimony of Joe L. Outlaw, Agricultural and Food Policy Center, Texas A&M University, before the Senate Committee on Energy and Natural Resources, June 12, 2008.

whereas adjustments in industries with longer production cycles, such as beef, can take a period of several years."

Reasons for Higher Feed Costs[7]

Strong Economic Growth in Developing Countries. Rising commodity prices, including for feed, are rooted in a confluence of factors, some with short-term impacts, but others longer-term. An example of the latter is steadily increasing world population, boosted by robust growth in purchasing power, especially in developing countries such as China and India. Analysts believe this trend could mean a sustained increase in global demand for more and different kinds of food — including, as household incomes improve, higher-value foods like meat and dairy products. This dietary shift in turn implies increased demand for animal feeds.

Weather-Related Crop Shortfalls. Global grain production declined in 2005 and 2006, cutting into existing stocks and reducing exportable supplies. Crops here as well as in Australia, Canada, the European Union, Eastern Europe, and some countries of the former Soviet Union were reduced mainly due to poor weather conditions. On the other hand, weather may have had less of an impact on corn than on food crops like wheat and rice: world corn output was expected to increase from 712 million metric tons (MMT) in 2006/2007 to 790 MMT in each of 2007/2008 and 2008/2009. In the United States, the world's leading producer, 2008 production was estimated by USDA at 12.1 billion bushels, approximately 1 billion bushels below the previous year. Among the reasons cited were spring rains and flooding followed by a dry August in the Corn Belt.[8]

Surging U.S. Exports. Since January 2002, the U.S. dollar has declined in value relative to the currency of U.S. export competitors (e.g., Canada, Australia, or the EU) and grain importing nations (e.g., Japan, Taiwan, etc.), effectively making U.S. grain exports cheaper and, therefore, more competitive, despite the rise in per-bushel prices. In terms of year-to-year export volumes, U.S. corn exports are projected up nearly 18% to a record of almost 2.5 billion bushels in the 2007/2008 marketing year.

Other Contributing Factors. Former USDA Chief Economist Keith Collins cited a number of other factors that have contributed to feed (and other food) price increases. These include higher demand for feed caused by increasing numbers of fed animals in the United States; a two-year rise in corn production costs from $186 per acre in 2005 to nearly $230 in 2007 due mainly to higher energy-related (fertilizer and fuel) costs; and actions by a number of foreign governments to insulate their own markets from high commodity prices, such as limiting exports of their own agricultural commodities and subsidizing domestic production and/or consumption.[9]

---

[7] Adapted in part from CRS Report RL34474, *High Agricultural Commodity Prices: What Are the Issues?*, where additional details, and (unless noted) sources may be found.

[8] USDA, ERS, *Feed Outlook*, September 16, 2008.

[9] Collins, Keith, *The Role of Biofuels and Other Factors in Increasing Farm and Food Prices*, June 19, 2008, study conducted in support of a review by Kraft Foods Global, Inc.

CRS-4

Government Biofuels Policy. Perhaps the most widely-debated factor has been government biofuels policy. Current U.S. biofuel production is almost entirely corn-based ethanol — nearly 6.5 billion gallons of corn-ethanol were produced in 2007, compared with an estimated 450 million gallons of primarily soybean-based biodiesel. The Energy Independence and Security Act of 2007 (EISA; P.L. 110-140) extended and substantially expanded the existing Renewable Fuel Standard (RFS), a usage requirement mandating that an increasing volume of biofuels be blended with conventional fuels. The RFS mandates the use of at least 9 billion gallons of biofuel in U.S. fuel supplies in 2009, more than doubling to 20.5 billion gallons by 2015 and to 36 billion gallons by 2022. The U.S. biofuels sector is also supported by a tax credit of 51 cents for every gallon of corn-based ethanol blended in the U.S. fuel supply (to drop to 45 cents in 2009), and an import tariff of 54 cents per gallon of imported ethanol. In addition, several federally-subsidized grant and loan programs assist biofuels research and infrastructure development.

Of the 160 U.S. ethanol plants — with total capacity now in excess of 9 billion gallons per year — only about a dozen use feedstocks other than corn (or milo or barley, also feed grains).[10] USDA estimates that in crop year 2007/2008, about 23% of the U.S. corn crop will have been used to produce ethanol; this share is projected to grow to more than 33% in 2008/2009. Although high corn prices have slowed plans for expanded capacity, plants already on line are expected to operate so long as they can cover their variable costs. The runup in corn demand, which is expected to be long-lasting, has directly sparked substantially higher corn prices to bid available supplies away from other uses like livestock feed, according to a number of agricultural economists.

Corn ethanol byproducts — now mainly distiller's grains from dry-mill ethanol plants but also corn gluten feed and corn gluten meal — can be fed by livestock producers to help offset higher feed corn prices. According to the National Corn Growers Association, the equivalent of an additional 1 billion bushels of feed will be available in the coming year through ethanol byproducts.[11] Because of the cost and difficulty of drying and shipping, distiller's grains are best utilized by feeders closest to ethanol plants. Research indicates that distiller's grains should constitute no more than 35-40% of feedlot cattle rations, 10%-20% of dairy cow rations, and 10% for hogs, broiler chickens, and turkeys, according to Anderson.

Reviewing several different studies and economic models, Collins concluded that implied changes in the price of corn due to its use in ethanol might range from 25% to 60%. A 2008 study by the Food and Agricultural Policy Research Institute (FAPRI) attempts to measure the pure and joint price effects of the U.S. biofuels RFS and the tax credits, suggesting that joint implementation of both the RFS and the tax credit supports corn prices by about 20%. A substantial portion of corn price effects are likely transmitted to the soybean market via competition for land, primarily in the Corn Belt where soybeans and corn are both widely grown. A study by the Center for Agricultural Research and Development (CARD) found that, jointly, the RFS and tax credit supported corn prices by 16%. FAPRI and CARD found the impacts of the tax incentives to be highly dependent on petroleum prices; higher petroleum prices substitute for government

[10] Source: Renewable Fuels Association, June 19, 2008, at [http://www.ethanolrfa.org/].

[11] "Figures Reveal How Livestock Benefit From Ethanol," *The PigSite*, accessed August 18, 2008, at [http://www.thepigsite.com/].

incentives and diminish their relative impact on corn prices. Neither study included the effects of grants and subsidized loans available for biofuels research and infrastructure.

Outlook and Options

The U.S. response to recent developments in grain markets is of keen interest to U.S. meat, dairy and poultry producers, who have long been wary of any government policies that can raise feed prices. Congressional committees in 2008 have held a number of hearings on the causes and impacts of rising commodity costs. The House Agriculture Committee convened six hearings between mid-May and mid-September 2008 on commodity futures trading and related matters, including proposals to increase oversight by the Commodity Futures Trading Commission of futures markets, where, some critics argue, excessive speculation added significantly to the 2008 price spikes in oil and other commodities; others counter that futures traders simply react to market conditions. Economists seem to agree that U.S. government initiatives are unlikely to have much impact on such price drivers as foreign population and income growth and concurrent demand for commodities, weather-related production shortfalls, and inflation generally, at least in the short term. However, a number of options have been offered to address other perceived reasons for high feed prices.

Reduce Domestic Ethanol Incentives. Several bills propose to reduce ethanol production incentives by repealing, lowering, or freezing the RFS, the biofuels tax credit, and/or the ethanol import tariff, such as H.R. 5911, H.R. 5986, H.R. 6137, H.R. 6183, H.R. 6324, S. 3031, and S. 3080. Outside of Congress, attention focused on an April 25, 2008 letter to the U.S. Environmental Protection Agency (EPA) from Texas Governor Rick Perry, requesting a waiver of 50% of the 2008 RFS for grain-derived ethanol, now at 9 billion gallons. The letter cited EPA's authority, under Sec. 211(o) of the Clean Air Act, to approve such a waiver if the RFS requirement would severely harm the economy of a state, region, or the country. The EPA, in August, denied the request, stating that its analysis "found no compelling evidence that the RFS mandate is causing severe economic harm during the time period specified by Texas."[12] Opponents have argued that an EPA waiver, or reductions in the financial incentives proposed in the various bills, would not have much of an immediate impact on corn prices (see FAPRI and CARD studies, above). At any rate, other factors such as global grain shortfalls, strong U.S. exports, and higher oil prices played a much larger role in high feed costs, some have continued to assert.

Use Conservation Land. Some have argued that as much as one-third of the nearly 35 million acres (April 2008) in USDA's conservation reserve program (CRP) could be returned to agricultural production without environmental harm. The 2008 farm bill (P.L. 110-246) caps national enrollment in the CRP, which is intended to keep environmentally sensitive lands out of production for 10-15 years, at 32 million acres. What about incentives like easing producer penalties who want to end their CRP contracts early, discouraging re-enrollment or renewal of existing contracts, permitting temporary use of more CRP acres for grazing, cutting hay, and/or raising other feedstuffs? Opponents of opening conservation land assert that the loss of environmental benefits will

---

[12] "EPA Keeps Biofuels Levels in Place after Considering Texas' Request," August 7, 2008 press release, which can be found, along with a fact sheet on the EPA findings and the August 13, 2008 *Federal Register* notice, at [http://www.epa.gov/otaq/renewablefuels/].

far outweigh any gains to animal agriculture, in part because such lands have always been considered marginally productive in the first place. Also, much of this marginal CRP acreage previously was more suitable for (and planted in) wheat, not corn or soybeans.

USDA on May 27, 2008 announced the potential eligibility of more than 24 million acres enrolled in CRP for hay and forage in 2008 to help livestock producers hit by higher feed prices. USDA set a number of limitations on use of CRP acreage to minimize impacts on nesting birds and on the most environmentally sensitive lands. The National Wildlife Federation sued the Department, arguing that it did not prepare an environmental impact statement on the new policy. In July, a U.S. District Court issued a temporary restraining order (TRO) and then a permanent injunction suspending USDA's CRP opening, except for applicants that had been approved by or had applied prior to the TRO, or who had invested at least $4,500 toward haying or grazing equipment and preparation prior to the TRO. Where the application was approved prior to the TRO, haying and grazing must be completed by November 10. For subsequent approvals, haying and grazing must end by September 30 and October 15, respectively.[13]

Financial Assistance for Livestock Producers. Over the past decade alone, many of the nearly annual supplemental appropriations measures containing emergency agricultural aid have included money for animal agriculture — amounting to at least $3.2 billion in total, often to cover the cost of high feed prices or lost pasture use.[14] Separately, Section 32 of the Act of August 24, 1935 is a permanent appropriation available to USDA for a variety of purposes, and it has been tapped not only to pay for disaster losses, but also to address economic and market problems.[15] Thus, some argue, there is policy precedent for providing direct payments to livestock and poultry producers to help them cope with sharply rising feed costs. On the other hand, both the 2008 farm bill (P.L. 110-246) and the FY2008 USDA appropriation (part of P.L. 110-161), contain language seeking to constrain Section 32 spending, or at least the Secretary's flexibility to use it.

Other Options. Some have urged U.S. officials to push harder for the elimination of various foreign barriers that obstruct international trade and distort market price signals to producers, including tariffs, unjustified technical and phytosanitary standards, export controls, and various domestic agricultural subsidies. Further, it is argued, investments should be increased, particularly in developing countries, in agricultural research and education to increase productivity, in transportation and marketing infrastructure, and in other types of alternative energy production here and globally. These efforts could all help to ease the upward pressure on feed grain and other commodity prices. Others counter that regardless of their merits, these types of solutions would offer little if any short-term relief to the financial problems currently facing livestock producers.

---

[13] See CRS Report RS21613, *Conservation Reserve Program: Status and Current Issues*, by Tadlock Cowan. The order does not affect a July 2008 USDA announcement to permit grazing only on CRP land in designated flood-stricken areas.

[14] The 2008 farm bill (P.L. 110-246) creates a new Agricultural Disaster Relief "Trust Fund" for crop years 2008-2011. Three of the five new programs under which payments could be made are relate to livestock. See CRS Report RL34207, *Crop Insurance and Disaster Assistance in the 2008 Farm Bill*, and CRS Report RL33958, *Animal Agriculture: 2008 Farm Bill Issues.*

[15] See CRS Report RL34081, *Farm and Food Support Under USDA's Section 32 Program.*

Order Code RL34191

CRS Report for Congress

Ethanol and Other Biofuels: Potential for U.S.-Brazil Energy Cooperation

September 27, 2007

Clare Ribando Seelke
Analyst in Latin American Affairs
Foreign Affairs, Defense, and Trade Division

Brent D. Yacobucci
Specialist in Environmental and Energy Policy
Resources, Science, and Industry Division

Prepared for Members and Committees of Congress

Ethanol and Other Biofuels: Potential for U.S.-Brazil Energy Cooperation

Summary

In the past several years, high oil and gas prices, instability in many oil-producing countries, and concerns about global climate change have heightened interest in ethanol and other biofuels as alternatives to petroleum products. Reducing oil dependency is a goal shared by the United States and many countries in Latin America and the Caribbean, a region composed primarily of energy-importing countries. In the region, Brazil stands out as an example of a country that has become a net exporter of energy, partially by increasing its production and use of sugar-based ethanol.

On March 9, 2007, the United States and Brazil, which together produce almost 70% of the world's ethanol, signed a Memorandum of Understanding (MOU) to promote greater cooperation on ethanol and other biofuels in the Western Hemisphere. The countries agreed to (1) advance research and development bilaterally, (2) help build domestic biofuels industries in third countries, and (3) work multilaterally to advance the global development of biofuels.

Many analysts maintain that the United States would benefit from having more energy producers in the region, while Brazil stands to further its goal of developing ethanol into a globally traded commodity. In addition to these economic benefits, some analysts think that an ethanol partnership with Brazil could help improve the U.S. image in Latin America and lessen the influence of oil-rich Venezuela under Hugo Chávez. However, obstacles to increased U.S.-Brazil cooperation on biofuels exist, including current U.S. tariffs on most Brazilian ethanol imports.

While some Members of Congress support greater hemispheric cooperation on biofuels development, others are wary of any cooperative efforts that might negatively affect U.S. ethanol producers. The Energy Diplomacy and Security Act of 2007, S. 193 (Lugar), approved by the Senate Foreign Relations Committee on March 28, 2007, would increase hemispheric cooperation on energy. S. 1007 (Lugar), the United States-Brazil Energy Cooperation Pact of 2007, calls for the same hemispheric cooperation groups as S. 193, and directs the Secretary of State to work with Brazil and other Western Hemisphere countries to develop biofuels partnerships. H.Res. 651 (Engel) recognizes and supports the importance of the U.S.-Brazil MOU on biofuels. In the 109th Congress, legislation was introduced that would have eliminated current tariffs on foreign ethanol, but in December 2006, Congress voted to extend the ethanol tariffs through December 31, 2008 (P.L. 109-432). In the 110th Congress, S. 1106 (Thune) would extend those tariffs through 2011, and H.R. 196 (Pomeroy) would make the tariffs permanent.

This report discusses the opportunities and barriers related to increasing U.S. cooperation with other countries in the hemisphere on biofuels development, focusing on the U.S.-Brazil agreement. This report may be updated. For more information, see CRS Report RL33290, *Fuel Ethanol: Background and Public Policy Issues*, and CRS Report RL33693, *Latin America: Energy Supply, Political Developments, and U.S. Policy Approaches.*

Contents

List of Figures

Ethanol and Other Biofuels: Potential for U.S.-Brazil Energy Cooperation

Introduction

Recent high oil and gas prices and concerns about global climate change have heightened interest in ethanol and other biofuels as alternatives to petroleum products. The Western Hemisphere produces more than 80% of the world's biofuels, led by Brazil (producing sugar-based ethanol) and the United States (producing corn-based ethanol).[1] Some have argued that increasing biofuels production in Latin America could bolster energy security, reduce greenhouse gas emissions, and promote rural development in the region. Others are concerned about the huge investment outlays and government subsidies needed to build up nascent biofuels industries. Skeptics also worry about the potential negative effects that increased biofuels production may have on the environment, labor conditions, and food prices in the region.

The United States and Brazil, the world's largest ethanol producers, have recently agreed to work together to promote the use and production of ethanol and other biofuels throughout the Western hemisphere. Increasing U.S.-Brazil energy cooperation was a top agenda issue when President Bush visited Brazil and when President Lula visited Camp David earlier this year. On March 9, 2007, the two countries signed an agreement to (1) advance research and development bilaterally, (2) help build domestic biofuels industries in third countries, and (3) work multilaterally to advance the global development of biofuels.[2]

Many Bush Administration officials and Members of Congress note that the new biofuels partnership with Brazil may help improve the U.S. image in Latin America and diminish the influence of President Chávez in the region. In the past few months, the U.S.-Brazil biofuels agreement has already had a significant political effect in Latin America. "Ethanol diplomacy" appears to be helping Brazil reassert regional leadership relative to oil-rich Venezuela under Hugo Chávez.[3]

---

[1] Inter-American Development Bank (IDB), *A Blueprint for Green Energy in the Americas*, April 2007, p. 7, available at [http://www.iadb.org/biofuels].

[2] "Memorandum of Understanding Between the United States and Brazil to Advance Cooperation on Biofuels," U.S. Department of State, Office of the Spokesman, March 9, 2007, available at [http://www.state.gov/r/pa/prs/ps/2007/mar/81607.htm].

[3] "Energy Summit was Stage for Oblique Regional Leadership Contest," *Latin News Weekly Report*, April 19, 2007; "Chávez, Lula Promote Competing Visions," *Miami Herald*, August 10, 2007.

Increasing biofuels cooperation with Brazil and other countries in Latin America may prompt challenges to existing U.S. trade, energy, and agriculture policies. For example, U.S. tariffs on foreign ethanol imports may prove to be an obstacle to U.S.-Brazil energy cooperation. In the 109th Congress, legislation was introduced that would have eliminated current tariffs on foreign ethanol. While some Members support ending the ethanol tariffs, other Members of Congress support further extensions of the ethanol import tariffs. Some have also proposed using tariff duties collected on foreign ethanol imports to fund advanced ethanol research and production within the United States.

This report examines the opportunities and barriers related to increasing U.S. cooperation with other countries in the hemisphere on biofuels development, focusing on the U.S.-Brazil agreement. It provides background information on Western hemisphere energy challenges, the ethanol industries in Brazil and the United States, and the biofuels potential in the region. It then raises a number of policy issues that Congress may choose to consider related to bolstering the development of ethanol and other biofuels in Latin America.

Biofuels: A Definition

The term *biofuels* generally refers to motor fuels produced from agricultural commodities or other biological materials such as agricultural and municipal wastes. The most widely used biofuel is ethanol (both in the United States and worldwide), an alcohol usually produced from the fermentation and distillation of sugar- or starch-based crops such as sugarcane or corn.[4] Ethanol can also be produced from cellulose-based materials (such as perennial grasses and trees), although the technology to produce cellulosic ethanol is in its infancy, and no commercial-scale cellulosic ethanol plants have been constructed.[5] Other biofuels include biodiesel and other renewable diesel fuel substitutes produced from vegetable oils or animal fats (such as soybean or palm oil), and butanol produced from various biological feedstocks.[6]

Biofuels have several potential benefits relative to petroleum-based fuels. First, the use of biofuels can reduce emissions of some pollutants relative to gasoline or diesel fuel. Second, most biofuels lead to lower emissions of greenhouse gases than petroleum fuels — some can lead to substantial greenhouse gas reductions. Third, biofuels can be produced domestically to displace some petroleum that would

---

[4] Ethanol can also be produced from petrochemical feedstocks, but such fuel is generally considered a fossil fuel, as opposed to a biofuel. For more information on ethanol, see CRS Report RL33290, *Fuel Ethanol: Background and Public Policy Issues*, by Brent D. Yacobucci.

[5] However, interest is growing. In February 2007, the U.S. Department of Energy announced $385 million in grants to construct the first six commercial-scale cellulosic ethanol plants in the United States. It is expected that these plants will be operational between 2009 and 2011.

[6] For more information on biodiesel and other renewable fuels, see CRS Report RL32712, *Agriculture-Based Renewable Energy Production*, by Brent D. Yacobucci.

otherwise be imported. Finally, some biofuels can reduce overall fossil energy consumption, given that much of the energy needed to grow the feedstock plant material is supplied by the sun.

There are also many potential drawbacks to biofuels. First, in nearly all cases, biofuels are more expensive to produce than petroleum fuels. Second, infrastructure limitations can lead to even higher costs for biofuels than for conventional fuels. Third, biofuels have their own potential environmental drawbacks, including increased emissions of some pollutants and the potential for increased greenhouse gas emissions (in some cases, depending on the particular biofuel) when the entire production process is taken into account.

While both the United States and Brazil — as well as many other countries — are studying and investing in many different biofuels, this report focuses on ethanol supply and use for several reasons: (1) ethanol production and consumption far exceeds that of other biofuels;[7] (2) ethanol can be (and is) used as a direct blending component with gasoline, and gasoline engines are dominant in passenger automobiles and light trucks;[8] and (3) current mandates in both Brazil and the United States favor ethanol use, either directly or indirectly.

Biofuels: A Potential Solution to Latin America's Oil Dependency?[9]

While oil and gas producers such as Venezuela, Mexico, Argentina, Bolivia, Colombia, Ecuador, and Trinidad, and Tobago are net energy exporters, most Latin American and Caribbean nations are net energy importers. Countries that rely on oil as their primary energy source are particularly vulnerable to increases in global oil prices. In many Caribbean island nations, oil accounts for more than 90% of total energy consumed. In Central America, oil dependency ranges from an estimated 51% in Costa Rica to 73%-78% in El Salvador, Honduras, and Panama, and to 84-85% in Guatemala and Nicaragua.[10]

Many Latin American countries experienced dramatic increases in their energy bills after oil price hikes began in 2005, straining government budgets in Central

---

[7] In 2006, the United States produced roughly 5.4 billion gallons of ethanol, as opposed to roughly 200 million gallons of biodiesel. The production of all other biofuels was even less.

[8] Biodiesel and other renewable diesel fuels must be used in diesel engines, which play a larger role in transporting goods. In those countries where diesel engines play a larger role in the light-duty vehicle market, renewable diesel fuels could play a larger role.

[9] For background on energy challenges facing countries in the Western Hemisphere, see CRS Report RL33693, *Latin America: Energy Supply, Political Developments, and U.S. Policy Approaches*, by Mark P. Sullivan, Clare M. Ribando, and Nelson Olhero.

[10] U.S. Department of Energy, March 2, 2006, "Latin America and High Oil Prices: Economic and Political Impact," *Latin American Special Report*, October 2006; House International Relations Committee, Hearing on "Western Hemisphere Energy Security," Testimony by Karen A. Harbert, Assistant Secretary for Policy and International Affairs.

America and the Caribbean. To fill a clear need and attempt to extend his influence in the region, Venezuelan President Hugo Chávez has offered oil on preferential terms in a program known as PetroCaribe launched in 2005.[11] His government has also reached preferential energy agreements with South American countries such as Argentina, Bolivia, and Ecuador. In December 2005, Mexico, perhaps in an attempt to act as a countervailing force to Venezuela in the region, initiated the Meso-American Energy Integration Program (PIEM) with Central America, the Dominican Republic, and Colombia.[12] As part of the plan, Mexico will supply the bulk of the crude oil to be processed by a new oil refinery in either Guatemala or Panama, which will be used primarily to satisfy Central America's energy needs.

Given declining oil production levels in both Venezuela and Mexico and the unstable political environment in Bolivia, a major natural gas producer in the region, some analysts have pointed to biofuels as a potential solution to Latin America's petroleum dependency. Producing biofuels would, some argue, allow oil-importing countries in the region to reduce energy costs and the need for imported oil. The Western Hemisphere produces more than 80% of the world's biofuels, led by Brazil (producing ethanol from sugar) and the United States (producing ethanol from corn). Biofuels proponents argue that the climate, surplus of arable land, and excess production of sugarcane and other potential biofuels crops make parts of Latin America ideally suited for an expanded biofuels industry.[13] The potential is greatest in tropical countries that have high crop yields and low costs for land and labor, characteristics that are present in several countries in Central America and the Caribbean.[14]

An April 2007 study by the Inter-American Development Bank (IDB), *A Blueprint for Green Energy in the Americas*, reports that some Latin American and Caribbean countries have shown great interest and promise in the development of biofuels. Beyond Brazil, which has been the leader in ethanol development and production, the study also highlights several other countries with potential for biofuels development. The study also suggests ways the IDB could support the development of biofuels production in the region, including support for a biofuels development fund, the development of regulatory frameworks, and research and development.[15] Regional highlights from the IDB study include the following:

- **Central America:** The bulk of ethanol production in Central America involves reprocessing hydrous ethanol from Brazil or the European Union (EU) for export to the United States. The IDB

[11] For more information, see CRS Report RL34288, *Venezuela: Political Conditions and U.S. Policy*, by Mark P. Sullivan and Nelson Olhero.

[12] Mexico Challenges Venezuela in Petro-Diplomacy Game," *EFE News Service*, June 2, 2006.

[13] IDB report, April 2007.

[14] Worldwatch Institute, "Biofuels for Transportation," Washington, D.C., June 7, 2006; Masami Kojima and Todd Johnson, "Potential for Biofuels for Transport in Developing Countries," World Bank, February 2006.

[15] IDB report, April 2007.

study asserts that while the sugarcane harvesting season in Central America is shorter than in Brazil, Costa Rica, El Salvador, and Guatemala have efficient sugar industries and could produce significant sugar-based ethanol. Costa Rica and Guatemala, which house 44% of Cental America's ethanol processing factories, have the most well-developed policies in place for biofuels development. Under the Dominican Republic-Central America-United States Free Trade Agreement (CAFTA-DR), signatory countries continue to share duty-free access for some ethanol exports to the United States under conditions established by the Caribbean Basin Initiative (CBI)[16], but exports from Costa Rica and El Salvador enjoy specific allocations. In the future, CAFTA-DR could spur indigenous ethanol production in Central America.

- **Caribbean:** Within the Caribbean, Jamaica has exported the largest amount of ethanol to the United States under CBI, most of it reprocessed hydrous ethanol from Brazil. Trinidad and Tobago has an ethanol dehydration plant, but the largest ethanol plants in the Caribbean are located in Jamaica and the Dominican Republic. Beyond Jamaica and the Dominican Republic, Grenada has been identified as having future production potential for sugar-based ethanol. The United Nations is working with Brazil to provide technology transfer and technical assistance to examine the potential of a biodiesel industry in Haiti from jatropha, a drought-resistant shrub that can grow almost anywhere.[17]

- **South America:** Most biofuels production in South America currently goes to satisfy domestic consumption, with Colombia and Argentina possessing the largest government programs in support of biofuels development. Biofuels exports from non-CBI countries are constrained by tariff barriers in the United States and the EU. Should Colombia and/or Peru conclude free trade agreements with the United States, their biofuels export potentials could expand. Argentina, Colombia, and Peru have potential to further develop ethanol from sugarcane; Colombia and Ecuador to produce more biodiesel from palm oil; and Chile has potential for second-generation ethanol production from woodchips.

---

[16] Most ethanol imports to the United States are subject to a 2.5% ad valorem tariff and (more significantly) a 54-cent-per-gallon added duty. However, to promote development and stability in the Caribbean region and Central America, the Caribbean Basin Initiative (CBI) allows imports of most products from those countries, including ethanol, duty-free. While many of these products are produced in CBI countries, ethanol entering the United States under the CBI is generally produced elsewhere and reprocessed in CBI countries for export to the United States. Up to 7% of the U.S. ethanol market may be supplied duty-free by ethanol imports from CBI countries. For more information, see CRS Report RS21930, *Ethanol Imports and the Caribbean Basin Initiative*, by Brent D. Yacobucci.

[17] Johanna Mendelson Forman, "Jatropha and Peace-building in Haiti," Statement for the Record for the House Foreign Affairs Committee, Subcommittee on the Western Hemisphere, March 19, 2007.

Brazil and the United States: Hemispheric Leaders in Ethanol Production

Brazil and the United States are by far the world's largest ethanol producers. In 2006, the United States was the largest producer of ethanol, with almost 4.9 billion gallons, followed closely by Brazil, with 4.5 billion gallons; together, the two countries produced 69% of ethanol in the world. Prior to discussing the U.S.-Brazil Memorandum of Understanding on biofuels, this section provides background information on ethanol production and usage in the two countries.

Ethanol Production Process

U.S. ethanol is generally produced and consumed in the Midwest, close to where the corn feedstock is produced.[18] In Brazil, São Paulo state is the key area for sugarcane and ethanol production, where integrated sugar plantations and mills produce refined sugar and fuel ethanol. Regardless of the feedstock, the main steps to ethanol production from sugar- or starch-based crops are as follows:

- the feedstock (e.g., corn or sugar) is processed to separate fermentable sugars;
- yeast is added to ferment the sugars;
- the resulting alcohol is distilled, resulting in "wet" or hydrous ethanol; and
- for use in gasoline, the distilled alcohol is dehydrated to remove any remaining water, resulting in "dry" or anhydrous ethanol.

To produce ethanol from cellulosic materials, considerably more physical and chemical/enzymatic processing is necessary to separate fermentable sugars that can then be converted to ethanol. This additional processing adds significant costs, making cellulosic ethanol currently uncompetitive with corn- or sugar-based ethanol.

U.S. Ethanol Industry and Market

The United States is currently the largest producer and consumer of fuel ethanol in the world. In 2006, the United States consumed roughly 5.4 billion gallons of fuel ethanol. Most of that ethanol (4.9 billion gallons) was produced in the Midwest from corn. A smaller amount was produced domestically from other feedstocks (e.g., sorghum) or imported directly or indirectly from Brazil. Over 99% of U.S. ethanol is blended into gasoline. Ethanol is blended into gasoline at up to the 10% level (E10), although much is blended at lower levels (5% to 7%). Roughly one-third of U.S. gasoline contains some ethanol. A small amount of ethanol is blended in purer

---

[18] When ethanol is shipped long distances, fuel costs are higher, as the fuel cannot be transported by pipeline and must be shipped by rail, truck, or barge. This added transport cost can be as much as 5 to 10 cents per gallon.

form as E85 (a blend of 85% ethanol and 15% gasoline), which can be used in flexible fuel vehicles (FFVs) specifically designed for its use.[19]

Various federal and state incentives promote the production and use of ethanol in the United States.[20] Since the 1970s, ethanol blended into gasoline (to make "gasohol") has been eligible for tax incentives of various forms. Currently, each gallon of pure ethanol blended into gasoline earns the blender a tax credit of 51 cents per gallon. Additional tax incentives exist for small producers. Further, as part of the Energy Policy Act of 2005 (P.L. 109-58), Congress established a Renewable Fuel Standard (RFS). Each year, the RFS requires a certain amount of renewable fuel be blended into gasoline. For 2007, the mandate is 4.7 billion gallons. The vast majority of this mandate will be met using ethanol. The mandate increases annually and will reach 7.5 billion gallons in 2012.

U.S. ethanol production capacity has grown rapidly in recent years, and is expected to grow even faster in the next few years. This rapid growth has generated upward pressure on corn demand and corn prices (as well as production of co-products such as animal feed), while lowering wholesale ethanol prices.

Brazil's Ethanol Industry and Market

Until 2004, Brazil was the largest producer of ethanol in the world. Since then, the United States has moved ahead of Brazil in annual production. Brazilian ethanol is produced almost exclusively from sugar cane. Brazilian ethanol plants tend to be integrated with sugar plantations and sugar mills. Depending on market forces, these plants have the capacity to shift some production from sugar to ethanol, or vice-versa.

Brazilian government support for ethanol began in 1975 when a presidential decree established the Brazilian National Alcohol Program ("Proalcool"). Originally established mainly as a way to support the Brazilian sugar industry from a collapse in sugar prices, Proalcool set a production target of 3 billion liters (some 0.8 billion gallons) in 1980.[21] The second phase of the program, established in 1979 in response to the OPEC oil embargo, made Proalcool explicitly into an energy policy and further expanded the production goal to 10.7 billion liters (2.8 billion gallons) by 1985.[22] In 2006, Brazil produced roughly 16.5 billion liters (4.4 billion gallons).

---

[19] It should be noted that while FFVs in the United States are designed to use any mixture of ethanol and gasoline up to E85, these vehicles are generally optimized to run on conventional gasoline and thus may not achieve some of the potential efficiency benefits possible with higher-level ethanol blends.

[20] For more information on federal biofuels incentives, see CRS Report RL33572, *Biofuels Incentives: A Summary of Federal Programs*, by Brent D. Yacobucci.

[21] F. Joseph Demetrius, *Brazil's National Alcohol Program: Technology and Development in an Authoritarian Regime*, New York, 1990, p. 11.

[22] Demetrius, Op. cit. p. 44.

Figure 1. Annual Ethanol Production in Brazil and the United States

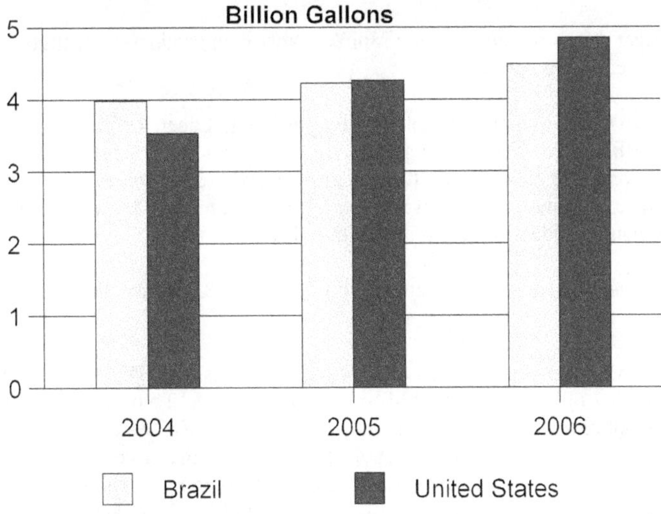

Source: Renewable Fuels Association, *Industry Statistics.* Accessed September 14, 2007, at [http://www.ethanolrfa.org/industry/statistics/].

To promote the goals of the Program, the Brazilian government has employed various policies through the years. These include requiring Petrobras, the state-owned oil company, to purchase a set amount of ethanol; tying the pump price of a liter of ethanol to a percentage of the price of gasoline (originally 59%, later increased to 80%); and requiring Brazilian automakers to produce dedicated ethanol vehicles that could run only on 100% ethanol.[23]

Currently, the Brazilian ethanol industry is thriving, and many of the requirements and policies from Proalcool have been eliminated. However, one key policy remains: all gasoline sold in Brazil must contain at least 20%-25% ethanol. Because of this mandate, as well as a large number of flexible fuel vehicles (FFVs) that can run on any blend of ethanol and gasoline, ethanol represents 40% or more of Brazilian gasoline demand.

---

[23] Demetrius, op. cit.; Jose R. Moreira and Jose Goldemberg, "The Alcohol Program," *Energy Policy*, April 1999, vol. 27, no. 4, pp. 229-245; Jose Goldemberg, "Brazilian Alcohol Program: An Overview," *Energy for Sustainable Development*, May 1994, vol. 1, no. 1, pp. 17-22.

Brazilian Ethanol: A Timeline

1973-1974 — Arab oil embargo.

1975 — Sugar prices plummet after a rapid expansion in production brought on by earlier historically high prices.

November 1975 — Proalcool established by presidential decree. The program set a target of 3 billion liters (0.8 billion gallons) of ethanol by 1980, largely blended into gasohol. To promote the industry, the government required Petrobras, the state-owned oil company, to purchase ethanol at a guaranteed price and blend it into gasoline. Brazilian ethanol production less than 1 billion liters per year.

1979 — Second oil shock; annual ethanol production at 3.7 billion liters (980 billion gallons).

July 1979 — Phase 2 of the Proalcool program is established. The production goal was expanded to 10.7 billion liters (2.8 billion gallons) in 1985. To promote this goal, several new policies were established: (1) the installation of100% ethanol pumps (as opposed to gasohol) was mandated at fueling stations; (2) the price of a liter of ethanol was pegged at 59% of the per-liter price of gasoline; (3) the government provided low-interest loans to agribusinesses to produce ethanol; and (4) the government and the Brazilian auto industry reached an agreement whereby the auto industry would produce a majority of new cars and light trucks as dedicated ethanol vehicles.

1980 — Neat (pure) ethanol sales permitted, sales of pure alcohol vehicles begins. Annual ethanol production at 5.5 billion liters (1.5 billion gallons).

1985-1995 — Annual ethanol production at roughly 11 billion liters (3 billion gallons).

1986 — Alcohol vehicle sales peak at roughly 90% of new light vehicle sales.

1989-1990 — Rise in ethanol prices, decrease in petroleum prices, and elimination of some ethanol subsidies lead to a drop in ethanol supply. Alcohol vehicle sales drop substantially.

1998 — Pure alcohol vehicle sales discontinued.

1999 — Ethanol prices liberalized. Annual ethanol production at roughly 11 billion liters (3 billion gallons).

2003 — Flexible fuel vehicle (FFV) sales begin.

2006 — Annual ethanol production at 17 billion liters (4.5 billion gallons). FFV sales represent roughly 90% of new vehicle sales by end of year.

Sources: Michael Barzelay, *The Politicized Market Economy: Alcohol in Brazil's Energy Strategy*, Berkeley, 1986; Christoph Berg, *World Fuel Ethanol Analysis and Outlook*, Kent, UK, 2004; Christoph Berg, *World Ethanol Production 2001*, Kent, UK, 2001; F. Joseph Demetrius, *Brazil's National Alcohol Program: Technology and Development in an Authoritarian Regime*, New York, 1990; Renewable Fuels Association, [http://www.ethanolrfa.org]; Harry Rothman, Rod Greenshields, and Francisco Roillo Calle, *The Alcohol Economy*, London, 1983.

CRS-10

The Role of Ethanol and Gasoline in the United States and Brazil

The United States consumed 5.4 billion gallons of ethanol in 2006, or roughly 4% of U.S. gasoline demand by volume. Nearly all of this fuel was consumed as gasohol at the 10% level or lower. A much smaller amount was consumed as E85 in FFVs. In contrast, Brazil's roughly 4 billion gallons of ethanol consumption in 2006 represented roughly half of Brazilian passenger vehicle fuel supply,[24] by volume. Because ethanol has a lower energy content than gasoline, in terms of energy, ethanol represents roughly 40% of Brazil's passenger vehicle fuel supply.

The Brazilian transportation sector is considerably smaller than the U.S. transportation sector, and diesel fuel plays a much larger role in motor vehicle fuel demand (including heavy trucks). While diesel fuel represents roughly a quarter of total U.S. highway fuel consumption,[25] in Brazil, diesel fuel consumption represents nearly two-thirds of all motor fuel. Therefore, while ethanol in Brazil displaces nearly half of all passenger vehicle fuel consumption, it represents a smaller percentage of total highway fuel consumption — perhaps 20% by volume and 14% in terms of energy.

Figure 2. Fuel Consumption in the United States and Brazil (billion gallons)

Sources: Energy Information Administration, *International Energy Annual 2004*, Washington, June 2006; Renewable Fuels Association, *Industry Statistics*, [http://www.ethanolrfa.org/industry/statistics], accessed July 20, 2007.

---

[24] Passenger vehicle fuel in Brazil consists mostly of gasoline and ethanol.

[25] Stacy C. Davis and Susan W. Diegel, Oak Ridge National Laboratory, *Transportation Energy Data Book: 26th Edition*, Oak Ridge, TN, 2007.

CRS-11

Figure 3. Vehicles Per 1,000 People in the United States and Brazil

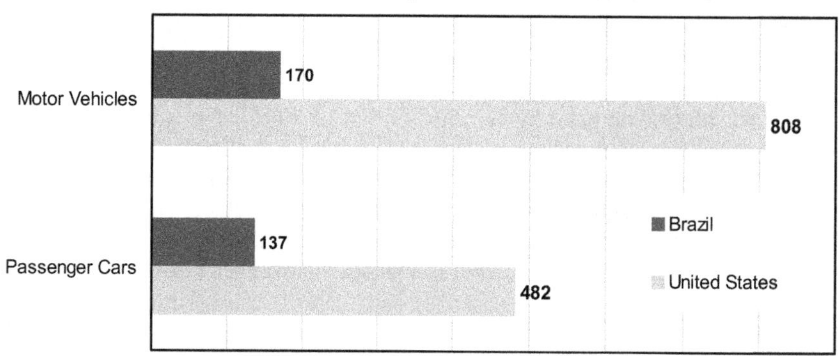

Source: The World Bank, *World Development Indicators 2006*, April 2006.

Figure 4. Ethanol as a Share of Fuel Demand in the United States and Brazil

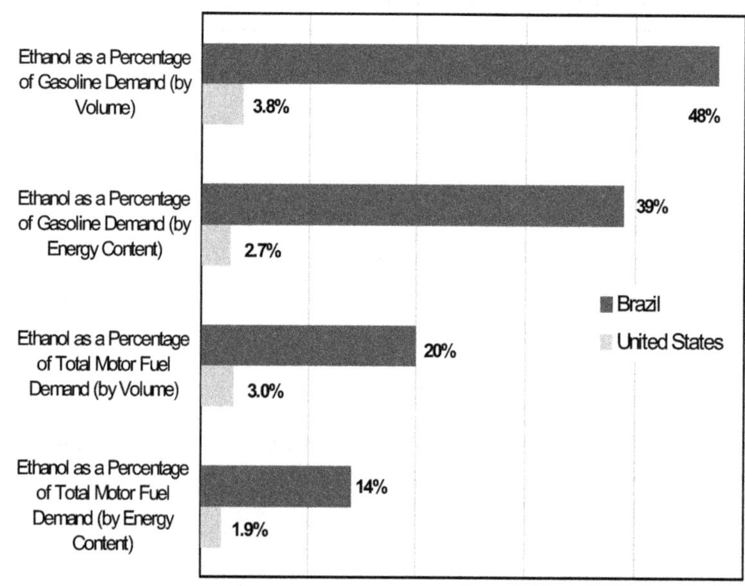

Sources: Energy Information Administration, *International Energy Annual 2004*, Washington, June 2006; International Energy Agency, *World Energy Outlook 2006*, Paris, 2006.

U.S. Ethanol Imports

Because of such high domestic demand, most Brazilian ethanol is produced for domestic consumption. However, U.S. imports of Brazilian ethanol have increased in recent years, especially in times of tight domestic supply. For example, in the spring of 2006, high U.S. ethanol demand, fueled by the phase-out of MTBE[26] (a competing gasoline blending component) led to a rapid rise in ethanol blending by gasoline suppliers that could not be met with domestically produced ethanol. As U.S. ethanol production capacity is growing rapidly — from roughly 5.5 billion gallons at the end of 2006, to roughly 6.2 billion gallons in July 2007, and an expected 11 to 12 billion gallons by the end of 2008[27] — whether it will remain profitable to import ethanol directly from Brazil is an open question.

Until recently, most Brazilian ethanol was imported into the United States through Caribbean Basin Initiative (CBI) countries in order to avoid import duties. (See section below on "Import Tariffs and Duties".) However, when U.S. ethanol prices are high relative to Brazilian production costs, it may be advantageous to import Brazilian ethanol directly, regardless of the tariff, as happened in the spring of 2006. **Figure 5** shows U.S. ethanol imports over the past eight years. Import data through the end of May 2007 suggest that 2007 imports will fall below 2006 levels but will remain high relative to previous years — perhaps 450 million gallons total.[28]

---

[26] Ethanol and MTBE have been used in the United States to extend gasoline stocks, increase the octane rating of gasoline, and to add oxygen to the fuel to meet clean air standards. For more information on MTBE, see CRS Report RL32787, *MTBE in Gasoline: Clean Air and Drinking Water Issues*, by James McCarthy and Mary Tiemann.

[27] Renewable Fuels Association, *Industry Statistics*, available at [http://www.ethanolrfa.org/industry/statistics/], accessed: July 18, 2007.

[28] USITC, *U.S. Imports of Fuel Ethanol, by Source, by Month: January 2007 through May 2007.*

Figure 5. Annual Ethanol Imports to the United States

Million Gallons Per Year

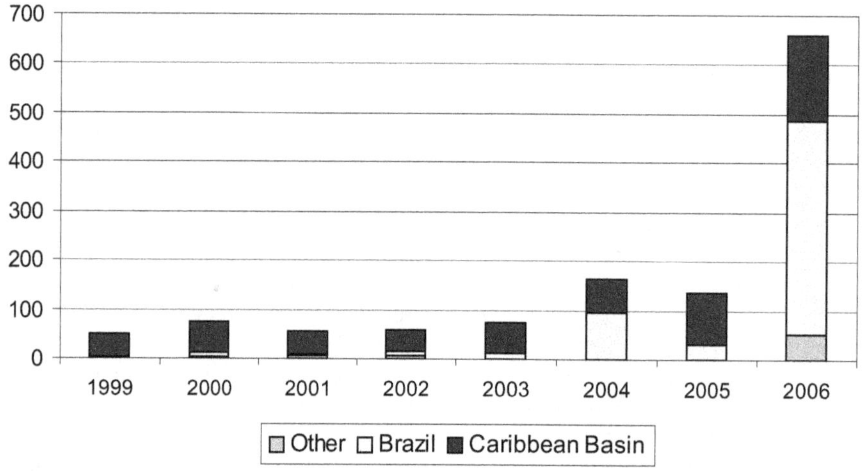

Sources: U.S. International Trade Commission (USITC), *Interactive Tariff and Trade DataWeb*, at [http://dataweb.usitc.gov], accessed March 9, 2006, and USITC, *U.S. Imports of Fuel Ethanol, by Source 1996-2006*, updated April 10, 2007.

U.S.-Brazilian Memorandum of Understanding on Biofuels

On March 9, 2007, the United States and Brazil signed a Memorandum of Understanding (MOU) to promote greater cooperation on ethanol and biofuels in the Western hemisphere. The agreement involves (1) technology-sharing between the United States and Brazil, (2) conducting feasability studies and providing technical assistance to build domestic biofuels industries in third countries, and (3) working multilaterally to advance the global development of biofuels. The first countries targeted for U.S.-Brazilian assistance are the Dominican Republic, El Salvador, Haiti, and St. Kitts and Nevis.[29]

Since March 2007, the United States and Brazil have moved forward on all three facets of the agreement.[30] Specific actions have included the following:

---

[29] "Memorandum of Understanding Between the United States and Brazil to Advance Cooperation on Biofuels," U.S. Department of State, Office of the Spokesman, March 9, 2007, available at [http://www.state.gov/r/pa/prs/ps/2007/mar/81607.htm]; "Joint Statement on the Occasion of the Visit by President Luiz Inácio Lula da Silva to Camp David," White House Press Release, March 31, 2007, available at [http://www.state.gov/p/wha/rls/prsrl/07/q1/82519.htm].

[30] "Advancing Cooperation on Biofuels: U.S.-Brazil Steering Group Meets August 20 in (continued...)

- **Bilateral:** Several high-level visits have taken place aimed at boosting bilateral cooperation on biofuels. A team of Brazilian scientists visited the U.S. Department of Energy and Department of Agriculture Laboratories in mid-September. A group of U.S. scientists are scheduled to visit Brazil in November. Officials from both countries are exploring bilateral professorial and graduate student exchanges.

- **Third-country efforts:** The U.S. and Brazilian governments are working with the Organization of American States (OAS), IDB, and the U.N. Foundation (UNF) to conduct feasibility studies in Haiti, the Dominican Republic, and El Salvador. The feasibility study on St. Kitts and Nevis has been completed. Officials from these four countries visited the United States in August to attend a biofuels conference.

- **Global efforts.** The United States and Brazil are working with other members of the International Biofuels Forum (IBF) to make biofuels standards and codes more uniform by the end of 2007. IBF members include Brazil, the United States, the European Union, China, India, and South Africa.

Some argue that the U.S.-Brazil agreement could provide the impetus needed to develop a viable biofuels industry in Latin America, a region with a comparative advantage in biofuels production. Other observers are less sure. They are concerned about the huge investment outlays that governments would have to make to ramp up biofuels production. The IDB estimates that at least $200 billion in new investments would have to be made for biofuels to provide even 5% of the region's transport energy by 2020.[31] Skeptics question whether countries that lack the type of enabling environment that Brazil possesses — infrastructure, research and extension services, technology, educated workforce, and credit market — should lend their support to biofuels before those items are in place.[32] On the domestic front, some analysts worry that increasing biofuels cooperation with Brazil and other countries in Latin America may prompt challenges to existing U.S. trade, energy, and agriculture policies.

Policy Considerations

Import Tariffs and Duties

Because of lower production costs and/or government incentives, ethanol prices in Brazil are generally significantly lower than in the United States. To offset the

[30] (...continued)
Brasilia," U.S. Department of State, Office of the Spokesman, August 22, 2007.

[31] IDB report, April 2007.

[32] Kojima and Johnson, 2006.

CRS-15

U.S. tax incentives that all ethanol (imported or domestic) receives, most imports are subject to a 2.5% ad valorem tariff, plus an added duty of $0.54 per gallon. This duty effectively negates the tax incentives for covered imports and has been a significant barrier to ethanol imports when U.S. domestic prices are low.

However, under certain conditions ethanol imports from Caribbean Basin Initiative (CBI) countries are granted tariff/duty-free status, even if the ethanol was actually produced in a non-CBI country. In this particular case, the CBI countries participate only in the final step of the production process — dehydration, after which the ethanol is shipped to the United States.[33] Up to 7% of the U.S. ethanol market may be supplied duty-free by ethanol dehydrated in CBI countries.

As shown in **Figure 2**, until recently, most U.S. ethanol imports came through the CBI. Whereas previously most imported ethanol imported was produced in Europe and dehydrated in CBI, now most CBI ethanol is produced in Brazil. As part of the Dominican Republic-Central American Free Trade Agreement (CAFTA-DR), Costa Rica and El Salvador are granted specific allocations within the 7% quota.

Because CBI ethanol is actually produced in countries subject to the duties, some stakeholders view this treatment of ethanol from the CBI as a "loophole" to avoid these duties. Proponents of the CBI provisions argue that the dehydration of ethanol promotes economic development in CBI countries, even if those countries are not using local feedstocks.

In addition to the concerns over imports of duty-free ethanol from CBI countries, there is also growing concern that a large portion of ethanol otherwise subject to the duties is being imported duty-free through a "manufacturing drawback."[34] If a manufacturer imports an intermediate product then exports the finished product or a similar product, that manufacturer may be eligible for a refund (drawback) of up to 99% of the duties paid. There are special provisions for the production of petroleum derivatives.[35] In the case of fuel ethanol, the imported ethanol is used as a blending component in gasoline. Jet fuel (containing no ethanol, but considered a "like commodity" to the finished gasoline) is exported to qualify for the drawback in lieu of finished gasoline containing the originally imported ethanol.[36] Some critics estimate that as much as 75% or more of the duties were eligible for the drawback in 2006. Therefore, critics question the effectiveness of the ethanol duties and the CBI exemption.

---

[33] Ethanol currently comes from four CBI countries: Costa Rica, Jamaica, El Salvador, and Trinidad and Tobago. CBI imports represent a significant percentage of U.S. fuel ethanol imports. For more information on ethanol imports from CBI countries, see CRS Report RS21930, *Ethanol Imports and the Caribbean Basin Initiative*, by Brent D. Yacobucci.

[34] For more information on drawbacks, see U.S. Customs Service, *Drawback: A Refund for Certain Exports*, Washington, February 2002.

[35] 19 U.S.C. 1313(p).

[36] Peter Rhode, "Senate Finance May Take Up Drawback Loophole As Part Of Energy Bill," *EnergyWashington Week*, April 18, 2007.

Proponents of the domestic ethanol industry argue that foreign ethanol producers receive benefits and incentives from their home countries, and that U.S. duties on imported ethanol should be strengthened. Further, they argue, limiting imports and promoting the domestic industry furthers U.S. national security and lowers our dependence on energy imports. Opponents of the duties argue that elimination of the duties would allow us to further diversify our energy supply and move toward more environmentally sound transportation fuels. Further, they argue that importing ethanol from Brazil is preferable to importing petroleum from less stable parts of the world.

On December 20, 2006, President Bush signed the Tax Relief and Health Care Act of 2006 (P.L. 109-432). Among other provisions, the act extended the duty on imported ethanol through December 31, 2008.

Energy Bill and Farm Bill Considerations

The U.S. Congress is currently considering new energy legislation, as well as a new Farm Bill. Both pieces of legislation could affect U.S. and foreign biofuels production.

Energy Bill. The Senate-passed energy bill (H.R. 6) would expand and extend the existing Renewable Fuel Standard (RFS). The RFS currently mandates the use of 7.5 billion gallons of renewable fuel in gasoline by 2012. The Senate bill would increase the mandate to 13.2 billion gallons in 2012 and 36 billion gallons in 2022. Further, by 2022, the bill would mandate the use of 21 billion gallons of "advanced biofuels," defined as "fuel derived from renewable biomass other than corn starch." Such a mandate would mean a significant increase in the role of non-corn-based ethanol. If enacted, the new RFS could mean a significant increase in demand for Brazilian sugar-based ethanol, especially if costs remain high for cellulosic ethanol and other non-corn biofuels.

Farm Bill. The previous farm bill, the Farm Security and Rural Investment Act of 2002 (P.L. 107-171), established an energy title (Title IX). Among other provisions, Title IX contained sections to promote the research and development of new biofuels, as well as to expand the production of existing biofuels. Those programs expired at the end of FY2006. The Administration's 2007 proposed farm bill energy title (IX) provides $1.6 billion in new funding for basic and applied research, as well as to share the risk associated with developing and commercializing a new technology through loan and loan guarantee programs. The primary focus of USDA's proposed new funding is the development of cellulosic ethanol production.

On July 27, 2007, the House approved its version of the Farm Bill, H.R. 2419. Among other provisions, the bill proposes a total of $3.2 billion in new funding for Title IX energy provisions over five years, including $1.4 billion for production incentive payments on new biofuels production. A key departure from current farm bill-related energy provisions is that most new funding is directed away from corn-starch-based ethanol production and towards cellulosic-based biofuels production or to new as-yet-undeveloped technologies with some agricultural linkage.[37]

Food vs. Fuel Debate

Because most ethanol is produced from food crops (either corn or sugar), there is concern that increasing biofuels production could lead to higher food prices.[38] In the case of corn, most corn in the United States is used for animal feed. Higher feed costs ultimately lead to higher prices for poultry, hogs, and cattle. The price of corn is also related to the price of other competing foodstuffs, such as grains and soybeans. In 2006, the expansion of corn-based ethanol production led to a sharp rise in corn prices, which some predicted would lead to higher U.S. and world food prices. In fact, the futures contract for March 2007 corn on the Chicago Board of Trade rose from $2.50 per bushel in September 2006 to a record high of over $4.37 per bushel in February 2007. Some analysts predict that if high oil prices continue, increases in global biofuels production may push corn prices up by 20% by 2010 and 41% by 2020.[39] On the other hand, commodity prices are dependent on many other factors besides the demand for biofuels crops. Corn prices have fallen slightly in the past few months since growers began predicting record crop yields for 2007, while wheat prices have soared because of weather-related production problems in many countries.[40]

Since most U.S. households do not spend a large percentage of their budgets on food, they may be able to absorb any increases in food prices that result from increasing biofuels production, at least in the short run. However, that is not likely to be the case for low-income households in the United States or for households in Latin America, the poorest of which often spend more than half of their household incomes on food. In January 2007, Mexico faced widespread strikes and social unrest as rising corn prices, fueled by the demand for corn-based ethanol, led to a 30% rise in the cost of corn tortillas, a basic dietary staple. Through Latin America, inflation rates have increased as a result of higher food prices, which have been attributed to increased demand for production of ethanol and other biofuels.[41]

Critics of biofuels argue that unless new technology is developed to produce biofuels from cellulosic materials, increasing biofuels production will lead to higher global food prices, which will in turn result in hunger and malnutrition in many

---

[37] For more information, see CRS Report RL34130, *Renewable Energy Policy in the 2007 Farm Bill*, by Randy Schnepf.

[38] This section was drawn from CRS Report RL33928, *Ethanol and Biofuels: Agriculture, Infrastructure, and Market Constraints Related to Expanded Production*, by Brent D. Yacobucci and Randy Schnepf.

[39] C. Ford Runge and Benjamin Senauer, "How Biofuels Could Starve the Poor," *Foreign Affairs*, May/June 2007.

[40] K.T. Arasu, "Wheat Eclipses Corn as Investors New Darling," *Reuters,* September 6, 2007.

[41] "Guatemala: Economic: Inflation: Recent Developments," *Global Insight*, August 27. 2007.

developing countries. Ethanol proponents dispute those predictions. They maintain that the availability of arable land, especially in Latin America, will allow plenty of space for biofuels production without encroaching upon other crops. In Brazil, for example, less than 9% of the country's total planted area is dedicated to sugar. They further argue that the food-versus-fuel debate may be more applicable in the case of corn than in the case of sugar, as recent expansion in ethanol production from sugarcane has not significantly affected global sugar prices.[42]

Environmental Concerns

While there are significant potential benefits from biofuels in terms of reduced petroleum consumption and reduced air pollution, there are also potential environmental drawbacks. These include the potential for increased greenhouse gas emissions, higher levels of surface water contamination, and increased pressure on land and water resources.

Greenhouse Gas Emissions. One of the key environmental concerns over biofuels is their effect on overall greenhouse gas emissions. Depending on the production process, biofuels can either lead to a net increase or decrease in greenhouse gas emissions throughout the fuel cycle relative to petroleum fuels. Because ethanol contains carbon, combustion of the fuel necessarily results in emissions of carbon dioxide (CO_2), the primary greenhouse gas. Further, greenhouse gases are emitted through the production and use of nitrogen-based fertilizers used in corn production, as well from fuels used in the operation of farm equipment and vehicles to transport feedstocks and finished products. However, because photosynthesis (the process by which plants convert light into chemical energy) requires absorption of CO_2, the growth cycle of the feedstock crop can serve, to some extent, as a "sink" to absorb some fuel-cycle greenhouse emissions.

Recent studies on energy consumption and greenhouse gas emissions have concluded that corn-based ethanol results in 13% to 22% lower greenhouse gas emissions relative to gasoline.[43] Ethanol from other feedstocks can lead to even lower greenhouse gas emissions. For example, sugar requires far less fertilizer to produce than corn, and less processing is necessary to prepare the feedstock for fermentation. Net greenhouse gas emissions from sugarcane-based ethanol could be as much as 56% lower than gasoline, and cellulosic ethanol could reduce emissions by 90% relative to gasoline.[44]

Ethanol production processes themselves can also lead to air quality concerns. For example, without proper emissions controls, ethanol plants produce emissions

---

[42] K.T. Arasu, "Ethanol Boom Won't Threaten Food Supply," *Reuters*, June 4, 2007.

[43] Alexander E. Farrell, Richard J. Plevin, Brian T. Turner, Andrew D. Jones, Michael O'Hare, and Daniel M. Kammen, "Ethanol Can Contribute to Energy and Environmental Goals," *Science*, January 27, 2006, pp. 506-508; U.S. Environmental Protection Agency (EPA), *Greenhouse Gas Impacts of Expanded Renewable and Alternative Fuels Use*, April 2007.

[44] EPA. Op. cit.

of volatile organic compounds (hydrocarbons) that can lead to negative health effects and can contribute to ozone (smog) formation. Further, burning of sugarcane fields before manual clearing can increase fuel cycle pollutant and greenhouse gas emissions. Mechanized harvesting without burning can improve the emissions profile of sugarcane ethanol, but greater mechanization would likely come at the cost of fewer jobs for cane cutters.

Because of the favorable emissions profile of ethanol from non-corn feedstocks, there is growing interest in moving U.S. biofuels consumption away from corn ethanol to either imported sugarcane ethanol or domestically produced cellulosic ethanol.

Water Contamination. Another key environmental concern surrounding ethanol is its effects on water quality. In the United States, corn requires a significant amount of chemical inputs. Runoff from fertilizers and pesticides finds its way into streams and other surface waters, potentially leading to algae blooms and other problems.

In Brazil, a key problem has been the discharge of nutrient-rich waste from ethanol production — "vinasse" — directly into streams or indirectly through soil contamination. Although legislation has been introduced to address this problem, lax enforcement of environmental standards in Brazil mean that pollution from ethanol production will likely continue to be a problem.[45]

Water Consumption. In addition to concerns over water quality, water consumption may also become an environmental concern, especially in the United States. Currently, Brazilian sugar and U.S. corn production do not require large amounts of water inputs. However, as feedstock production, especially U.S. corn, expands into drier areas, more water may be needed, putting additional pressure on already stretched water resources.

Land Use/Soil Quality. Concerns have also been raised about the effects of agricultural production for biofuels on land resources. For example, in the United States, corn has generally been rotated with soybeans to promote soil quality. However, as corn production for ethanol expands, much of the land that had been in rotation is shifting away from soybean production. This could lead lower concentrations of soil nutrients, increasing the need for fertilizers and other chemical inputs. And while cellulosic biofuels in general appear more sustainable, some concerns have also been raised about their sustainability, especially if environmentally sensitive areas (e.g., Conservation Reserve Program land) are used for bioenergy production.

In Brazil, concerns focus on protection of habitats in the cerrado (Brazilian savanna) and the Amazon rain forest. Expansion of sugarcane planting has led to

---

[45] Jose Moreira. "Water Use and Impacts Due Ethanol Production in Brazil," *International Water Management Institute and Food and Agricultural Organization Conference: Linkages between Energy and Water Management for Agriculture in Developing Countries*, Hyderabad, India, January 29-30, 2007.

rapid depletion of wooded areas of the cerrado. Further, as ethanol expands into existing pastureland, cattle-breeding has been displaced into the cerrado and the Amazon.[46] Increased demand for soybeans (both for food and as a feedstock for biodiesel) has added pressure to expand soybean production into the Amazon.

Labor Issues

Some analysts see expanding the region's biofuels industry as a way to create jobs and promote rural development in Latin America. In most countries in the region, biofuels production is a labor-intensive process that creates jobs in agriculture, manufacturing, and transport. UNICA, the São Paulo Sugarcane Association, estimates that roughly 1 million direct jobs and 3 million indirect jobs have been created in Brazil as a result of the country's biofuels industry. Another study asserts that some 2,333 jobs are created in Brazil for every 1 million tons of sugarcane harvested. Most of those jobs are as sugarcane cutters. Ethanol proponents estimate that some 14,000 jobs could be created in Central America through increased production of E10 (10% ethanol) fuel. Similar forecasts have been made for other countries and subregions in Latin America and the Caribbean.[47]

Analysts agree that biofuels production generates jobs, but some question the number and quality of those jobs.[48] Skeptics argue that because biofuels production often displaces other existing agricultural activities, net job gains may be minimal. They also assert that unless governments make a concerted effort to ensure that small-scale producers have a role to play in biofuels production, large agribusinesses will continue to dominate the biofuels industry in Latin America. Finally, most analysts acknowledge that as biofuels production becomes increasingly mechanized, a development that brings efficiency and environmental benefits, less agricultural jobs are going to be generated. As was stated above, there is likely to be a tradeoff between increased employment and more environmentally benign practices.

Skeptics of using biofuels to promote rural development also question the quality of most jobs created by the biofuels industry, particularly in countries producing ethanol from sugarcane, where jobs are often low-paying, hazardous work as seasonal cane cutters. The Brazilian government has acknowledged that there

---

[46] Isaias de Carvalho Macedo, editor, *Sugar Cane's Energy*, Sao Paulo, September 2005, pp. 133.

[47] Presentation by Alfred Szwarc, Senior Advisor to the São Paulo Sugarcane Association (UNICA), at the Corporación Andina de Fomento Conference on Trade and Investment in the Americas, September 5, 2007; "Issue Paper on Biofuels Development in Latin America and the Caribbean," prepared for the IDB, 2006; Luiz Augusto Horta Noguierz, "Perspectivas Para las Biocombustibles en América Central," Incentivos a las Renovables y Biocombustibles, Alianza en Energía y Ambiente para Centroamérica, El Salvador, February 2006.

[48] For a discussion of these problems, see Worldwatch Institute, June 2006, and Kojima and Johnson, World Bank, February 2006.

have been instances of forced labor on some sugarcane properties in Brazil, particularly in the northeast region of the country.[49]

Biofuels and Geo-politics in Latin America

The U.S.-Brazil MOU on biofuels, the agreement is also intended to have a political effect in the region. Many Bush Administration officials and Members of Congress note that the new biofuels partnership with Brazil may help improve the U.S. image in Latin America and diminish the influence of President Chávez in the region. The United States has increasingly regarded Brazil as a significant power, especially in its role as a stabilizing force and skillful interlocutor in Latin America. U.S. officials tend to describe Brazil as an amicable partner governed by a moderate leftist government and have responded positively to Brazil's efforts to reassert its regional leadership, which has recently been challenged by the rise of oil-rich Hugo Chávez in Venezuela.[50]

In recent months, Brazil has increasingly used so-called "biofuels diplomacy" as a diplomatic and economic tool to raise its profile in Latin America and throughout the world. President Chávez, recognizing that increasing biofuels production and usage in Latin America could diminish his regional influence, quickly attacked the Brazil-U.S. biofuels agreement, stating that it would raise food prices and hurt the poor. In early April 2007, despite Chávez's criticisms of ethanol, one of his allies, President Correa of Ecuador, signed a biofuels production agreement with Brazil. Some Members of the Morales government in Bolivia are also supportive of biofuels production. In mid-April 2007, Chávez was forced to backtrack on his initial opposition to all biofuels production in the region while attending South America's first regional energy summit. Competition between Brazil and Venezuela for leadership in the region has accelerated in the past few months. In August 2007, as President Lula took a six-day tour of Mexico, Central America and the Caribbean to promote biofuels production agreements, President Chávez visited Argentina, Bolivia, Ecuador and Uruguay, where he signed a series of oil and gas agreements.[51]

Congressional Action

In the past two years, there has been significant congressional interest in issues related to energy security. Some of that interest has focused on how to ensure that countries in the Western Hemisphere, which currently supply about half of U.S.

---

[49] "Brazil Labor Minister Calls Sugarcane Labor 'Degrading,'" *Dow Jones Commodity Service*, June 11, 2007.

[50] Monte Reel, "U.S. Seeks Partnership with Brazil on Ethanol; Countering Oil-rich Venezuela is Part of Aim," *Washington Post*, February 8, 2007.

[51] "Brazil Gets Tough on Energy," *Latin American Regional Report*, April 2007; "Venezuela, Brazil at Odds Over Ethanol as Chávez Hosts South American Energy Summit," *AP*, April 16, 2007; Marifeli Perez-Stable, "Marching to Different Drummers; Lula, Chavez," *Miami Herald*, August 17, 2007.

imports of crude oil and petroleum products, remain reliable sources of energy for the United States. Another area of interest has been to promote cooperation among Latin American countries, which are divided between net energy exporting and importing nations, to ensure that enough clean, affordable, and reliable energy sources are exploited to support regional growth and development. Members have cited Brazil as an example of a country that has successfully reduced its reliance on foreign oil by using alternative energies. In addition to the importance of following Brazil's example in the field of biofuels development, some Members have cited the importance of U.S. engagement in regional efforts to develop biofuels and other renewable energies.

On September 19, 2007, the House Western Hemisphere Subcommittee held a hearing on "U.S.-Brazil Relations" during which Chairmen Eliot Engel and many of the witnesses cited biofuels cooperation as a primary example of the expanding strategic relationship between the United States and Brazil. They discussed how the U.S.-Brazil MOU on biofuels may encourage both countries to work together to advance their national, regional, and international interests. Despite this potential for increasing U.S.-Brazil collaboration on biofuels, one witness warned that this unique opportunity may be lost if the countries are unable to resolve the underlying agricultural disputes that divide them, such as current U.S. subsidies and tariffs that protect corn-based ethanol producers.[52]

Legislation

In the 109th Congress, Members were somewhat divided over whether to keep the current 2.5% ad valorem tariff and added duty of $0.54 per gallon on foreign ethanol imports in place. Legislation was introduced that would have eliminated the two duties on foreign ethanol: H.R. 5170 (Shadegg) and S. 2760 (Feinstein), the Ethanol Tax Relief Act of 2006. However, in December 2006, Congress voted to extend the duties on foreign ethanol through December 31, 2008 (P.L. 109-432). In the 110th Congress, S. 1106 (Thune) would extend those tariffs through 2011, and H.R. 196 (Pomeroy) would make the tariffs permanent.

Several legislative initiatives in the 110th Congress would increase hemispheric cooperation on energy issues, including biofuels development and distribution. S. 193 (Lugar), the Energy Diplomacy and Security Act of 2007, calls for the establishment of a regional-based ministerial forum known as the Hemisphere Energy Cooperation Forum that would, among its many activities, be involved in developing an Energy Sustainability Initiative to promote the development, distribution, and commercialization of renewable fuels in the region. The bill also calls for the establishment of a Hemisphere Energy Industry Group to increase public-private partnerships, foster private investment, and enable countries to devise energy agendas on various topics, including the development and deployment of biofuels. The Senate Foreign Relations Committee reported favorably on the bill on April 12, 2007, without amendment (S.Rept. 110-54).

---

[52] Testimony of Paulo Sotero, Director of the Brazil Institute of the Woodrow Wilson International Center for Scholars, before the House Western Hemisphere Subcommittee, September 19, 2007.

CRS-23

Another initiative, S. 1007 (Lugar), the United States-Brazil Energy Cooperation Pact of 2007, calls for the same cooperation groups as S. 193, and directs the Secretary of State to work with Brazil and other Western Hemisphere countries to develop partnerships to accelerate the development of biofuels production, research, and infrastructure. The bill was introduced on March 28, 2007, and referred to the Senate Foreign Relations Committee.

H.Res. 651 (Engel), introduced on September 19, 2007, recognizes the warm friendship and expanding relationship that exists between the United States and Brazil, commends Brazil for reducing its dependency on oil by using alternative energies, and recognizes the importance of the March 9, 2007, United States-Brazil Memorandum of Understanding (MOU) on biofuels cooperation.

Outlook

Rising demand for ethanol and other biofuels has sharpened attention on whether the United States and Brazil, the leaders in biofuels production, should increase cooperation, share technology, and work to expand the global biofuels market. Of the three pillars of the U.S.-Brazil MOU on biofuels, progress on the first (technology-sharing) and third (working multilaterally to advance biofuels) pillars are likely to occur most quickly. In the short to medium term, collaborative research and development activities may yield the largest potential benefit for both countries, particularly if they are able to hasten the development of cellulosic ethanol technology. Producing ethanol from dedicated energy crops and waste products may allay many of the environmental and food-versus-fuel concerns that are drawbacks of producing ethanol from food crops like sugar or corn. Both countries also stand to benefit from working together on the global front to establish consistent ethanol standards and codes, a crucial step in the process for ethanol to become a globally traded energy commodity.

While some analysts believe the U.S.-Brazil agreement may be enough to spur viable biofuels markets in "third countries" (pillar two), those efforts may not be feasible. First, governments may lack the resources or political will to make the huge investment outlays necessary to develop their biofuels industries. Second, many countries lack the arable land necessary to develop biofuels without encroaching on traditional agricultural lands. A third concern with increasing sugar-based biofuels production is that the sugar industries of many countries in the Caribbean (including the Dominican Republic) are struggling because of high labor costs and efficiency problems. Fourth, as previously mentioned, there are serious labor and environmental concerns about rapidly increasing biofuels production.

In the next few months, results from U.S.-Brazil feasibility studies for Haiti, the Dominican Republic, and El Salvador are expected to be completed. The St. Kitts study has already determined that although producing biofuels for transport would not be feasible there, bio-electricity could be generated for domestic use. The results of the other feasibility studies and the willingness of the governments of each of those countries to embrace biofuels development are likely to affect the selection of a second round of countries to receive U.S.-Brazil technical assistance. While U.S.

officials are eager to expand third-country initiatives into South America, Brazilian officials have reportedly been reluctant to give the United States a foothold into its sphere of influence.

While some Members of Congress have been supportive of energy cooperation efforts like the U.S.-Brazil MOU, others might not support any initiatives that they feel will adversely affect U.S. corn-based ethanol producers. Indeed, the U.S.-Brazil MOU does not address two key issues that many Brazilians feel are significant obstacles to expanding bilateral and regional biofuels cooperation, namely the current subsidies and tariffs that protect U.S. corn-based ethanol producers. Since many Members strongly favor extending the current subsidy programs for corn producers and tariffs on foreign ethanol, these issues may be obstacles to maintaining expanded U.S.-Brazil biofuels cooperation. In addition, Members who feel that Brazil's positions on agricultural trade during the failed Free Trade Area of the Americas (FTAA) and in the World Trade Organization (WTO) negotiations have adversely affected U.S. interests may also be opposed to the MOU on biofuels. On the other hand, some may see energy cooperation as an issue on which a positive U.S.-Brazil agenda can be based, presenting a unique opportunity to overcome past trade disputes.

Order Code RS21930
Updated March 18, 2008

CRS Report for Congress

Ethanol Imports and the Caribbean Basin Initiative

Brent D. Yacobucci
Specialist in Energy and Environmental Policy
Resources, Science, and Industry Division

Summary

Fuel ethanol consumption has grown significantly in the past several years, and it will continue to grow with the establishment of a renewable fuel standard (RFS) in the Energy Policy Act of 2005 (P.L. 109-58) and the expansion of that RFS in the Energy Independence and Security Act of 2007 (P.L. 110-140). This standard requires U.S. transportation fuels to contain a minimum amount of renewable fuel, including ethanol.

Most of the U.S. market is supplied by domestic refiners producing ethanol from American corn. However, imports play a small but growing role in the U.S. market. One reason for the relatively small role is a 2.5% ad valorem tariff and (more significantly) a 54-cent-per-gallon added duty on imported ethanol. These duties offset an economic incentive of 51 cents per gallon for the use of ethanol in gasoline. However, to promote development and stability in the Caribbean region and Central America, the Caribbean Basin Initiative (CBI) allows the imports of most products, including ethanol, duty-free. While many of these products are produced in CBI countries, ethanol entering the United States under the CBI is generally produced elsewhere and reprocessed in CBI countries for export to the United States. The U.S.-Central America Free Trade Agreement (CAFTA) would maintain this duty-free treatment and set specific allocations for imports from Costa Rica and El Salvador. Duty-free treatment of CBI ethanol has raised concerns, especially as the market for ethanol has the potential for dramatic expansion under P.L. 109-58 and P.L. 110-140.

In the United States, fuel ethanol is largely domestically produced. A value-added product of agricultural commodities, mainly corn, it is used as a gasoline additive and as an alternative to gasoline. To promote its use, ethanol-blended gasoline is granted a significant tax incentive. However, this incentive does not recognize point of origin, and there is a duty on most imported fuel ethanol to offset the exemption. But a limited amount of ethanol may be imported under the Caribbean Basin Initiative (CBI) duty-free, even if most of the steps in the production process were completed in other countries. This duty-free import of ethanol has raised concerns, especially as U.S. demand for ethanol has been growing. Further, duty-free imports from these countries, especially

Congressional Research Service ⁂ The Library of Congress
Prepared for Members and Committees of Congress

Costa Rica and El Salvador, have played a role in the development of the U.S.-Central America Free Trade Agreement (CAFTA).

Fuel Ethanol

Ethanol is an alcohol fuel produced from the fermentation of simple sugars.[1] Most ethanol in the United States is produced from corn. In other countries, sugarcane or other plants are common feedstocks. In the United States, the increased demand for corn leads to higher revenues for U.S. corn farmers. Ethanol is usually blended in gasoline (a mixture called "gasohol") to increase octane, improve combustion, and extend gasoline stocks. Currently, about 3% to 5% of total U.S. gasoline demand is actually met by ethanol, and roughly half of U.S. gasoline contains some ethanol.

U.S. ethanol is generally produced and consumed in the Midwest, close to where the corn feedstock is produced. The main steps to ethanol production are as follows:

- The feedstock (e.g., corn) is processed to separate fermentable sugars.
- Yeast is added to ferment the sugars.
- The resulting alcohol is distilled.
- Finally, the distilled alcohol is dehydrated to remove any remaining water.

This final step — dehydration — is at the heart of the issue over ethanol imports from the CBI, as discussed below.

Ethanol Imports

According to the United States International Trade Commission, the majority of all fuel ethanol imports to the United States came through CBI countries between 1999 and 2003 (see **Figure 1**).[2] In 2004, imports from Brazil to the United States grew dramatically, but in 2005, CBI imports again represented more than half of all U.S. ethanol imports. With an increase in ethanol demand in 2006 due to voluntary elimination of MTBE — a competitor for ethanol in gasoline blending — imports grew dramatically, roughly quadrupling imports in any previous year.[3] Most of this increase was in direct imports from Brazil. Historically, imports have played a relatively small role in the U.S. ethanol market. Total ethanol consumption in 2005 was approximately 3.9 billion gallons, whereas imports totaled 135 million gallons, or about 4%. Imports from the CBI totaled approximately 2.6%. In 2006, total imports represented roughly 13% of the 5.0 billion gallons consumed in 2006; ethanol from CBI countries represented

---

[1] For more information on ethanol, see CRS Report RL30369, *Fuel Ethanol: Background and Public Policy Issues*, by Brent D. Yacobucci.

[2] It should be noted that between 1999 and 2003, Saudi Arabia was the largest exporter to the United States of ethanol. However, this ethanol is synthetic (produced from fossil fuels) and does not qualify for the tax incentives for ethanol-blended fuel. Therefore, ethanol from Saudi Arabia is used as an industrial feedstock and is subject to different tariff treatment than fuel ethanol.

[3] For more information on the MTBE phaseout, see CRS Report RL31361, *"Boutique Fuels" and Reformulated Gasoline: Harmonization of Fuel Standards*, by Brent D. Yacobucci.

roughly 3.4%. In 2007, total imports represented roughly 6% of U.S. consumption (6.8 billion gallons); ethanol from CBI countries represented roughly 3.6%.

Figure 1. Annual Ethanol Imports to the United States

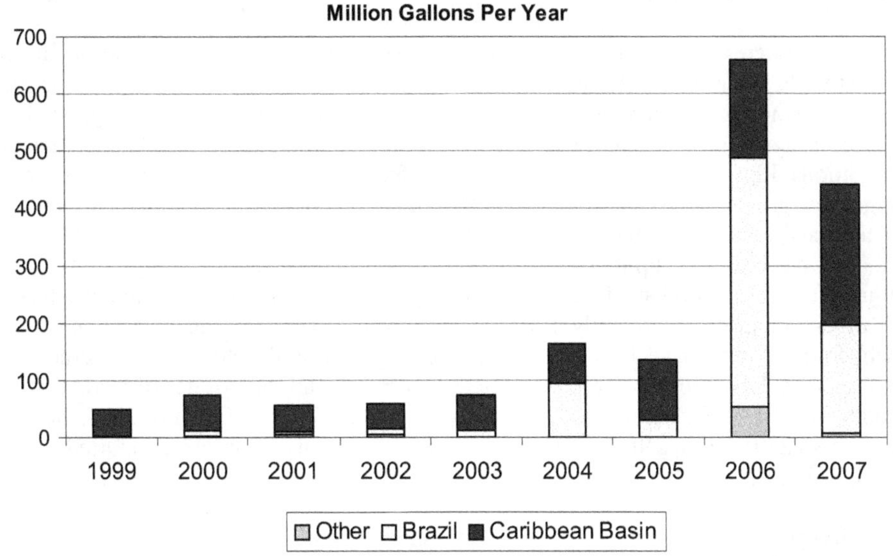

Million Gallons Per Year

700
600
500
400
300
200
100
0

1999 2000 2001 2002 2003 2004 2005 2006 2007

Other Brazil Caribbean Basin

Source: U.S. International Trade Commission (USITC), *Interactive Tariff and Trade DataWeb*, at [http://dataweb.usitc.gov], accessed March 9, 2006, and USITC, *U.S. Imports of Fuel Ethanol, by Source 1996-2007*, updated February 2008.

One reason for limited imports — even though, in some cases, production costs for ethanol in foreign countries are significantly lower than in the United States — is a most-favored-nation tariff of 2.5% and an added duty of 54 cents per gallon.[4] In many cases, this tariff negates lower production costs in other countries. For example, by some estimates, Brazilian production costs have been roughly 50% lower than in the United States.[5] A key motivation for the establishment of the tariff was to offset a tax incentive for ethanol-blended gasoline ("gasohol").[6] This incentive is currently valued at 51 cents per gallon of pure ethanol used in blending. Unless imports enter the United States duty-free, the tariff effectively negates the incentive for those imports. With U.S. wholesale ethanol prices ranging from roughly $1.50 to $2.50 per gallon for most of the time between January 2006 to March 2008, the tariff has presented a significant barrier to

---

[4] Technically, the tariff is 14.27 cents per liter, which is equal to 54 cents per gallon.

[5] "NCGA's Adams Addresses World Energy Crisis at ACE Meeting," *NCGA News*, August 16, 2004; Kevin Diaz, "Cargill Takes Heat Over Ethanol Import Plan," *Star Tribune*, July 2, 2004.

[6] U.S. General Accounting Office, *Fuel Ethanol: Imports from Caribbean Basin Initiative Countries*, April 1989. For more information on the excise tax exemption, see CRS Report 98-435, *Alcohol Fuels Tax Incentives*, by Salvatore Lazzari.

CRS-4

imports.[7] However, during the voluntary phaseout of MTBE, there was a significant spike in wholesale prices between April 2006 and September 2006, with wholesale prices nearing $6.00 per gallon in some markets during the summer of 2006.[8] This runup in prices significantly improved the profitability of importing ethanol, regardless of the duty.

Ethanol and the CBI

As Congress noted in the Customs and Trade Act of 1990, the Caribbean Basin Initiative (CBI) was established in 1983 to promote "a stable political and economic climate in the Caribbean region."[9] As part of the initiative, duty-free status is granted to a large array of products from beneficiary countries, including fuel ethanol under certain conditions. If produced from at least 50% local feedstocks (e.g., ethanol produced from sugarcane grown in the CBI beneficiary countries), ethanol may be imported duty-free.[10] If the local feedstock content is lower, limitations apply on the quantity of duty-free ethanol. Nevertheless, up to 7% of the U.S. market may be supplied duty-free by CBI ethanol containing no local feedstock.[11] In this case, hydrous ("wet") ethanol produced in other countries, historically Brazil or European countries, can be shipped to a dehydration plant in a CBI country for reprocessing.[12] After the ethanol is dehydrated, it is imported duty-free into the United States. Currently, imports of dehydrated ethanol under the CBI are far below the 7% cap (approximately 3% in 2006). For 2006, the cap was about 270 million gallons,[13] whereas about 170 million gallons were imported under the CBI in that year.[14]

Dehydration plants are currently operating in Jamaica, Costa Rica, El Salvador, Trinidad and Tobago, and the U.S. Virgin Islands.[15] Jamaica and Costa Rica were the two largest exporters of fuel ethanol to the United States from 1999 to 2003. (In 2004 and 2006, direct imports from Brazil exceeded imports from all other countries combined.)[16]

---

[7] Chemical Week Associates, "Octane Week Price Report," *Octane Week*, various issues, January 2006 to March 2006, and Chicago Board of Trade, *Ethanol Derivatives, updated through January, 2008*, Chicago, February 13, 2008.

[8] Chicago Board of Trade, *op. cit.*

[9] P.L. 101-382, §202; 19 U.S.C. 2701 note: congressional findings.

[10] P.L. 99-514, §423; 19 U.S.C. 2703 note: ethyl alcohol and mixtures thereof for fuel use.

[11] Ibid.

[12] U.S. House of Representatives, Committee on Ways and Means, *Hearing on Fuel Ethanol Imports from Caribbean Basin Initiative Countries*, April 25, 1989.

[13] *69 Federal Register 76956.*

[14] The quota for a given year is calculated based on 7% of U.S. consumption in the preceding year. Therefore, as U.S. consumption is growing, the quota represents somewhat less than 7% of total U.S. consumption in that year.

[15] Petrojam, Ltd., *Petrojam Ethanol Limited - Alcohol Sources.* U.S. International Trade Commission (USITC), *U.S. Imports of Fuel Ethanol, by Source 1996-2007*, February, 2008

[16] USITC, *Interactive Tariff and Trade DataWeb*, at [http://dataweb.usitc.gov]. March 9, 2006.

Despite criticisms in the United States, new dehydration facilities began production in Trinidad and Tobago in 2005[17] and the U.S. Virgin Islands in 2007.

Duty-free ethanol imports have also played a role in discussions regarding the U.S.-Central America Free Trade Agreement (CAFTA).[18] Under this agreement signed by the Bush Administration and the participating countries, specific allocations (of the 7% duty-free cap for CBI ethanol) are set aside for Costa Rica and El Salvador. These allocations effectively limit the amount of fuel that other CBI countries can import duty-free. Costa Rica's allocation is 31 million gallons per year, while El Salvador was granted an initial allocation of approximately 6.6 million gallons per year, increasing by roughly 1.3 million gallons in each subsequent year. However, El Salvador's allocation may not exceed 10% of the total CBI allocation (or 0.7% of the U.S. market). The agreement was signed on May 28, 2004. Congress approved the agreement in 2005, and implementing legislation was signed by President Bush on August 2, 2005 (P.L. 109-53). As both countries exceeded their allocations in 2005, 2006, and 2007, the ultimate effects of the allocations is unclear.

Growing U.S. Ethanol Market

The U.S. ethanol market has grown dramatically over the past several years. Between 1990 and 2007, U.S. ethanol consumption increased from about 900 million gallons per year to 6.8 billion gallons per year. Much of this growth has resulted from Clean Air Act requirements that gasoline in areas with the worst ozone pollution contain an oxygenate, such as ethanol, and the establishment of a renewable fuel standard (RFS) in the Energy Policy Act of 2005 (P.L. 109-58). The RFS required that gasoline sold in the United States contain a renewable fuel, such as ethanol. The mandate required 4.0 billion gallons of renewable fuel in 2006, increasing to 7.5 billion gallons in 2012. The Energy Independence and Security Act of 2007 (P.L.110-140) expanded the RFS to 9.0 billion gallons in 2008, increasing to 36 billion gallons in 2022. In addition, the expanded RFS specifically requires the use of an increasing amount of "advanced biofuels" — biofuels produced from feedstocks other than corn starch (including sugar cane ethanol). While domestic producers anticipate greater demand for their product under the RFS, they are also concerned that duty-free ethanol imports through the CBI could dramatically increase, to their detriment.

Duty Drawback

In addition to the concerns over imports of duty-free ethanol from CBI countries, there is growing concern that a large portion of ethanol otherwise subject to the duties is being imported duty-free through a "manufacturing drawback."[19] If a manufacturer imports an intermediate product then exports the finished product or a similar product,

---

[17] This project has received particular scrutiny from some critics because its construction was financed through a loan insured by the U.S. Export-Import Bank.

[18] For more information on CAFTA, see CRS Report RL31870, *The Dominican Republic-Central America-United States Free Trade Agreement (CAFTA-DR)*, by J. F. Hornbeck.

[19] For more information on drawbacks, see U.S. Customs Service, *Drawback: A Refund for Certain Exports*, Washington, February 2002.

that manufacturer may be eligible for a refund (drawback) of up to 99% of the duties paid. There are special provisions for the production of petroleum derivatives.[20] In the case of fuel ethanol, the imported ethanol is used as a blending component in gasoline, and jet fuel (considered a like commodity) is exported to qualify for the drawback.[21] Some critics estimate that as much as 75% or more of the duties were eligible for the drawback in 2006. Therefore, critics question the effectiveness of the ethanol duties and the CBI exemption.

Congressional Action

Some Members of Congress have expressed concern over duty-free imports of dehydrated ethanol that originates in Brazil or other countries. Therefore, there is growing interest from some Members of Congress to eliminate the CBI exemption and/or modify the manufacturing drawback for petroleum products.

Although some stakeholders are concerned over increased ethanol imports and their effect on the U.S. industry, others believe that tariffs on imported ethanol should be eliminated entirely. They argue that increased use of ethanol, regardless of its origin, would further displace gasoline consumption. They also argue that inexpensive imported ethanol would help mitigate any fuel price increases from the renewable fuels standard.

Conclusion

With growing demand for ethanol, there is increased interest in foreign imports. Because ethanol from CBI countries is granted duty-free status, there is the possibility that imports of dehydrated ethanol will grow because of this avenue provided in the law. While CBI countries have not yet reached their quota for ethanol refined in other countries and dehydrated in the Caribbean, CBI imports have increased over the past few years, and may exceed the quota in future years. CBI imports have the potential to increase significantly over the next few years, especially as the domestic market grows under the renewable fuels standard. In addition, the manufacturing drawback could provide another avenue for duty-free ethanol imports directly from Brazil and other countries.

Low-cost ethanol imports could have an advantage over domestically produced ethanol, which could affect the U.S. ethanol industry and American corn growers. However, the U.S. ethanol industry has grown significantly in the past several years, and will likely continue to grow regardless of the level of imports.

---

[20] 19 U.S.C. 1313(p).

[21] Peter Rhode, "Senate Finance May Take Up Drawback Loophole As Part Of Energy Bill," *EnergyWashington Week*, April 18, 2007.

Regulatory Announcement

EPA Lifecycle Analysis of Greenhouse Gas Emissions from Renewable Fuels

Background

As part of revisions to the National Renewable Fuel Standard program (commonly known as the RFS program) as mandated in the Energy Independence and Security Act of 2007 (EISA), EPA has analyzed lifecycle greenhouse gas (GHG) emissions from increased renewable fuels use. EISA established eligibility requirements for renewable fuels, including the first U.S. mandatory lifecycle GHG reduction thresholds, which determine compliance with four renewable fuel categories. The regulatory purpose of EPA's lifecycle GHG emissions analysis is therefore to determine whether renewable fuels produced under varying conditions meet the GHG thresholds for the different categories of renewable fuel. Determining compliance with the thresholds requires a comprehensive evaluation of renewable fuels, as well as of gasoline and diesel, on the basis of their lifecycle emissions.

EISA defines lifecycle GHG emissions as follows:

> The term 'lifecycle greenhouse gas emissions' means the aggregate quantity of greenhouse gas emissions (including direct emissions and significant indirect emissions such as significant emissions from land use changes), as determined by the Administrator, related to the full fuel lifecycle, including all stages of fuel and feedstock production and distribution, from feedstock generation or extraction through the distribution and delivery and use of the finished fuel to the ultimate consumer, where the mass values for all greenhouse gases are adjusted to account for their relative global warming potential.[1]

EISA established specific lifecycle GHG emission thresholds for each of four types of renewable fuels, requiring a percentage improvement compared to lifecycle GHG emissions for gasoline or diesel (whichever is being replaced by the renewable fuel) sold or distributed as transportation fuel in 2005. EISA required a 20% reduction in lifecycle GHG emissions for any renewable fuel produced at new facilities (those

[1] Clean Air Act Section 211(o)(1)

Office of Transportation and Air Quality
EPA-420-F-10-006
February 2010

Regulatory Announcement

constructed after enactment), a 50% reduction in order to be classified as biomass-based diesel or advanced biofuel, and a 60% reduction in order to be classified as cellulosic biofuel.

Threshold Determinations

EPA is making threshold determinations based on a methodology that includes an analysis of the full lifecycle of various fuels, including emissions from international land-use changes resulting from increased biofuel demand. EPA has used the best available models for this purpose, and has incorporated many modifications to its proposed approach based on comments from the public, a formal peer review, and developing science. EPA has also quantified the uncertainty associated with significant components of its analyses, including important factors affecting GHG emissions associated with international land use change. EPA is confident that its modeling of GHG emissions associated with international land use is comprehensive and provides a reasonable and scientifically robust basis for making threshold determinations. Based on this analysis, EPA is determining that:

- Ethanol produced from corn starch at a new natural gas, biomass, or biogas fired facility (or expanded capacity from such a facility) using advanced efficient technologies (ones that we expect will be most typical of new production facilities) will meet the 20% GHG emission reduction threshold compared to the 2005 gasoline baseline.
- Biobutanol from corn starch also meets the 20% threshold.
- Biodiesel and renewable diesel from soy oil or waste oils, fats, and greases will meet the 50% GHG threshold for biomass-based diesel compared to the 2005 petroleum diesel baseline.
- Biodiesel and renewable diesel produced from algal oils will also comply with the 50% threshold should they reach commercial production.
- Ethanol from sugarcane complies with the applicable 50% reduction threshold for advanced biofuels.
- For cellulosic ethanol and cellulosic diesel, the pathways modeled in our analysis (for feedstock and production technology) would comply with the 60% GHG reduction threshold for cellulosic biofuel.
- Determinations for additional fuels and fuel pathways can be found in Section V of the preamble.

In addition to finalizing threshold compliance determinations for pathways that we specifically modeled, as shown above, in some cases our technical judgment indicates that other pathways are likely to be similar enough that we can extend these determinations. These include fuels that are produced from five categories of feedstocks similar to those already modeled and which are expected to have less or no indirect land use change:

1. Crop residues such as corn stover, wheat straw, rice straw, citrus residue
2. Forest material including eligible forest thinnings and solid residue remaining from forest product production
3. Secondary annual crops planted on existing crop land such as winter cover crops
4. Separated food and yard waste including biogenic waste from food processing
5. Perennial grasses including switchgrass and miscanthus

Regulatory Announcement

Threshold determinations for certain other pathways were not possible at this time because sufficient modeling or data is not yet available. In some of these cases, we recognize that while a renewable fuel is already being produced from an alternative feedstock and we have the data needed for analysis, we did not have sufficient time to complete the necessary lifecycle GHG impact assessment for this final rule. EPA anticipates modeling grain sorghum ethanol, woody pulp ethanol, and palm oil biodiesel after this final rule and including the determinations in a rulemaking within 6 months.

For other fuels, we are establishing a process whereby a biofuel producer or importer can petition the Agency to also consider whether a fuel pathway would be eligible for use in complying with an EISA standard. EPA will use the data supplied in the petition to evaluate whether the information for the fuel pathway, combined with information developed in this rulemaking for other fuel pathways, is sufficient to allow EPA to determine whether the new fuel pathway qualifies. EPA will process these petitions as expeditiously as possible, taking into consideration that some fuel pathways are closer to the commercial production stage than others.

Our Analysis

In order to calculate the lifecycle GHG emissions of various fuels, EPA utilized models that take into account energy and emissions inputs for fuel and feedstock production, distribution, and use, as well as economic models that predict changes in agricultural markets. In developing this analysis, the Agency employed a collaborative, transparent, and science-based approach. Through technical outreach, the peer review process, and the public comment period, EPA received and reviewed a significant amount of data, studies, and information on our proposed lifecycle analysis approach. We incorporated a number of new, updated, and peer-reviewed data sources in our final rulemaking analysis, including better satellite data for tracking land use changes and improved assessments of N2O impacts from agriculture.

We also performed dozens of new modeling runs, uncertainty analyses, and sensitivity analyses which are leading to greater confidence in our results. We have updated our analyses in conjunction with, and based on, advice from experts from government, academia, industry, and not for profit institutions.

The new studies, data, and analysis performed for the final rulemaking impacted the lifecycle GHG results for biofuels in a number of different ways. In some cases, updates caused the modeled analysis of lifecycle GHG emissions from biofuels to increase, while other updates caused the modeled emissions to be reduced. Overall, the revisions since our proposed rule have led to a reduction in modeled lifecycle GHG emissions as compared to the values in the proposal. For example, for corn ethanol the final rule analysis found less overall indirect land use change (less land needed), thereby improving the lifecycle GHG performance of corn ethanol. The main reasons for this decrease are:

- Based on new studies that show the rate of improvement in crop yields as a function of price, crop yields are now modeled to increase in response to higher crop prices. When higher crop yields are used in the models, less land is needed domestically and globally for crops as biofuels expand.

3

Regulatory Announcement

- New research available since the proposal indicates that distillers grains and solubles (DGS), a corn ethanol production co-product, is more efficient as an animal feed (meaning less corn is needed for animal feed) than we had assumed in the proposal. Therefore, in our analyses for the final rule, domestic corn demand and exports are not impacted as much by increased biofuel production as they were in the proposal analysis.
- Improved satellite data allowed us to more finely assess the types of land converted when international land use changes occur, and this more precise assessment led to a lowering of modeled GHG impacts. Based on previous satellite data, the proposal assumed cropland expansion onto grassland would require an amount of pasture to be replaced through deforestation. For the final rulemaking analysis we incorporated improved satellite data, as well as improved economic modeling of pasture demand, and found that pasture is also likely to expand onto existing grasslands. This reduced the GHG emissions associated with an amount of land use change.

Next Steps/Future Work

While EPA is using its current lifecycle assessments to inform regulatory determinations in this final rule, as required by EISA, we also recognize that as the state of scientific knowledge continues to evolve in this area, the lifecycle GHG assessments for a variety of fuel pathways will continue to be enhanced. Therefore, the Agency is committing to further reassess these determinations and lifecycle estimates. As part of this ongoing effort, we will ask for the expert advice of the National Academy of Sciences, as well as other experts, and incorporate their advice and any updated information we receive into a new assessment of the lifecycle GHG emissions performance of the biofuels being evaluated in this final rule. EPA will request that the National Academy of Sciences evaluate the approach taken in this rule, the underlying science of lifecycle assessment, and in particular indirect land use change, and make recommendations for subsequent lifecycle GHG assessments on this subject. This new assessment could result in new determinations of threshold compliance compared to those included in this rule. These would apply to future production from plants that are constructed after each subsequent rule incorporating a revised lifecycle assessment methodology.

Additional detail on the different components of EPA's lifecycle analysis can be found in the preamble and the Regulatory Impact Analysis that accompany the Final Rule.

For More Information

For more information on this proposal, please contact EPA's Office of Transportation and Air Quality, Assessment and Standards Division information line at:

U.S. Environmental Protection Agency
Office of Transportation and Air Quality
2000 Traverwood Drive
Ann Arbor, MI 48105
Voicemail: (734) 214-4636
E-mail: asdinfo@epa.gov

Or visit: www.epa.gov/otaq/renewablefuels/index.htm

4

Resources from TheCapitol.Net

Live Training

<www.CapitolHillTraining.com>

- Capitol Hill Workshop
  <www.CapitolHillWorkshop.com>
- Understanding Congressional Budgeting and Appropriations
  <www.CongressionalBudgeting.com>
- Advanced Federal Budget Process
  <www.BudgetProcess.com>
- The President's Budget
  <www.PresidentsBudget.com>
- Understanding the Regulatory Process: Working with Federal Regulatory Agencies
  <www.RegulatoryProcess.com>
- Drafting Effective Federal Legislation and Amendments
  <www.DraftingLegislation.com>

Capitol Learning Audio Courses™

<www.CapitolLearning.com>

- Congress and Its Role in Policymaking
  ISBN: 158733061X
- Understanding the Regulatory Process Series
  ISBN: 1587331398
- Authorizations and Appropriations in a Nutshell
  ISBN: 1587330296

Other Resources

Government

- Biofuels in the US Transportation Sector, Energy Information Administration, Department of Energy <*http://www.eia.doe.gov/oiaf/analysispaper/biomass.html*>
- Ethanol, Energy Information Administration, Department of Energy <*http://www.eia.doe.gov/oiaf/ethanol3.html*>
- Alternative Fuels and Advanced Vehicles Data Center: Ethanol, Department of Energy <*http://www.afdc.energy.gov/afdc/ethanol/*>
- Ethanol Myths and Facts, Office of Energy Efficiency and Renewable Energy, Department of Energy <*http://www1.eere.energy.gov/biomass/ethanol_myths_facts.html*>
- Ethanol and a Changing Agricultural Landscape, Economic Research Service, Department of Agriculture <*http://www.ers.usda.gov/Publications/ERR86/*>
- Ethanol Expansion in the United States: How Will the Agricultural Sector Adjust? Economic Research Service, Department of Agriculture <*http://www.ers.usda.gov/Publications/FDS/2007/05May/FDS07D01/*>
- National Renewable Energy Laboratory (NREL) <*http://www.nrel.gov*>

Associations, Coalitions, News

- American Coalition for Ethanol <*http://www.ethanol.org*>
- Biodiesel Magazine <*http://www.biodieselmagazine.com*>
- Biofuels Digest <*http://www.biofuelsdigest.com*>
- BioFuels Journal <*http://www.biofuelsjournal.com*>
- Brazilian Sugarcane Industry Association (UNICA) <*http://english.unica.com.br*>
- Canadian Renewable Fuels Association <*http://www.greenfuels.org*>
- Clean Fuels Development Coalition (CFDC) <*http://www.cleanfuelsdc.org*>
- Ethanol: ATTRA—National Sustainable Agriculture Information Service <*http://attra.ncat.org/farm_energy/ethanol.html*>
- Ethanol Producers And Consumers (EPAC) <*http://www.ethanolmt.org*>
- Ethanol Promotion and Information Council <*http://www.drivingethanol.org*>
- EthanolMarket.com <*http://www.ethanolmarket.com*>
- European Bioethanol Fuels Association <*http://www.ebio.org*>
- Growth Energy <*http://www.growthenergy.org*>

- Industrial Ethanol Association (IEA) *<http://www.industrial-ethanol.org>*
- International Ethanol Trade Association (IETHA) *<http://www.ietha.org>*
- National Biodiesel Board *<http://www.biodiesel.org>*
- National Corn Growers Association *<http://www.ncga.com>*
- National Ethanol Vehicle Coalition *<http://www.e85fuel.com>*
- Renewable Fuels Association (RFA) *<http://www.ethanolrfa.org>*
- Sugar Association *<http://www.sugar.org>*
- University of Illinois Center for Advanced BioEnergy Research *<http://www.bioenergy.uiuc.edu>*
- US Beet Sugar Association *<http://www.beetsugar.org>*

Books

- *Alcohol Can Be a Gas!: Fueling an Ethanol Revolution for the 21st Century*, ISBN 0979043778
- *Anaerobic Biotechnology for Bioenergy Production: Principles and Applications*, ISBN 0813823463
- *Beet-Sugar Handbook*, ISBN 0471763470
- *Beyond Oil and Gas: The Methanol Economy*, ISBN 3527324224
- *Biodiesel America: How to Achieve Energy Security, Free America from Middle-east Oil Dependence And Make Money Growing Fuel*, ISBN 0970722745
- *Biodiesel: A Realistic Fuel Alternative for Diesel Engines*, ISBN 1846289947
- *Biodiesel Basics and Beyond: A Comprehensive Guide to Production and Use for the Home and Farm*, ISBN 0973323337
- *Biodiesel: Growing a New Energy Economy*, Second Edition, ISBN 1933392967
- *Biodiesel Power: The Passion, the People, and the Politics of the Next Renewable Fuel*, ISBN 0865715416
- *Bioenergy*, ISBN 1555814786
- *Biofuels*, ISBN 047002674X
- *Biofuels (Advances in Biochemical Engineering/Biotechnology)*, ISBN 3540736506
- *Biofuels: Biotechnology, Chemistry, and Sustainable Development*, ISBN 1420051245
- *Biofuels Engineering Process Technology*, ISBN 0071487492

- *Biofuels for Road Transport: A Seed to Wheel Perspective*, ISBN 1848821379
- *Biofuels for Transport: Global Potential and Implications for Energy and Agriculture*, ISBN 1844074226
- *Biofuels: Implications for the Feed Industry*, ISBN 9086860435
- *Biofuels Refining and Performance*, ISBN 0071489703
- *Biomass and Alternate Fuel Systems: An Engineering and Economic Guide*, ISBN 0470410280
- *Biomass for Renewable Energy, Fuels, and Chemicals*, ISBN 0124109500
- *Biorefineries—Industrial Processes and Products: Status Quo and Future Directions*, ISBN 3527310274
- *Biorenewable Resources: Engineering New Products from Agriculture*, ISBN 0813822637
- *Careers in Renewable Energy: Get a Green Energy Job*, ISBN 097737243X
- *Catalysis for Sustainable Energy Production*, ISBN 3527320954
- *Clean Money: Picking Winners in the Green Tech Boom*, ISBN 0470283564
- *Do It Yourself Guide to Biodiesel: Your Alternative Fuel Solution for Saving Money, Reducing Oil Dependency, and Helping the Planet*, ISBN 1569756244
- *Energy from Biomass: A Review of Combustion and Gasification Technologies*, ISBN 0821343351
- *Energy Plant Species: Their Use and Impact on Environment and Development*, ISBN 1873936753
- *Energy Storage: A Nontechnical Guide*, ISBN 159370027X
- *Ethanol and Biofuels: Production, Standards and Potential*, ISBN 1606922246
- *Fast Pyrolysis of Biomass: A Handbook*, ISBN 1872691072
- *From the Fryer to the Fuel Tank: The Complete Guide to Using Vegetable Oil as an Alternative Fuel*, ISBN 0970722702
- *Genetic Improvement of Bioenergy Crops*, ISBN 1441924221
- *Handbook of Alternative Fuel Technologies*, ISBN 0824740696
- *Handbook of Plant-Based Biofuels*, ISBN 1560221755
- *Handbook of Sugar Refining: A Manual for the Design and Operation of Sugar Refining Facilities*, ISBN 0471183571
- *Handbook on Bioethanol: Production and Utilization*, ISBN 1560325534
- *Introduction to Agricultural Engineering Technology: A Problem Solving Approach*, ISBN 0387369139

- *Introduction to Chemicals from Biomass*, ISBN 0470058056
- *Miscanthus Bioenergy*, ISBN 3836493314
- *Practical Fermentation Technology*, ISBN 0470014342
- *Process Synthesis for Fuel Ethanol Production*, ISBN 1439815976
- *Pyrolysis and Gasification of Biomass and Waste*, ISBN 1872691773
- *Renewable Energy*, ISBN 0199261784
- *Renewable Energy Policy*, ISBN 0595312187
- *Renewable Energy: Technology, Economics and Environment*, ISBN 3540709479
- *Run Your Diesel Vehicle on Biofuels: A Do-It-Yourself Manual*, ISBN 0071600434
- *Sourcebook of Methods of Analysis for Biomass and Biomass*, ISBN 1851665277
- *Sustainable Ethanol: Biofuels, Biorefineries, Cellulosic Biomass, Flex-fuel Vehicles, and Sustainable Farming for Energy Independence*, ISBN 0978629302
- *SVO: Powering Your Vehicle With Straight Vegetable Oil*, ISBN 0865716129
- *Synthetic Fuels*, ISBN 0486449777
- *Synthetic Fuels Handbook: Properties, Process, and Performance*, ISBN 007149023X
- *The Alcohol Fuel Handbook*, ISBN 0741406462
- *The Biofuel Delusion: The Fallacy of Large Scale Agro-Biofuel Production*, ISBN 1844076814
- *The Biomass Assessment Handbook: Bioenergy for a Sustainable Environment*, ISBN 1844075265
- *The Brilliance of Bioenergy: In Business and Practice*, ISBN 190291628X
- *The Clean Tech Revolution: The Next Big Growth and Investment Opportunity*, ISBN 006089623X
- *The Handbook of Biomass Combustion and Co-firing*, ISBN 1844072495
- *Water Implications of Biofuels Production in the United States*, ISBN 030911361X

Video and Movies

- 2008 Global Conference: Innovator: Kevin Walsh on Financing the Future of Renewable Energy, ASIN B001HL01DM
- 2008 Global Conference: The Race to the Finish: Next Gen Biofuels, ASIN B001HL01QY
- 60 Minutes—The Ethanol Solution (May 7, 2006), ASIN B000FOTA5C
- A Crude Awakening—The Oil Crash (2006), ASIN B000PY52IG
- Alcohol Can Be a Gas! Fueling an Ethanol Revolution for the 21st Century, ASIN B001A-LYGZ8
- Food, Inc., ASIN B0027BOL4G
- Future Fuels 1: Consumer's Guide to Alternative Energy, ASIN 1607433346
- H. Douglas Lightfoot's Nobody's Fuel—energy supply is more important than climate change, ASIN B000PHX2CG
- King Corn, ASIN B001EP8EOY
- Modern Marvels: Environmental Tech (2007), ASIN B0019KAQBS
- Modern Marvels: Renewable Energy (2006), ASIN B0019KAQB8
- Modern Marvels—Sugar, ASIN B000FBFYZU
- Science NOW 2009: Episode 6: Algae Fuel, ASIN B002TCRQ3G
- The Big Energy Gamble (2009), ASIN B001PUTN2K
- The Future of Food, ASIN B000V5IOWK

About TheCapitol.Net

We help you understand Washington and Congress.™

For over 30 years, TheCapitol.Net and its predecessor, Congressional Quarterly Executive Conferences, have been training professionals from government, military, business, and NGOs on the dynamics and operations of the legislative and executive branches and how to work with them.

Instruction includes topics on the legislative and budget process, congressional operations, public and foreign policy development, advocacy and media training, business etiquette and writing. All training includes course materials.

TheCapitol.Net encompasses a dynamic team of more than 150 faculty members and authors, all of whom are independent subject matter experts and veterans in their fields. Faculty and authors include senior government executives, former Members of Congress, Hill and agency staff, editors and journalists, lobbyists, lawyers, nonprofit executives and scholars.

We've worked with hundreds of clients across the country to develop and produce a wide variety of custom, on-site training. All courses, seminars and workshops can be tailored to align with your organization's educational objectives and presented on-site at your location.

Our practitioner books and publications are written by leading subject matter experts.

TheCapitol.Net has more than 2,000 clients representing congressional offices, federal and state agencies, military branches, corporations, associations, news media and NGOs nationwide.

Our blog: Hobnob Blog—hit or miss ... give or take ... this or that ...

TheCapitol.Net is on Yelp.

Our recommended provider of government training in Brazil is PATRI/EDUCARE <www.patri.com>

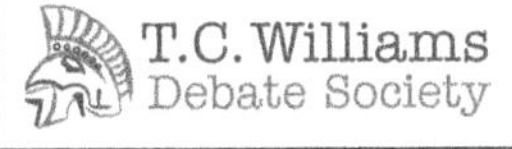

TheCapitol.Net supports the T.C. Williams Debate Society, Scholarship Fund of Alexandria, and Sunlight Foundation

TheCapitol.Net

Non-partisan training and publications that show how Washington works.™

PO Box 25706, Alexandria, VA 22313-5706 703-739-3790 www.TheCapitol.Net

www.ingramcontent.com/pod-product-compliance
Lightning Source LLC
Chambersburg PA
CBHW081457200326
41518CB00015B/2286